Earth Changes and the Human-Cosmic Connection

Earth Changes and the Human-Cosmic Connection

Pierre Lescaudron with Laura Knight-Jadczyk

Red Pill Press 2021

Second edition (black and white)
ISBN 978-1-7349074-3-8
Red Pill Press (www.redpillpress.com)

Contents

Acknowledgements

To be honest, when I read a book I usually skip the acknowledgment section, which doesn't seem to be much more than a compulsory but boring list of names and thanks.

However, while writing this book I saw how important the help from others was and I would like very much for you to know who helped transforming what was barely a dream a few years ago into the tangible reality you're holding in your hands right now.

In order to prevent you from skipping this section, I tried to make it short and entertaining. Also, if you read it to the end you will have a little surprise ...

First I want to thank Laura Knight-Jadczyk, a.k.a. 'Read this book'.
Through her (almost too numerous!) recommended readings, the articles she has written, the ideas she presented and the discussions we shared, Laura contributed tremendously to the realization of this book. In all fairness, many of the developed concepts (especially the good ones) are not my ideas but hers.

Arkadiusz Jadczyk, a.k.a. 'Where are the equations?'
Ark is a renowned mathematician and physicist. His language, even his whole world, is made up of equations and numbers of which this book is almost totally devoid. Despite this unforgivable flaw (from a conscientious scientist perspective), Ark had the patience to go through some of the craziest ideas described in this book and check their scientific feasibility.

Neil Bradley, a.k.a. 'Did you read this article?'
Some people obsess about cleanliness or money, Neil obsesses about collecting, editing and reading articles about all kinds of topics relating to the subject of this book (the above doesn't mean that he's not clean or rich). Consequently, many quotes and sources come from his precious discoveries. Also, Neil proofread the whole book, substantially improving its clarity and fluidity. It was not an easy job.

Joseph Quinn, a.k.a. 'When your only tool is a hammer, all you see is nails'.
Indeed, after spending quite some time studying electric phenomena associated with earthly and cosmic events, I tended to see electricity everywhere. That's a fairly standard specialization bias that, fortunately, was challenged. Joe also has a keen interest in weather and climate and provided some very interesting ideas relating to those topics.

Scott Ogrin, a.k.a. 'What is an electron?'
In times of overly daring speculations, Scott knew how to bring out the painful reality check. Sadly but truly, despite tons of theories, we really don't know much about our universe and some very basic but fundamental concepts like electrons, gravity, electricity. Scott was wise enough to repeatedly point out those limitations.

I also want to thank Myriam Kieffer who skillfully designed the cover of this book, astrophysicist Amokrane Berdja who helped calculating Nemesis parameters and Amelia Martin who shared Photoshop techniques and contributed to some of the illustrations you'll find in this book. A special thanks to Juliana Barembuem and Michael Franzl for creating the layout and printing files.

Last but not least, a warm thank you to all the editors of the information website *Signs of the Times*[1] who have built over the years an invaluable knowledge resource.

And now, here is, as promised, the previously mentioned surprise, namely the proposed titles that didn't make the cut.

As you can imagine, finding a title is quite a challenge. You want it to be short, catchy, never used before and meaningful. So, to find the ideal candidate you have to go through many trials and errors (especially errors). I don't know if the selected title is perfect but I'm sure some of the others were much worse.

The presumptuous 'Why science is wrong'

The latinizing 'Sol, Nemesis, Gaia et Homo'

The apocalyptic 'OMG we're all going to die'

The enigmatic 'Sixth extinction and the human cosmic connection'

The messianic 'The Mandate of Heaven'

The extensive 'The approaching Sun's companion and its cometary swarm, a human cosmic connection'

The alarming 'Before it's too late'

The inclusive 'All and everything'

The catastrophist 'Sinkholes, chemtrails and climate changes, several effects of one single cause'

The provocative 'F*** you NASA'

The Hollywoodian 'Return of the Dark star'

The esoteric 'As above so below'

The Gallicizing 'The sky is falling on our heads'

The transcendental 'The celestial mandate'

The not-so New Agey 'You thought you could create your own reality'

[1] www.sott.net

Foreword by Laura Knight-Jadczyk

Earth Changes and the Human-Cosmic Connection: The Secret History of the World, Book Three is an attempt to answer certain questions that were left dangling at the end of my last book, *Comets and the Horns of Moses: The Secret History of the World, Book Two*, specifically this idea that there is a dynamic interaction between the actions of human beings and the cosmos.

That hypothesis emerged over time as I continued to research history with an eye open for environmental catastrophes affecting the beginning, development, and end of civilizations. What I noticed, as I organized my material into tables, was a distinct correlation: in one column, I would place the archaeological evidence for catastrophic physical collapse, in another, the geological and astronomical (back calculated) evidence, and in a third, the historical narrative of what was 'going on'. The results were compelling and an entire volume devoted to that tabular organization of the material is in preparation.

In *Comets and the Horns of Moses*, a short section was devoted to the topic of plasma physics and how this could provide a model for understanding what the ancients were reporting in respect of the Sky Gods and their dramatic encounters with the Earth and humanity, but there was only space to cover this briefly and more questions were raised by that short exposition that demanded answers. The main author of the present book, Pierre Lescaudron, was my fellow traveler through those topics and it was thanks to him that the plasma physics section made any sense at all; he read my fumbling attempts to articulate what I was perceiving in the historical record, and wrote the explanatory passages. So, it was only logical that he should carry through on that study and attempt to answer the questions that had emerged. His engineering education and knowledge base in these areas is superior to my own, and delightfully, his ability to explain in a simple way helped me enormously and I am certain that it will do the same for the reader.

Geophysicist Amos Nur points out that using archaeological evidence to infer what may have caused catastrophic destruction of cities and entire regions is actually a pretty simple idea, but one which most archaeologists, guided by historians, don't really like. Historians want to 'stick to the texts', ignoring the strong evidence of textual tampering in respect of 'what really happened', and the partnership with archeologists has been so long established that there is stiff opposition to bringing hard sciences to bear on their carefully constructed narratives and chronologies.

Nur's work[2] is an excellent introduction to the problem and deals with evidence of massive earthquake destruction. What he does not address are the associated celestial phenomena that invariable accompany the earthquakes in the records of the ancients. And certainly, what neither he nor individuals from other disciplines address are the strong signal correlations between human behavior on the mass scale and cosmic apparitions followed by cataclysmic destruction.

The problem, as becomes clear in the present book, is the intense pressure on historians, political

[2] Nur, A., *Apocalypse, Earthquakes, Archaeology, and the Wrath of God*, Princeton University Press, 2008

scientists, ecologists, archaeologists, to describe major disasters in history as being due to human causes. The most visible modern promoter of this view is Jared Diamond who, in his analysis of societies' collapse,[3] never even mentions earthquakes or volcanic eruptions, much less cosmic influences.

Earth Changes and the Human-Cosmic Connection will show you why politically controlled science needs to ascribe massive destruction to human causes rather than to the whims of Nature or the 'Anger of the Gods'. It also posits that there is a connection between human beings, Nature, the Cosmos and disasters on Earth and seeks to explain how this can be understood scientifically. Interestingly, the conclusion is that, yes, humans are probably one of the causes, but definitely not in the way that has been proposed hitherto. What is more, the very denial of this fact may contribute significantly to the level of devastation. The widespread scientific bias against recognizing the effects of cosmic influences on the history of humanity reflects a deep pathology in the human mass psyche.

When the political leaders of those ancient times began to form coalitions with the religious leaders in order to ideologically support their cravings for domination and control and self-aggrandizement, it was the beginning of the end. The result of the rampant imperialism manifested by the pathologically infiltrated Bronze Age civilizations was near total destruction of everything and everybody, though certainly, as we noted in the previous volume, such cosmo-planetary phase transitions may take place over a protracted (to us) period of time. But, at some point, things can – and do – change very, very fast, on a very large scale, even in a single day, as Plato suggested in his story of Atlantis. More terrifying still is that the same thing happened again about 1500 years ago, bringing the Roman Empire to an end.

Scientists and historians today may argue that there is insufficient evidence or that the theory cannot be tested – falsified – that cosmic disasters can change the course of human history or bring civilizations to their knees. They claim that the historical evidence says otherwise: societies respond to 'minor disasters' by rebuilding and carrying on as before – that's what their re-written histories tell them. But that is not really the case, as we will see from a couple of historical studies included in this book that include evidence that the historians exclude or ignore, and the more massive historical study coming in the next volume of the *Secret History* series will provide even more. There is abundant evidence that political instability, extended warfare, mass hysteria, are strongly correlated to earthquakes, climate change, and Celestial apparitions which may interact with our planet and societies far more frequently and destructively than modern historians acknowledge.

The purpose of the present book is to show how it can happen, that it HAS happened, and that we may very well experience it again rather soon. And it does it very well!

Laura Knight-Jadczyk

[3] Diamond, J., *Collapse: How Societies Choose to Fail or Succeed*, Penguin Books, 2011

Preface

I began writing the material that would become the basis for this book in October 2010. Initially it was supposed to be an article to provide some tentative explanations for the unusually heavy snowfall in the northern hemisphere before winter had even started. Nothing more.

So I read here and there about the role the jet stream probably plays in our planet's weather. This exposed me to the electrical dimension of weather events, and from there I began studying electricity's role in various natural and astrophysical phenomena ('Electric Universe Theory'). A few months later, the *Dot Connector Magazine*[4] published my article dealing with these specific topics, 'Jet stream: Is an ice age imminent?'

But, as is often the case, writing this first article brought more questions than answers. Why was the Sun's activity dropping so dramatically? What is the cause of solar cycles? What is the role played by the oceanic current? What is the effect of these factors on climate cycles and weather patterns? What hidden factors are involved?

I concluded my article by announcing my intention to write a follow-up addressing these questions. This second article would hopefully provide satisfactory answers that would wrap everything up; case closed. Famous last words!

As I began pulling on the threads of the Electric Universe theory, I realized that this exercise was going to take much longer and would be more complicated than I had expected. On top of that, a fundamental paradox emerged: the Sun was eerily quiet, while the main driver of solar activity, namely cometary activity (as I'll argue later on) was reaching levels never seen in the modern age.

This led me to consider existing ideas about our Sun's companion star ('The Nemesis Theory') and an accompanying (theorized) cometary swarm. To date, it remains the only solution I can think of that adequately explains and reconciles lasting reduced solar activity concomitant with increased cometary activity. We'll come to this in more detail further on.

In the course of my additional research, I stumbled upon the L'Aquila trial. In April 2009, a magnitude 6.3 earthquake struck the town of L'Aquila, Italy. Of course, earthquakes occur rather frequently in Italy, but this one was peculiar because it had been correctly predicted by a scientist named Gioacchino Giuliani. A few weeks before the catastrophe, Giuliani had driven around the Province of L'Aquila in his truck equipped with speakers, warning the local population of the impeding danger. After Giuliani made his prediction, but before the quake occurred, the local authorities launched a witch-hunt against him. Even Italian Prime Minister Berlusconi became directly involved, and finally Giuliani was condemned and *forced to remove the website displaying his scientific tests and predictions*. The mainstream dogma prevailed: earthquakes (like many other natural phenomena) are unpredictable.

But when the earthquake occurred almost exactly as Giuliani had predicted, everything changed.

[4] Issue #13, www.thedotconnector.org/mag/

Three hundred people were killed and the inhabitants of the area were understandably upset, because the authorities had reassured them that there was no danger. They were also very angry about what had happened to Giuliani, who had not only been right about his prediction, but had cared enough about the people to try to warn them.

The authorities found themselves in a decidedly uncomfortable situation and looked for a scapegoat. In October 2012, seven members of the Italian National Commission for the Forecast and Prevention of Major Risks were each *sentenced to six years' imprisonment* after being convicted of multiple manslaughter for downplaying the likelihood of a major earthquake during tremors felt six days prior to the event. In effect, they were condemned for failing to predict the L'Aquila earthquake!

The L'Aquila trial followed a pattern I had observed repeatedly while investigating the 'man-made global warming' scam. In both cases, we find the authorities covering up what is really happening: global cooling instead of global warming, and an increase in the number of earthquakes instead of constant frequency. In addition, the authorities were lying about the causes: cosmically induced cooling vs. man-made warming, cosmically induced quakes vs. unpredictable quakes.

Struck by this similarity, I began to wonder if this *modus operandi* might apply to other phenomena besides just climate change and earthquakes. As I eventually discovered, a number of 'new phenomena' are being covered up in a similar fashion. Noctilucent clouds, sinkholes, floods, record low temperatures, contrails forming in unusually icy lower atmosphere layers, strange sounds, fireballs, a weakening jet stream and a stalling Gulf Stream are all occurring more and more frequently around the globe, while the mainstream media consistently downplay their significance and write them off as totally unrelated or, worse, due to anthropogenic global warming.

After looking into these phenomena, it was clear to me that a hypothetical solar companion and accompanying cometary swarm approaching the solar system could explain these allegedly man-made and unrelated anomalies, which now appeared to be intimately related to each other and unlikely to be related to human activity. So, with all these discoveries and the evidence I had collected, it seemed that the case could be closed and the book finally ready.

However, before adding the finishing touches to the manuscript in preparation for publication, I spent some time helping Laura Knight-Jadczyk on her recently published book, *The Secret History of the World Vol. II: Comets and the Horns of Moses.* This book, and the forthcoming volumes in the series, present a fascinating thesis: that there is a positive correlation between the evil deeds committed by imperialistic powers and the occurrence of 'cosmic' catastrophes that end up destroying the very Empire that both causes them and covers them up (or blames them on scapegoats).

An incredible number of historical records, testimonies and archeological data have been compiled, analyzed and cross-referenced, showing that this human-cosmic feedback mechanism has occurred time and again throughout the entire history of humanity. This phenomenon is so obvious when organized in a statistical format, and so prevalent, that I think it should be considered the main driver of evolution, devolution and even human life on Earth.

Before these staggering discoveries, I had assumed that the Sun's companion and its accompanying cometary swarm, if they existed, were most likely following a predictable orbit solely ruled by mechanistic laws. It was a neat and simple conclusion, but the above-mentioned feedback mechanism opened a whole new can of worms: Can the violence and the lies spread by powerful elites somehow trigger cosmic reactions? Does human consciousness have any influence on matter as has been suggested by some physicists, occurring at the microcosmic level? How does the observer influence the observed event at the macrocosmic level? Does a group of individuals (the world population, for example) have more influence on an observed event when they all 'resonate' on the same belief frequency? What happens when those beliefs are lies, opposed to the reality? And finally, if all the above is correct, what can we do to alter the bleak future that such an idea suggests?

In a nutshell, the sequence of questions and discoveries described above is the outline of the book you're about to read.

I will begin by covering some basic ground on the Electric Universe theory in Part I, showing the forces at work in our solar system and beyond, including plasma dynamics, electric discharge, double-layer plasma sheaths, and how these phenomena apply at planetary, solar and galactic levels.

In Part II, I will focus on the Nemesis theory and how it fits into the Electric Universe model, connecting it with both the increase in cometary activity of recent years and the decrease in solar activity. I also introduce the topic of cyclical extinctions caused by encounters with comets following various cycles.

Part III moves closer to home, focusing on the effects we should expect to find on Earth if such a scenario is true, including a slowdown of the Earth's rotation, an increase in earthquakes, fireballs and meteors in our atmosphere, meteorite impacts on the ground, volcanic eruptions and sinkholes, global cooling, disruptions of the Gulf Stream and the jet streams, more frequent hurricanes, lightning and tornadoes, floods, droughts, strange and new types of clouds and anomalous sounds, crop failure and pole shifts. In fact, all of these phenomena are now on record, as I will show.

Finally, in Part IV, I take my analysis even closer to home, explaining the human-cosmic connection as it operates in human history. I reintroduce the forgotten concept of the Mandate of Heaven and describe how it was progressively erased from human understanding and history, focusing on the medieval Dark Ages when an enormous cover-up took place, engineered by religion. Then I describe the possible science behind this phenomenon, with data from quantum theory, information theory and psi research, finally summing it all up and suggesting what our future may have in store, and what we can do about it.

Now that you know the path we are going to follow, you can already foresee that we will address a broad variety of scientific fields. In particular, we will rely on several research fields, some of which are still considered relatively unorthodox, namely plasma physics, plasma cosmology and information theory. These might sound like big and complicated topics, but I assure you that they will be presented in a way that is informative and easy to understand.

The whole book should be accessible to any lay reader. For this reason, numerous graphics and charts will complement the written material, which will itself be divided into short, 'easily digestible' chapters. Throughout the course of the book, I will attempt to elaborate working hypotheses that might provide answers to the above-mentioned questions. These are of course just hypotheses. Testing them against the facts and observations should show us how reliable they are.

My goal with this book is simply to state those hypotheses, bring together some existing scientific results and observations collected by various researchers, establish some connections between seemingly unrelated facts and suggest some causation in supposedly independent phenomena. I hope it will inspire readers to discover more about these crucial and fascinating topics and to improve and expand our knowledge of them.

This book doesn't claim any kind of fully developed theory, groundbreaking discovery or immutable truth. For that I should have elaborated full mathematical equations, conducted lab experiments and followed a true scientific methodology. This is not the case. To be honest, I would need years to acquire the proper knowledge and conduct the necessary experiments, and from my limited perspective, I don't think we have that much time left!

So, there are probably mistakes in this book. But what I am sure of is that its content is much closer to reality than what is conveyed by mainstream science about our planet, our Sun and the cosmos. With all that said, let's jump right in, beginning with some history concerning mainstream cosmology and the emergence of plasma cosmology.

Part I.

Electricity and Plasma

Chapter 1.

The mainstream universe vs. the Electric Universe

According to conventional space science, the movements of the bodies within the solar system are exclusively ruled by gravitational laws. German astronomer Johannes Kepler established his three laws of planetary motion in the beginning of the 17th century,[5] at a time when electricity was virtually unknown. Benjamin Franklin conducted his famous kite experiment 150 years later.[6]

So, in the mainstream Book of Reality, gravity rules the solar system. Additionally, the Sun's output is mostly photon radiation; space is a perfect vacuum; earthquakes are due to plate tectonic movements; weather events are due to air temperature or pressure differences (or both); human activity and *cosmic* events are totally uncorrelated.

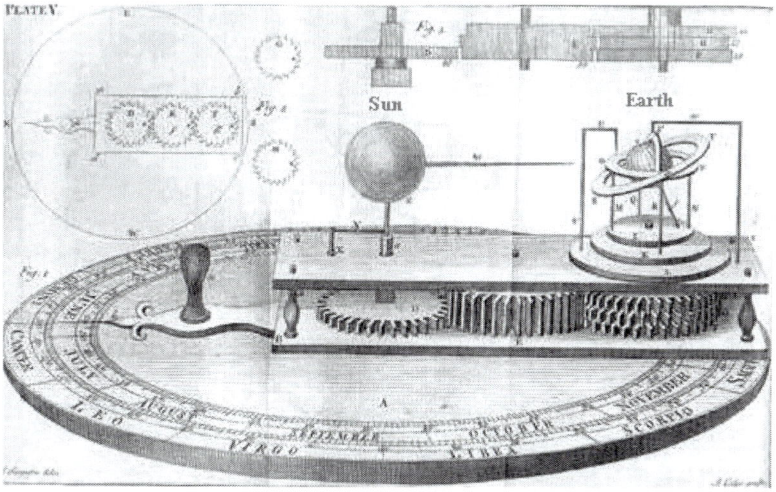

Figure 1: The astral movements by Newton. A mechanistic worldview where life is a linear system in which all events can be reduced to and explained as matter in motion. (© Unknown)

Over the last several centuries, science has produced an elaborate conceptual framework to rationalize and explain most natural phenomena (including many not mentioned in the short list above). According to mainstream science, we are living in a clockwork universe made of predictable events and undisturbed celestial movements. The solar system is a cosmic sanctuary, harmoniously animated by mechanical laws. Life on Earth flows like an uninterrupted, quiet, linear river, and evolution proceeds incrementally over thousands of generations in this relatively uneventful environment.

The problem is, this clockwork model of the universe has numerous inconsistencies and fails to explain several observable phenomena. In particular, during the past few years there has been an

[5] 'Kepler's laws of planetary motion', *Encyclopaedia Britannica.*
See: www.britannica.com/EBchecked/topic/315 260/Keplers-laws-of-planetary-motion

[6] 'A quick biography of Benjamin Franklin', *US History.* See: www.ushistory.org/franklin/info/index.htm

increase in unusual and extreme phenomena that don't conform to the dominant dogmas. When a new phenomenon of this type is observed, mainstream science simply attempts to force-fit the observed facts into the pre-existing theories, even when those theories are inadequate to account for the new phenomenon.

When twisting the theories or the facts is not enough, mainstream science comes up with convoluted explanations, paradoxical statements or theories so complicated or abstract that they can't even be understood or tested. In other cases, mainstream science just remains silent when confronted with an observation – an ugly fact – that clearly doesn't fit the dominant paradigm, or worse, threatens to shake its very foundations.

While the media and most of the scientific community were hammering this clockwork universe model into the people's minds, another group of prominent scientists could see the problems inherent in the conventional explanations and sought more valid models: something that could account for a wider range of facts. These scientists were attempting to understand the nature of the universe by acknowledging its main components: electricity and plasma.

Traces of this movement can be found in the late 19th century, when *Scientific American* published an article stating that Professor Zollner of Leipzig ascribed the 'self-luminosity' of comets to 'electrical excitement.' Zollner proposed that:

> ... the nuclei of comets, as masses, are subject to gravitation, while the vapors developed from them, which consist of very small particles, **yield to the action of the free electricity of the sun ...**

Then, regarding comet tails, the August 11th, 1882, issue of *English Mechanic and World of Science* included the following:

> There seems to be a rapidly growing feeling amongst physicists that both the self-light of comets and the phenomena of their tails belong to the order of **electrical phenomena**.

In 1896, *Nature* published an article stating:

> It has long been imagined that the phenomenon of **comet's tails are in some way due to a solar electrical repulsion**, and additional light is thrown on this subject by recent physical researches.

For decades the 'electric plasma' school pursued its research out of the media spotlight and without major research grants. Irving Langmuir,[7] winner of the Nobel Prize in Chemistry in 1932, Kristian Birkeland,[8] seven-time Nobel Prize nominee, and Hannes Alfven,[9] winner of the Nobel Prize in Physics in 1970, can be considered the founding fathers of Plasma Cosmology. The three of them conducted numerous laboratory experiments on plasma properties and extrapolated their results through scalability (we'll clarify this point soon in Chapter 6) to cosmological phenomena.

[7] 'Irving Langmuir – Biographical', *Nobel Prize.* See: www.nobelprize.org/nobel_prizes/chemistry/laureates/1932/langmuir-bio.html

[8] 'Kristian Birkeland', *Plasma Universe.* See: www.plasma-universe.com/Kristian_Birkeland

[9] 'Hannes Alfven – Biographical', *Nobel Prize.* See: www.nobelprize.org/nobel_prizes/physics/laureates/1970/alfven-bio.html

This is one difference between mainstream cosmology and plasma cosmology. The latter *can test its hypotheses*, and therefore, elaborate theories. In contrast, mainstream cosmology, with its 'big bang', dark matter, dark energy, matter-generating black holes, three-flavoured neutrinos and neutron stars, creates new *ad hoc* concepts that cannot be tested! These ideas should remain what they are: scientific speculation.

For example, dark matter has never been observed in the universe; it was proposed simply as a way to balance out equations that suggested the universe must have more matter than is readily observable. As retired professor of electrical engineering Donald E. Scott puts it, quoting Princeton University cosmologist Jim Peebles:

> 'It's an embarrassment that the dominant forms of matter in the universe are hypothetical.' I suggest that these dominant forms are in the category of particle known as **Fabricated Ad hoc Inventions Repeatedly Invoked in Efforts to Defend Untenable Scientific Theories** (FAIRIE DUST). [10]

Figure 2: Hannes Alfven, father of plasma cosmology and winner of the Nobel Prize in Physics, 1970. (© Nobel)

The point is not to discard Newton's or Kepler's laws (i.e. gravity) but to acknowledge that other forces are also at work, in particular electromagnetic forces. In some cases, gravitational forces are even negligible compared to electromagnetism, as we are going to see.

[10] Scott, D. E., *The Electric Sky*, p. 40

Chapter 2.

Electromagnetic forces vs. gravitational forces

As we've seen above, according to mainstream science, gravity is the main force controlling the behavior of celestial bodies. Usually, electromagnetic forces are considered nonexistent or negligible at best. However, *electromagnetic forces are stronger than gravitational forces* by a magnitude of 10^{39} (i.e. not 39 times stronger, but 1-followed-by-39-zeros times stronger), making electromagnetism the *de facto* 'driving force' in our universe.

The comparative strength of gravity and electricity was illustrated by the oil drop experiment designed by Robert Millikan,[11] winner of the Nobel Prize in Chemistry in 1923 (see diagram below). Millikan demonstrated that an oil drop charged with only one electron (thanks to X-ray ionization) can be lifted up by the electromagnetic force when exposed to a strong electric field.[12] Thus, the electromagnetic force exerted on one single electron can overcome the gravitational pull exerted by the whole planet on the oil droplet.

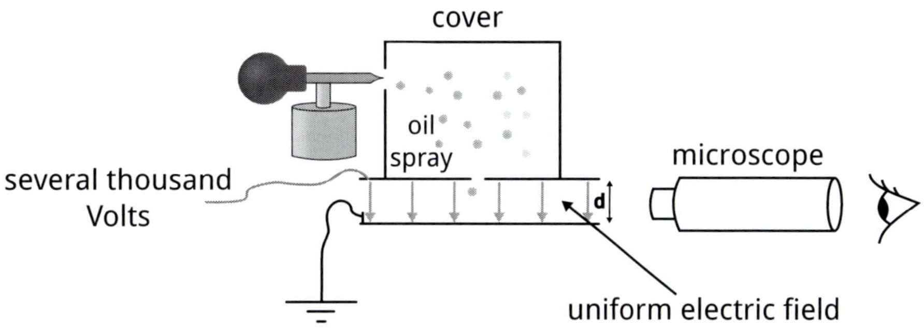

Figure 3: Schematic diagram of Millikan's experiment. The electric pull-up of one single electron balances the phenomenally weak (relatively speaking) gravitational force exerted by the whole planet on the oil drop. (© Theresa Knott)

To be precise, Millikan's pulverized oil droplets are much smaller than oil drops. Typically a droplet is 0.1 microns in radius[13] while a drop is about 1000 microns (1 millimeter). Since there are about 10^{21} atoms in one single drop of water,[14] one droplet contains roughly 10^{17} atoms. So Millikan showed that *the electromagnetic force exerted by one single electron could counteract the weight (i.e. the gravitational force) of 10^{17} atoms.*

The prevalence of electromagnetic magnetic forces over gravity is even more striking when distance increases:

---

[11] 'Robert Millikan – Biographical', *Nobel Prize.* See: www.nobelprize.org/nobel_prizes/physics/laureates/1923/millikan-bio.html

[12] For a droplet to levitate, the electric field has to be 32,100 volts. For more explanations see Chapter 26: 'Hurricanes, lightning and tornadoes'.

[13] Harrison, R. G., 'Atmospheric Electricity And Cloud Microphysics', p. 3

[14] 'How many atoms are in a single drop of water', *MadSci: Chemistry.* See: www.madsci.org/posts/archives/2000-10/971190308.Ch.r.html

> The strength of the magnetic field produced by an electric current (e.g., a cosmic-sized Birkeland current) **falls off inversely as the *first* power of the distance from the current. Both electrostatic and gravitational forces between stars fall off inversely as the *square* of the distance.**[15]

To use a concrete example, when electromagnetic forces are divided by 100 (10^2) because of the distance between two bodies, gravitational forces are divided by 10,000 (10^4) because of the same distance. While gravity might play a significant role along with electromagnetic forces *within* celestial bodies, for distant interactions *between* bodies (star-star, star-planet, star-comet, etc.) *gravitation is generally negligible* and electromagnetic forces become the main player.

[15] Scott, D. E., *The Electric Sky*, p. 44

Chapter 3.

What is plasma?

Before going any further, let's introduce one of the main characters of this book, namely 'plasma', or ionized gas. To understand the electric nature of the universe we first have to comprehend the nature of its prime constituent. Irving Langmuir coined the term 'plasma' after the resemblance of ionized gas to living blood cells. Indeed, the life-like properties of plasma[16] are quite unusual compared to other states of matter:

> At the Berkeley Radiation Laboratory, [David] Bohm began what was to become his landmark work on plasmas. A plasma is a gas containing a high density of electrons and positive ions, atoms that have a positive charge. To his amazement he found that **once they were in a plasma, electrons stopped behaving like individuals and started behaving as if they were part of a larger and interconnected whole. Although their individual movements appeared random, vast numbers of electrons were able to produce effects that were surprisingly well-organized.** Like some amoeboid creature, the plasma constantly regenerated itself and enclosed all impurities in a wall in the same way that a biological organism might encase a foreign substance in a cyst. So struck was Bohm by these organic qualities that he later remarked he'd frequently had the impression the electron sea was 'alive'.[17]

Alive or not, plasma is the most common phase of matter in the universe, both in terms of mass and volume. It constitutes 99% of the visible universe[18] and is therefore far more prevalent than the three other phases: solid, liquid, and gas. All stars are made of plasma, and even interstellar space is filled with plasma. The chart below shows that plasmas occur in all kinds of environments, temperatures and states of matter. Even metals are categorized as plasmas, since they are (solid) matter with **free dissociated electrons**[19] (see top left of figure 4).

Paradoxically, plasma, the most common state of matter, is also the least mentioned and least studied. While students of the hard sciences are taught in detail the properties of solids, liquids and gases, not much is said about plasma. So, let's try to do it some justice.

Plasma is matter (usually a gas, but it can also be a solid or a liquid) where a specific portion of the particles has been ionized. An ionized particle is a particle that has lost one or more of its electrons. So while a 'normal' gas is made of non-ionized particles, a plasma is made of dissociated positive particles and negative electrons.

---

[16] Alfred, J., 'Plasma life forms', *Unexplained mysteries*. See: www.unexplained-mysteries.com/column.php?id=111 062

[17] Talbot, M., *Holographic Model of the Universe*, p. 39

[18] Peratt, A. L., *Advances in Numerical Modeling of Astrophysical and Space Plasmas*, p. 98

[19] *Ibid.* p. 97

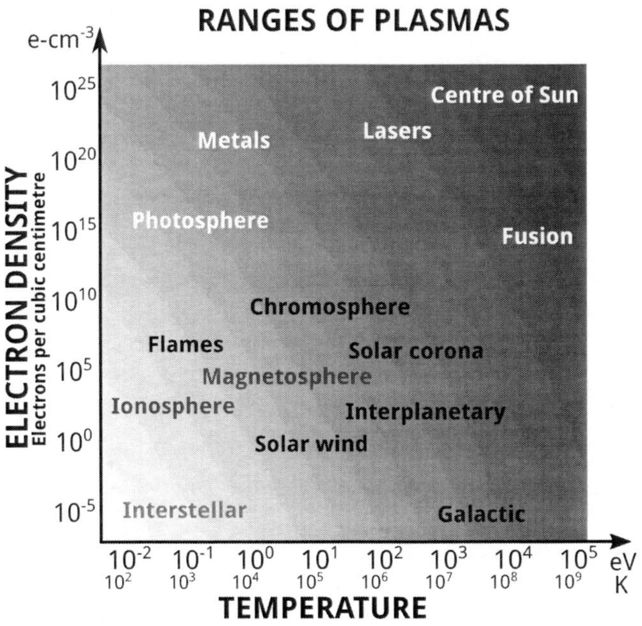

Figure 4: Various kinds of plasma displayed according to their temperature and electron density. (© Wikipedia Commons)

In the above definition, the term 'particle' can stand for either 'molecule' or 'atom'. Let's take the example of an atom of hydrogen (H) – as depicted in figure 5 – which is made of one proton (i.e. the nucleus) and one electron orbiting around the nucleus (see left part of figure 5). If the atom of hydrogen is ionized, the nucleus (proton) will be separated from the electron (see right side of figure 5).

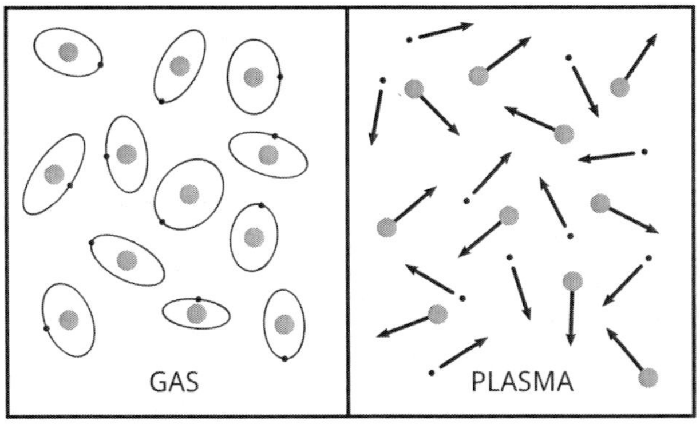

Figure 5: The difference between a gas and a plasma (hydrogen atom).

During ionization, an input of energy expels an electron from an existing atom. This results in one free electron (represented by the black dots in the right diagram) and one positively charged ion (represented by the grey ball in the right diagram). Now the charges are separated and the gas is ionized – it is a plasma.

Chapter 4.

Electric properties of plasmas

Plasmas exhibit very specific electric properties. They are not insulators (which provide very high resistance) like non-ionized gases, nor are they superconductors (which provide zero resistance), but they are **very good conductors**, even better than copper or gold. The typical electrical impedance of plasma is about 30 ohms.[20] For comparison, the typical impedance[21] of copper ranges between 300 and 600 ohms.[22]

Electric current, including plasma current, occurs between two bodies exhibiting different electric charges. In such a case, the body that is positively charged (i.e. likely to receive electrons to balance its charge) is called 'anode', and the body that is negatively charged (i.e. likely to give away electrons) is called 'cathode'.

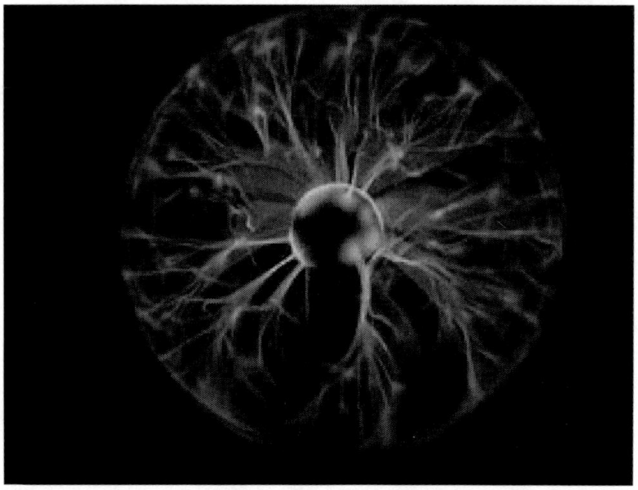

Figure 6: A plasma ball, or plasma globe, displaying electric discharges in gaseous plasma. The gas filling the ball is usually neon.

If the *charge difference* is strong enough – if the distance between the two electrodes (anode and cathode) is small enough and the gas in between is dense enough – then the gas will become ionized (i.e. charge separation will free up electrons). The charges will then start balancing between the two bodies by conducting electrons from the cathode to the anode, or positive ions from the anode to the cathode (or both). This phenomenon is very common. For example, it occurs in a fluorescent tube or in a plasma globe.

In a plasma globe, plasma filaments (i.e. electron and positive ion flows) extend from the central electrode to the outer glass electrode and attempt to balance the charge difference. Keep this plasma globe example in mind because it's a very good analogy for what occurs at stellar and even galactic scales.

[20] Demidov, B. A. *et al.*, 'Anomalous Resistance And Microwave Radiation From A Plasma In A Strong Electric Field', *Soviet Physics* (August 1965) 21(2)

[21] Impedance is a form of resistance. The higher the impedance, the lower the conductivity.

[22] Cebik, L. B., 'Some (Old) Notes on Home-Brew Parallel Transmission Lines', *QSL*. See: www.qsl.net/co8tw/openline.htm

Chapter 5.

Discharge modes

Plasma exhibits different modes of discharge depending on the *current density* (amperes per square meter) that crosses it. When the current is weak, the discharge occurs in **'dark mode'**, which means that no visible radiation or light is emitted. This is what happens with dark asteroids, dark stars or in interstellar space, for example (as we'll see when we discuss scalability in the next chapter). The plasma is simply traversed by a current too weak to make it glow.

When the current density increases, the plasma starts glowing. This is what occurs in a neon light, in a comet (i.e. a glowing asteroid) or in the Sun's corona (the bright 'atmosphere' of the Sun). This is called **'glow mode'**.

If the current increases even more, plasma enters the **'arc discharge mode'**, producing a sudden and violent electric discharge. This is what occurs in lightning or during arc-welding. It's also the phenomenon observed in the plasma globe mentioned above. It can also occur on comets leading to explosive brightening and/or break-up as happened famously to Comet Shoemaker-Levy,[23] and other comets before and since.

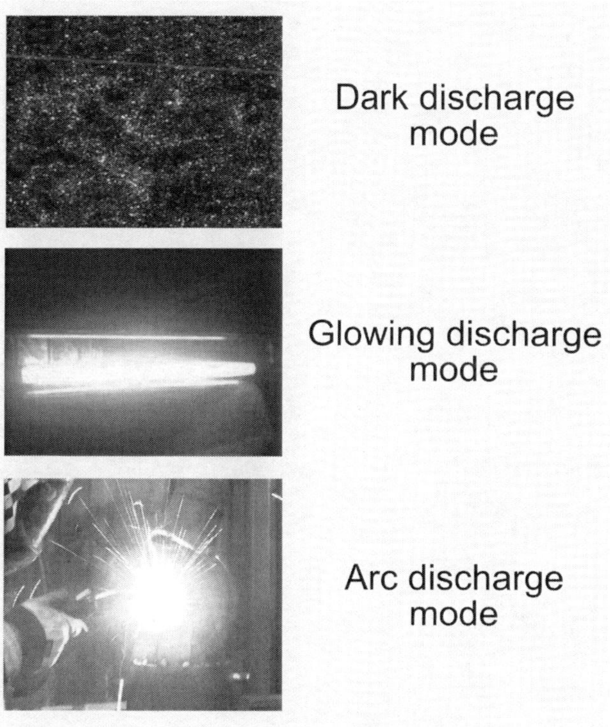

Figure 7: From top to bottom: dark discharge mode (interstellar space), glowing discharge mode (neon light), arc discharge mode (arc welding). (© Sott.net)

Figure 7 shows the three discharge modes exhibited by plasma.

In conclusion, plasma displays three modes of discharge depending on the current density that crosses it.

[23] More details about this event in Chapter 18: 'Comet or asteroid?'

Chapter 6.

The scalability of plasma

A very interesting property of plasma is its *scalability*. This means that plasma exhibits similar behaviors and properties no matter how big or small the example is: in the laboratory or in the cosmos. As a matter of fact, plasma occurs in a wide range of sizes beyond that of the atomic scale (i.e. an electron and its nucleus dissociated from each other). This range of sizes spans from about 10^{-10} meters in diameter to the galactic scale, which, in the case of our own Milky Way, is about 10^{20} meters in diameter.

This means that **plasma spans a magnitude of 30 orders** (10^{30}, or 10 followed by 30 zeroes) as far as size scale is concerned. Plasma exhibits similar properties along the whole of this range. The picture above illustrates the analogy between microscopic plasma (atomic scale) and macroscopic plasma (solar system scale).

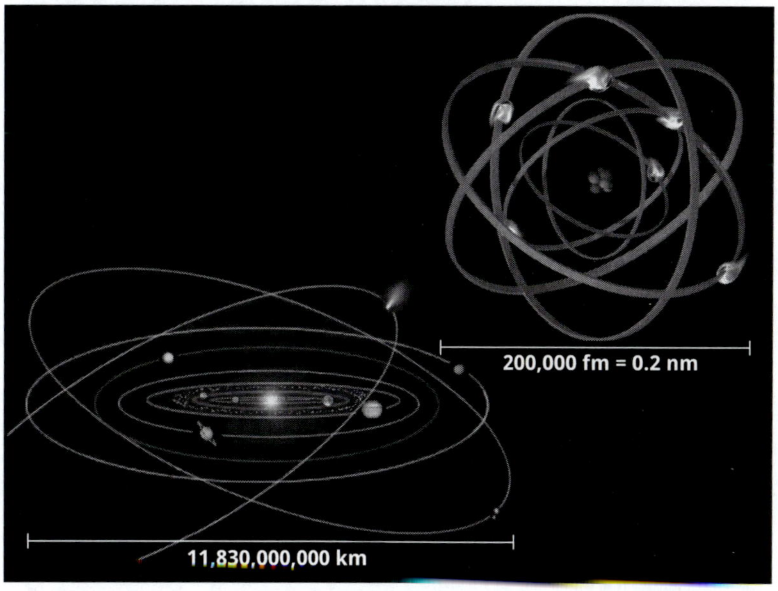

Figure 8: Artist's rendition of the solar system–atom analogy.

Because of this broad scalability, plasma cosmologists can make observations, state hypotheses, and most importantly, test them in their labs (i.e. small scale) and through 'similarity transformations' apply their results to plasmas that are many magnitudes larger (cosmic phenomena, for example). From this perspective, plasma lab experiments applied to cosmological events are not unlike testing aircraft or studying natural turbulent flow in wind tunnels with smaller-scale models, and then applying the results to real-sized objects.

This testability allows for making predictions and testing of theories with various concrete experiments, which is an essential aspect of the scientific method, as stated by philosopher of science Karl Popper:

> The criterion of the **scientific status of a theory is its falsifiability**, or refutability, or **testability**.

Chapter 7.

The 'insulating bubble'

Besides its very low electric resistance, another characteristic of plasma is its ability to create a sort of 'insulating bubble' around charged bodies:

> Irving Langmuir discovered that one of the most important properties of plasma is its **ability to electrically isolate one section of itself from another**. The isolating wall is made up of two closely spaced layers, one consisting of positive charges and the other of negative charges. Langmuir called it a double sheath. Today it is called a *double layer* (DL).
>
> Because plasmas are excellent conductors, there will not be a significant voltage drop across them while they are carrying current. **If there is a significant voltage difference between two locations within a plasma, a DL will form between them and most of the voltage difference will be contained in it.** In other words, a DL is where the strongest electric field in any plasma will be found. Plasmas have an almost magical ability to isolate themselves from foreign intruders.[24]

Note that when the electric potential of a celestial body becomes equal to the potential of the surrounding plasma, the insulating bubble disappears. This is the case with our Moon, which is devoid of any insulating bubble.[25]

Figure 9 shows a slice-through image of a charged body (left) and its surrounding DL (right). The DL exhibits **three specific sections**. The middle section accounts for most of the DL (volume wise). The electric potential of this thick region is relatively uniform, creating a very weak electric field. As a result, electric current is relatively limited; hence the DL's insulating properties. Most of the *electric potential difference* occurs at the extremities; where the DL approaches the charged body (left) and the surrounding space / plasma (right).

The **DL dramatically limits the discharge of the body**. Without the DL, the body's surface would be in direct contact with the surrounding space, which exhibits a much different electric potential. This potential difference would lead to a quick and total discharge of the body. From this perspective, celestial bodies are quite similar to spherical *capacitors*.

A capacitor,[26] also known as a condenser, is simply two electrodes separated by an insulator called 'dielectric' material. This material can be air, paper, wood, glass, etc. When the capacitor is connected to a source of electric current like a battery, the charges slowly build up in the electrodes

[24] Scott, D. E., *The Electric Sky*, p. 74

[25] Note that while our Moon is currently devoid of any magnetism, it was not always the case. The Moon's surface does bear remnant magnetism. The rocks returned to Earth by our Moon landings showed evidence of this magnetism. Unfortunately, the orientation of the rocks prior to removal was not recorded. See *ibid.*, p. 214

[26] Capacitor phenomena were first described in 1745 by German physicist Ewald Georg von Kleist.

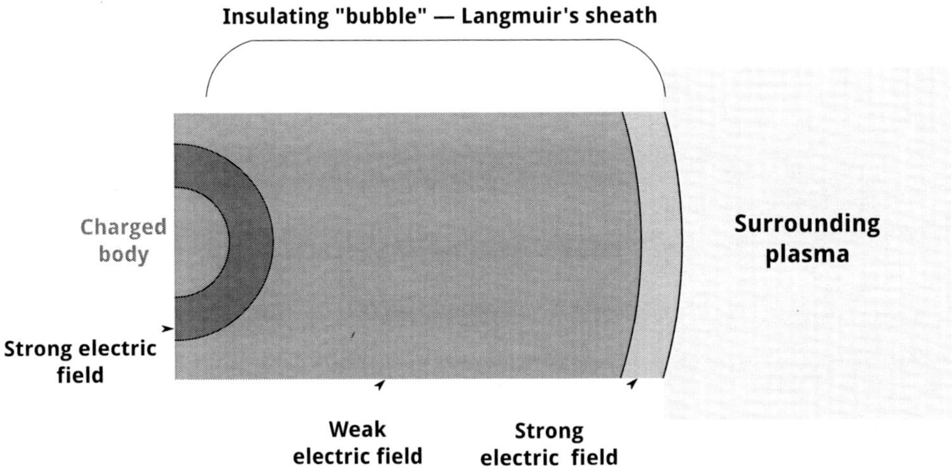

Figure 9: The insulating bubble surrounding an electrically charged object. (© Sott.net)

Figure 10: A 19th-century parallel plate capacitor. The air between the two plates plays the role of the dielectric. (© Wikimedia Commons)

and the electric field between the two electrodes increases. At this point there is hardly any current flowing between the electrodes. When the electric field reaches a critical value (known as 'breakdown voltage'), an electric discharge occurs and a strong current suddenly flows between the two electrodes. Then the slow charge build-up restarts.

The Taser or stun gun is a typical application of capacitors, where a 9-volt battery charges a capacitor for several seconds, after which it can deliver a very short high voltage electric discharge (up to 150,000 volts).[27]

If we think of a charged celestial body as a capacitor, then one electrode is the celestial body itself, the other electrode is the outer layer of the 'insulating bubble' or DL, and the capacitor's dielectric (the insulating material) is the plasma contained within the 'bubble', also known as the *plasmasphere*.[28]

The Sun has its own DL, namely the heliosphere. The Earth's DL is called the ionosphere. All planets of the solar system are embedded within the Sun's heliosphere while being surrounded by their own insulating bubbles. Figure 11 shows the Earth's DL (the ionosphere) embedded within the Sun's DL (heliosphere) and the heliopause, which is the outer layer of the Sun's 'bubble'.

While the existence of the electric nature of the Earth's ionosphere was recognized a long time ago, it was only in 1994 that the electric properties of the *heliosphere* were finally observed:

> Voyager 1 reached the 'edge of the solar system' in December 2004 in a region 94 times more distant from the Sun than the Earth. It is a region where it was expected that the 'supersonic solar wind' would suddenly slow down when it met the interstellar medium at what is called the 'termination shock'. This fluid-mechanical analog is vastly different from the electrical model, in which the Sun is a 'unipolar' corona discharge. That is, the Sun forms the physical anode (positively charged object) and the plasma itself performs the function of the cathode (negatively charged object). High voltage transmission lines on Earth undergo a similar 'coronal discharge' into the surrounding air, which forms a 'virtual cathode' (not an object, but a region of charge).
>
> In this model, almost the entire voltage difference between the Sun and its galactic environment, estimated to be tens of billions of volts, will be concentrated at the virtual cathode boundary with interstellar space. So it is only to be expected that the data from Voyager 1 is anomalous for the magnetohydrodynamic 'shock' model while it seems consistent with the electrical model. In the electrical model, Voyager 1 has entered the 'Faraday dark space' region of the solar discharge where the solar electric field is reversed. That would account for the pile-up of the protons in the solar wind and the steady increase in anomalous cosmic rays from a more distant region.[29]

As shown in figure 11, the heliosphere is not spherical but ellipsoidal in shape. This shape is due to the fact that the whole solar system, including the heliosphere, is orbiting around the core of our galaxy. The speed of the Sun relative to the center of the Milky Way is estimated at 220 km/s.[30]

[27] Harris, T., 'How stun guns work', *How Stuff Works*. See: electronics.howstuffworks.com/gadgets/other-gadgets/stun-gun3.htm

[28] In the rest of the book we will interchangeably use the words plasmasphere and magnetosphere. The plasmasphere is the ionized region around a celestial body, while the magnetosphere is the magnetized region around a celestial body. While these are often overlapping concepts, this is not always the case. For example, Venus has little if any magnetic field, but it does have a large plasmasphere. See: Scott, *op. cit.*, p.105

[29] Thornhill, W. & Talbott, D., *The Electric Universe*, pp. 44–45

[30] 'Spiral Galaxies', *University of Alberta*.

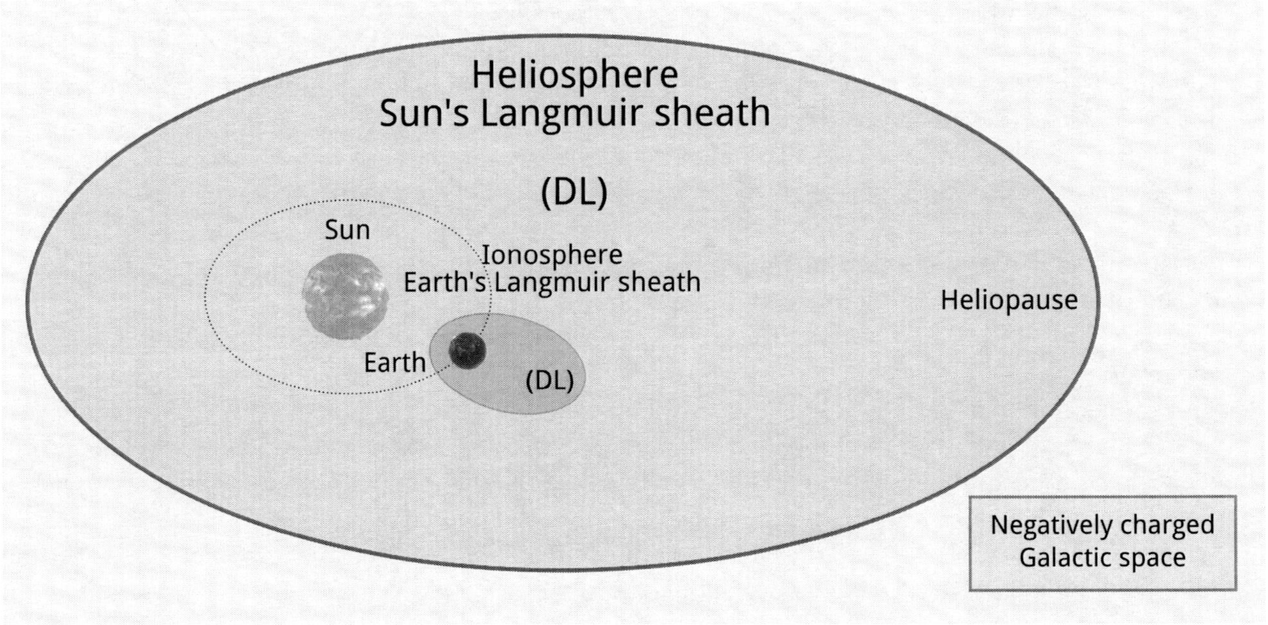

Figure 11: Depiction of the Earth and its insulating bubble (the ionosphere) embedded in the heliosphere, the Sun's own insulating bubble. Not to scale. (© Sott.net)

The solar system is travelling to the left; therefore the left side of the heliosphere is 'compressed',[31] while the right of the heliosphere is elongated.

For many years, the shape of the heliosphere was a controversial topic. Some scientists claimed it was a sphere with the Sun at its center, while plasma cosmologists claimed that it was an ellipsoid, like the Earth's ionosphere or a comet's coma, because of the very same electromagnetic phenomena at work (remember scalability). It's only recently that the ellipsoid shape of the heliosphere was confirmed by NASA's Interstellar Boundary Explorer.[32] (See figure 12.)

Note that, unlike the Sun and its planets, most moons in the solar system, including our Moon, don't have a DL or plasmasphere of their own. Their electric potential is equal to the electric potential of the surrounding space.

However there is at least one exception. Ganymede (see figure 13), Jupiter's 3rd moon, and the biggest moon in the solar system, does show the presence of a plasmasphere,[33] i.e. a DL. This might be due to the fact that Ganymede is a recently captured, electrically active body that interacts with Jupiter, the most electrically active planet in the solar system.

See: www.ualberta.ca/~pogosyan/teaching/ASTRO_122/lect24/lecture24.html

[31] In a similar way the Earth's insulating bubble is compressed on its sunward side because of the solar winds it is exposed to; it also exhibits an ellipsoidal shape. This idea is developed further in Chapter 28: 'Jet streams'.

[32] The Interstellar Boundary Explorer (IBEX) is a $169-million spacecraft owned by NASA. IBEX observations were made public in July 2013 and showed what is now called the 'heliotail'. See: www.nasa.gov/content/nasa-s-ibex-provides-first-view-of-the-solar-system-s-tail/#.UeQLO23oDGo

[33] Kivelson, M. G., *et al.*, 'Ganymede's magnetosphere: Magnetometer overview.' *Journal of Geophysical Research: Planets (1991–2012)* (1998) 103(E9): 19 963–19 972.

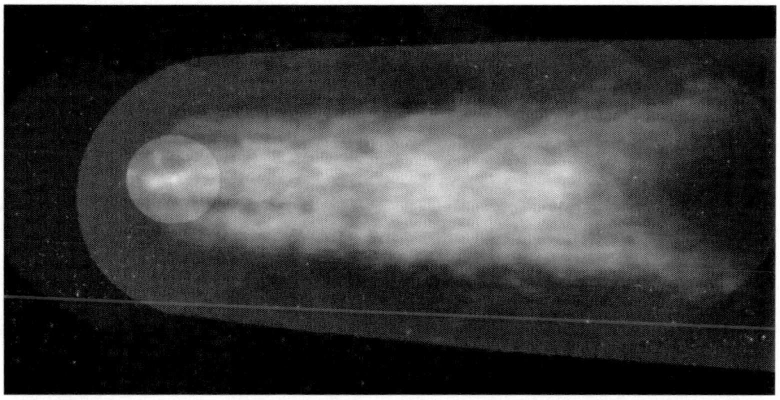

Figure 12: Representation of the Sun's ellipsoidal insulating 'bubble'. (© NASA / IBEX)

Figure 13: Jupiter's moon Ganymede. (© NASA)

Chapter 8.

Electric charge of the Sun, the Earth and their DLs

As noted above, most of the universe is made of plasma. This also applies to the solar system. Therefore, in this ionized environment, electric charges are almost everywhere. In this chapter we will try to better understand the *relative* electric charges of various celestial bodies (comets, moons, planets, stars, galaxies) in their cores, on their surfaces, and in their DLs (double layers).

Keep in mind that 'relative charge' does not mean 'absolute charge'. In other words, saying that A is more positively charged than B doesn't necessarily mean that the overall charge of A is positive in an absolute, universe-wide sense. It only means that it is more positive than B, or less negative than B with which it has a relationship or is 'relative to'.

Ultimately, these relative charges are what really matter because it is the difference in relative charges that causes electric currents, whatever the absolute charges (positive or negative) are. Since our objective here is to better understand the differences between surface charge, DL charge, core charge, etc., we will focus on relative charge.

As a rule of thumb, most celestial *bodies* have a charge that is negative overall,[34] and those bodies are generally surrounded by an even more negatively charged DL, which in turn is surrounded by an even more negatively charged galactic or interstellar plasma. Applying this to our Sun, it follows that **in the solar system, the Sun is the most positively charged body – relatively speaking – although its absolute charge is negative**, but less negative than the planets, the comets, the heliosphere, and galactic space that surround it. Therefore, planets and comets can be considered negatively charged bodies *relative* to the Sun.

Figure 14 is the same as figure 11, except that we have added the *relative* electric charges of the Earth, the Sun and their respective DLs.

At the level of the Sun, the relative electric charges are as follows: the core of the Sun is more positive than the surface of the Sun. The Sun (surface and core) is more positive than its 'bubble' (heliosphere), which encompasses the Earth and all the other planets of the solar system. The Sun and its heliosphere are more positive than the surrounding galactic plasma.

Concerning Earth, we can see that, like the Sun, its core is more positive than its surface. The Earth (surface and core) is more negative than its 'bubble' (ionosphere). The Earth and its ionosphere combined are more negatively charged than the surrounding plasma (heliospheric plasma).

In both cases the electric charges follow a gradient. For example, concerning the Sun: while moving away from the center of the Sun (from core to surface, heliosphere, heliopause, galactic

[34] De Grazia, A. & Milton, E., *Solaria Binaria*, p. 29

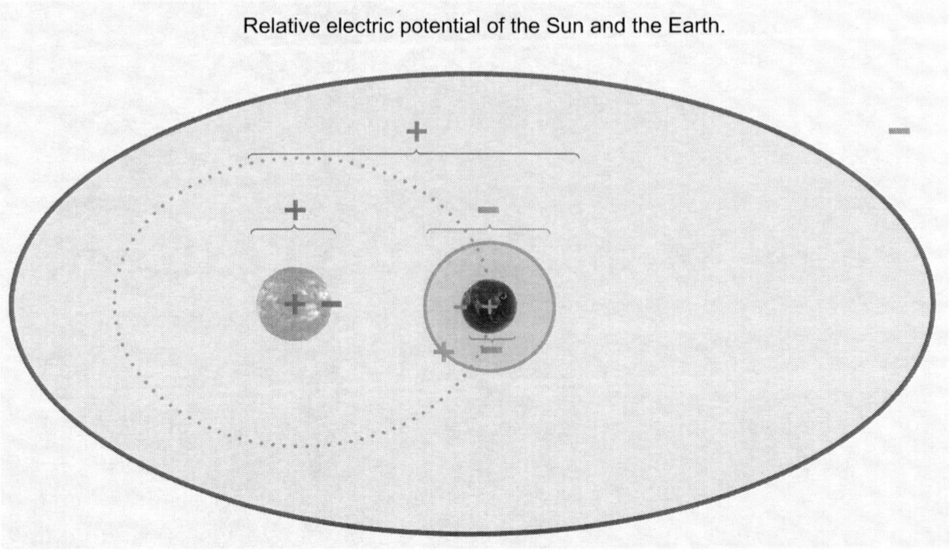

Figure 14: Relative electric charges between and within the Sun and the Earth (notice that a body can be ascribed a relative positive charge while its absolute charge is negative). (© Sott.net)

space) the electric charge becomes increasingly negative:

Electric potential of the Sun > electric potential of the heliosphere > electric potential of the galactic space.

Conversely, for the Earth, the charge becomes more and more positive as it moves away from the core:

Electric potential of the Earth < electric potential of the ionosphere < electric potential of the surrounding space.

Note, however, an apparent discrepancy: the Sun exhibits a (relatively) negative electric charge on its surface, while its overall charge is (relatively) positive.

The Sun acts as a generator. At the Sun's surface, positively charged protons are carried away by the 'solar winds'[35] towards the outer layer of the heliosphere, while electrons drift back and accumulate on the Sun's surface. These two factors explain the negative charge of the Sun's surface relative to its core.

In contrast, the Earth doesn't act as a generator. It is powered by the Sun, which maintains Earth's positively charged ionosphere. Since charges of opposite polarity attract each other, the positively charged ionosphere attracts electrons present on the Earth's surface, hence the negative electric potential of the Earth's surface relative to its core.

The local electric charges described above (surface and core charges) are average values (average surface charge, average core charge). However, the cores or the surfaces of celestial bodies do not exhibit the same electric charge everywhere. This means, for example, that although the surface

[35] The term 'solar wind' is misleading. Wind is considered a purely mechanical process describing a flow of air due to differences in air pressure. In contrast to air wind, solar wind is mostly *an electric process describing the flow of charged particles being driven by electrical fields.* Since solar wind is a popular term, we'll keep using it throughout the rest of this book, but keep in mind its limitations.

of Earth is overall more negative than the atmosphere, locally it can be more positive in particular regions. This can lead to various types of electric discharge phenomena.

Lightning is a local phenomenon that balances charges in the regions where it occurs. These local charge imbalances are the reason why we observe cloud-to-ground lightning (the most prevalent type where the ground is more positive than the clouds), but also ground-to-cloud lightning (when the ground is more negative than the clouds) and cloud-to-cloud lightning (when two clouds exhibit very different electric potentials).

These sudden and massive electric discharges enable a balancing of charge between two locations that exhibit strong local positive and/or negative charges. We'll cover this topic in more detail later.[36]

Figure 15: Cloud-to-cloud lightning balancing electric potentials between two atmospheric regions. (© Wikipedia)

Returning to the plasma globe analogy, when you touch the surface of the globe, a thick plasma filament forms between the central electrode and the area of the globe you are touching. In a similar manner, when a mosquito flies between two wires of a bug zapper, it triggers an electric discharge. In both cases the foreign object (finger, bug) increases the local conductivity, sets up a path of least (electrical) resistance and enables a discharge to occur through it.

While the **DL acts as an electric insulator, dampening the discharge between the body and the surrounding plasma, the DL is not a perfect insulator. As with any kind of capacitor, electric current still flows through the insulator, whether in the form of slow leakage current or more intense discharges**, hence the three types of discharges described above (dark, glow, and arc modes). *The presence of charged objects (e.g., comets, planets, etc.) in the DL of a celestial body is the main trigger of massive discharges.* We will address this point further below.

[36] See Chapter 26: 'Hurricanes, lightning and tornadoes'

Chapter 9.

External power sources of celestial bodies

The reason capacitors can repeatedly discharge, and yet maintain an electric potential difference between their anodes and cathodes, is because **they are plugged into an external source of energy**. So where does all the electricity in our solar system come from? University of Michigan astrophysicist Michael J. Longo has thoroughly studied more than 40,000 galaxies.[37] Following numerous steps in data treatment and analysis – I'll spare you the details of his calculations, which you can check in his paper – his conclusion was the following:

> The worrisome alignment of the equinoxes and ecliptic with the AE [Axis of the Ecliptic] is now seen as an accident due to the ecliptic's definition along *RA* [right ascensions] =180° and 0°, near the Galactic poles. It is not a sign of a serious foreground bias in the WMAP data. **All the alignments can be explained with a cosmic magnetic field that aligns electrons' cyclotron orbit axes and imprints its multipoles on the CMB.**[38]

In simpler terms, what Longo concluded was that *the rotation axes of galaxies are aligned along the same curve* and that this alignment could not be due to random chance. Along with Longo, Alfven,[39] Campanelli[40] and Schwarz[41] have all strongly suggested that the alignment of galactic rotation axes is due to *a giant ring of electric current.* Although the ultimate source of this giant electric ring circling the cosmic 'void' is still unknown, the alignment of the galaxies' rotation axes is indirect proof of its existence. Figure 16 depicts this alignment of galactic rotation axes on a ring-shaped intergalactic current (medium grey color).

This giant ring of Birkeland current (see next chapter: 'Interstellar plasma') may be the external **source of energy that powers galaxies**. It may also explain the 'handedness' addressed by Longo in his paper, by **electrically inducing a specific direction and speed of rotation in the galaxies** located along its current.[42] As shown in figure 17, *intra*galactic space seems to follow the same process as *inter*galactic space.

The galactic jet (central vertical jet in the center of figure 17) is a massive Birkeland current coming from the intergalactic current ring (the one described above that aligns galaxies with each other), which enters the galaxy through its core and spreads through the galactic arms. Just as

[37] Longo, M., 'Does the Universe Have a Handedness?', *arXiv:0812.3437 [astro-ph]* (2008)

[38] ccapp.osu.edu/workshops/GLCW8/glcw8/talks/mLongo.pdf

[39] Alfven, H., *et al.*, 'Interstellar clouds and the formation of stars', *Astrophysics and Space Science* (1978) 55(2): 487–509

[40] Campanelli, L., *et al.*, 'Ellipsoidal universe can solve the cosmic microwave background quadrupole problem', *Phys. Rev. Lett.* (2006) 97(13): 131 302

[41] Schwarz, D., 'Is the low-l microwave background cosmic?', *Astrophysics Journal* (24 November 24 2004)

[42] See Chapter 12: 'Homopolar motors'

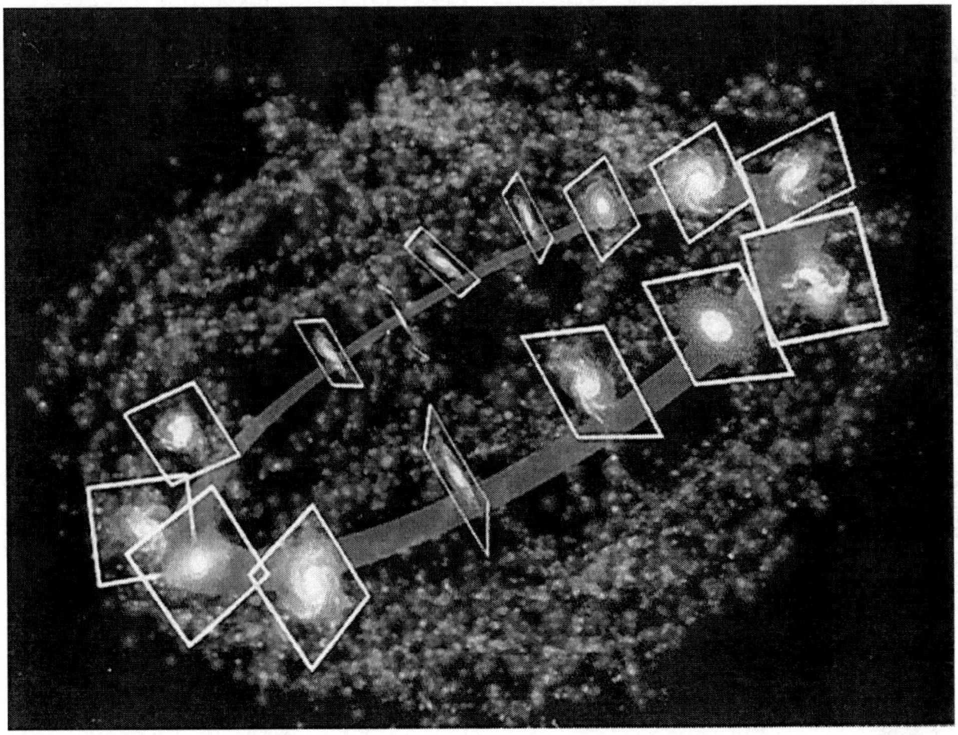

Figure 16: Galaxies located on the shells of the largest cosmic voids exhibit rotation axes aligned with the electric current that surrounds the voids. (© Trujillo, Carretero, Patiri)

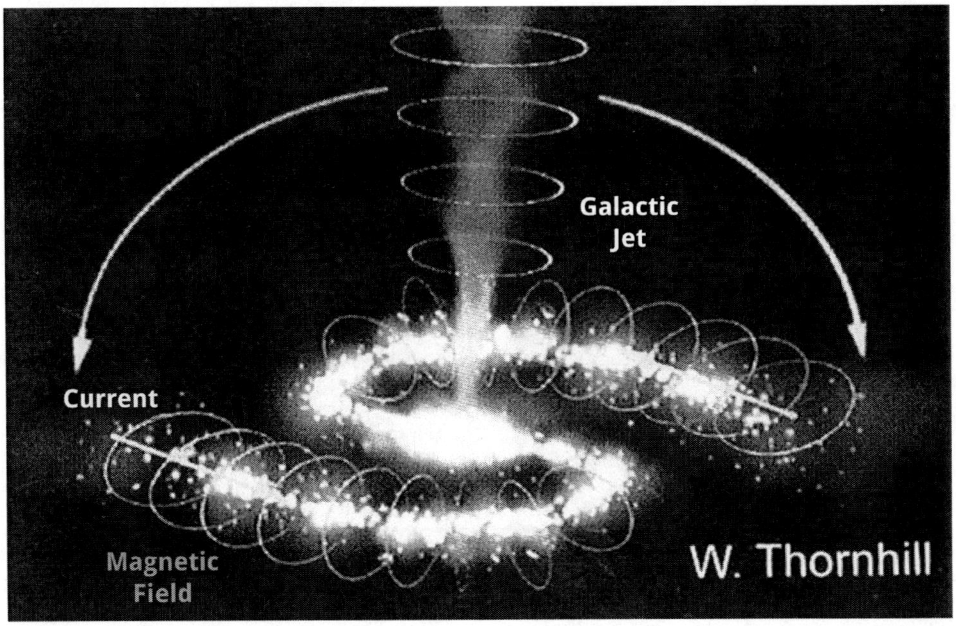

Figure 17: Electric currents entering the galactic core and traveling through galactic arms. (© W. Thornhill)

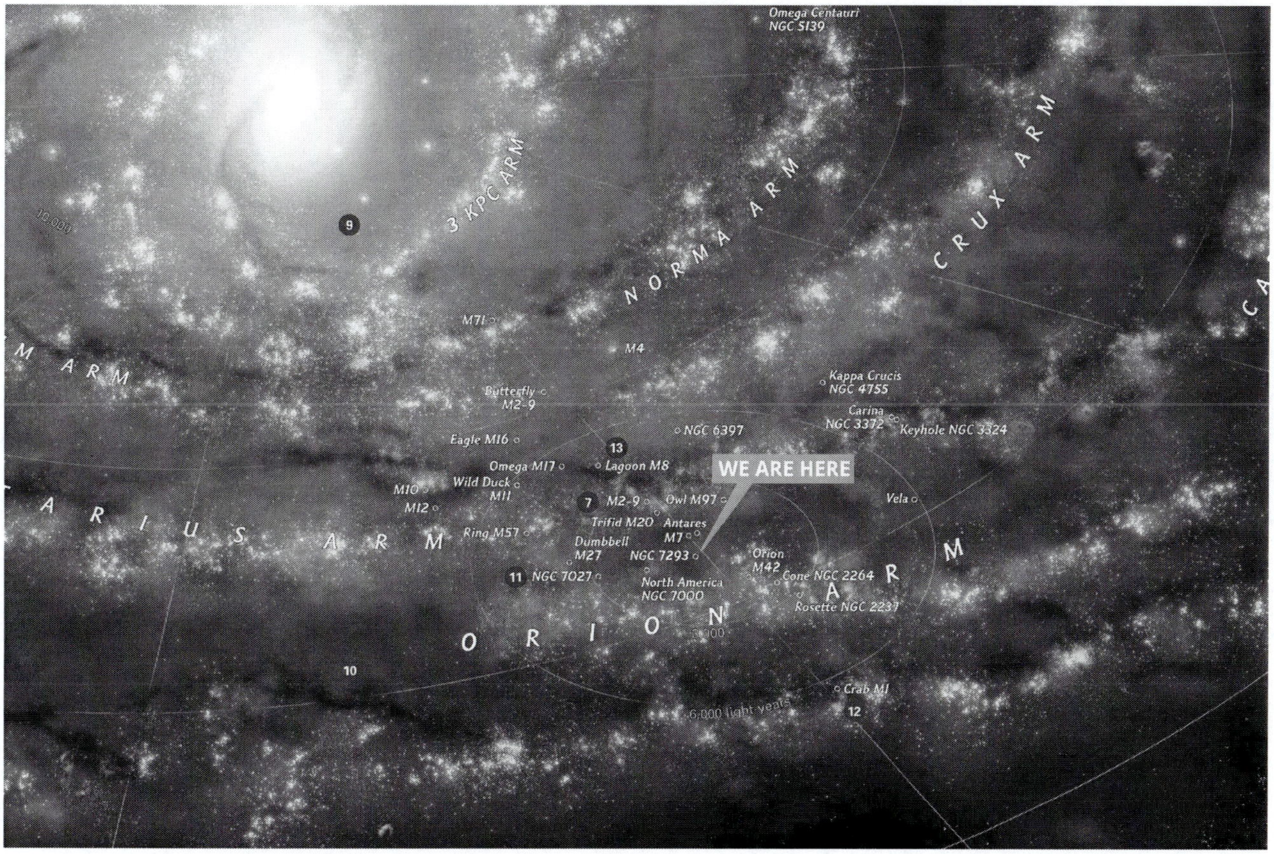

Figure 18: In the Milky Way, like in other galaxies, stars are aligned along the galactic arms. The caption reading "WE ARE HERE" shows the location of our solar system. (© National Geographic)

the current between galaxies may explain their alignment and same rotation direction, the current *within* galaxies may explain why most stars are located within galactic arms, i.e. along the galaxies' own electric current.

This is the case with our own galaxy, the Milky Way, where most stars are grouped along galactic arms as depicted in figure 18. It also applies to our solar system, which is embedded in one arm of the Milky Way, namely the Orion arm[43] (bottom right part of figure 18).

In addition to aligning galaxies along the 'intergalactic ring' and making them spin, as well as grouping the stars within galaxies along galactic arms, Birkeland currents may also be the external electric source that powers the stars themselves[44] and makes them spin. In their turn, stars may be the external electric source that powers planets and makes *them* spin.

These remain hypotheses only. Today the external electric source that powers celestial bodies is still unknown. Even plasma cosmologists have no definite answer to this question:

[43] Byrd, D., *'Which spiral arm of the Milky Way contains our Sun?'*, Earthsky. See: earthsky.org/space/does-our-sun-reside-in-a-spiral-arm-of-the-milky-way-galaxy

[44] Scott, D. E., *The Electric Sky*, p. 85

> Studies of magnetic fields in the spiral arms of galaxies show that electrical currents flow along the arms in the form of spiraling Birkeland filaments. **Ultimately, we don't know where the power comes from.**[45]

From the observations listed above, it seems that **celestial bodies are electrically powered by external electric sources in a cascading fashion**. The intergalactic void powers the galaxies, aligns them and makes them spin. In turn, the galaxies align the stars, power them and make them spin. Finally the stars power the planets and make them spin.

If this hypothesis is correct, it means that stars – including our Sun – do not 'run out of fuel' in the way that the mainstream solar hypotheses suggest.

So, for our solar system, the Sun powers the planets – including the Earth – and makes them spin. Notice also that all planets are located on the same plane (the plane of the ecliptic) just as most stars are located on the galactic plane.

In sum, the very same process seems to repeat itself at all different scales, like a fractal. A greater entity electrically powers a smaller one, while the latter powers an even smaller one. At every scale, we notice the presence of Birkeland currents crossing the space plasma, surrounding celestial bodies (galaxies, stars, and planets) and directing this aligning, orbiting and spinning cosmic ballet.

[45] Thornhill, W. & Talbott, D., *The Electric Universe*, p. 61

Chapter 10.

Interstellar plasma

Until recently space was considered totally empty, a perfect vacuum. This position is still widely held, although it is not entirely true, as I've suggested in previous chapters. **Space is not empty. It is filled with plasma.** This space plasma is mostly constituted of very light molecules: hydrogen and helium ions plus electrons, and its concentration is approximately one (ionized) particle per cubic centimeter.[46] For comparison, atmospheric air concentration is about 10^{13} particles per cubic centimeter.

Figure 19: A Birkeland current crossing 'empty' interstellar space.

As shown by the Birkeland current in figure 19, which crosses light years of 'empty' outer space, the very low concentration of space plasma doesn't prevent electric phenomena from occurring. Remember the Millikan experiment and how the electromagnetic force exerted by one single electron affected a wide portion of the space surrounding it? On cosmic scales, the electric properties of space plasma allow electric currents to flow between celestial bodies because such plasma is highly conductive. This enables electric exchanges between the surface of a celestial body and the outer layer of its DL, as well as anything within the DL.

According to Hannes Alfven and James McCanney, plasma in space is quasi-neutral or slightly positive. However, there is some controversy surrounding the electric charge of the solar wind. While the official position is that solar wind is electrically neutral, British mathematician and geophysicist Sydney Chapman stated as early as 1930 that it was comprised of positively charged plasma. More recently, physicist Luis Alvarez[47] argued that the solar wind exhibits an overall positive electric

[46] Tsytovich, V. N., *Elementary Physics of Complex Plasmas*, p. 7

[47] Luis Alvarez (1911–1988), University of Berkeley researcher, winner of the Nobel Prize in Physics, 1968.

charge.[48] Jean Martin Meunier[49] also states that the solar wind is not electrically neutral and demonstrates it as follows:

> The solar wind is electropositive in the whole; it contains much more protons h+, than electrons. Why? Because electrons are evacuated in the galactic space, with speeds of 10,000 to 300,000 km/s by the solar UV, X and gamma rays (Compton effect). Consequence: the solar wind (speed of 300 to 900 km/s) is a current of protons trying to re-equilibrate the loss of electrons.[50]

---

[48] Trower, W. P., *Luis Walter Alvarez – A biographical memoir*, p. 7

[49] Ex-CNRS researcher and ex-secretary of the French section of IAGA (International Association of Geomagnetism and Aeronomy).

[50] Nodon, A., *'Prévisions météo d'après les taches solaires'.* See: albert-nodon.e-monsite.com/pages/recherche-au-20-siecle/previsions-meteo-d-apres-les-taches-solaires/

Chapter 11.

Currents in plasma

Remember the plasma globe and the glowing filaments connecting the central electrode to the outer plastic layer of the globe? That's a typical plasma current discharge. But why do plasma currents take such a filamentary shape? To understand this phenomenon we have to remember our high school physics, in particular the lesson on electromagnetism and how an electromagnetic field is generated by a live electric wire.

In figure 20, we can see that the electric current in the wire (three grey horizontal arrows) generates magnetic loops (black circular arrows) that are perpendicular to the electric current. Similarly, the electric current in plasma will generate magnetic loops that are perpendicular to it. However, unlike a rigid copper wire, in most cases plasma is *fluid*. Because of this, **magnetic loops constrain the plasma discharges into the shape of filaments** (figure 21). Thus, the magnetic field gives a filamentary shape to the very current that generates it. In this sense, an electric current in plasma creates the magnetic field that then serves to channel or 'pinch' it.[51] Another way to say this is that the pinch is the compression of an electrically conducting filament by magnetic forces. These **filamentary plasma currents are also called Birkeland** currents.[52]

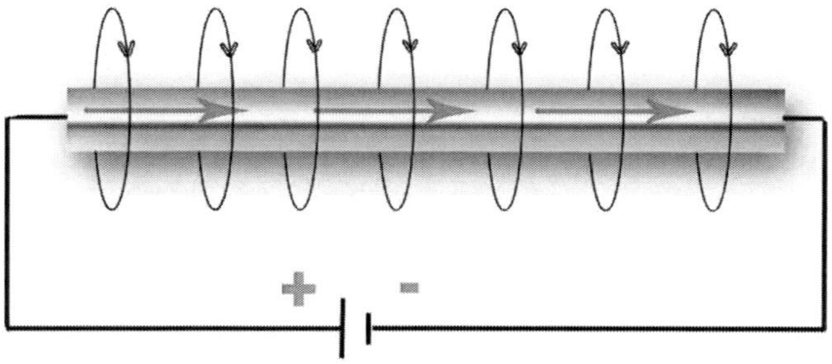

Figure 20: Magnetic field generated by an electrical current crossing a straight wire. (© physick.wikispaces.com)

[51] Pinches are created in the laboratory in equipment related to nuclear fusion. Pinches may also become unstable and generate radiation across the electromagnetic spectrum, including radio waves, x-rays and gamma rays, and also neutrons and synchrotron radiation. There are different kinds of pinches including theta pinch, the screw pinch and the Z-pinch. The name refers to the direction of the current in the devices, i.e., the Z-axis on a mathematical diagram. Any machine that causes a pinch effect due to current running in that direction is correctly referred to as a Z-pinch system, and this encompasses a wide variety of devices used for an equally wide variety of purposes including fusion power research. Pinches are used to generate X-rays, and they have applications to particle beams including particle beam weapons, and astrophysics.

[52] Kristian Birkeland had written in 1913 that what is now called the 'solar wind' generates currents in space that cause the auroras. Birkeland's theory was disputed at the time by the British geophysicist and mathematician Sydney Chapman, a senior figure in space physics, who argued the mainstream view that currents could not cross the vacuum

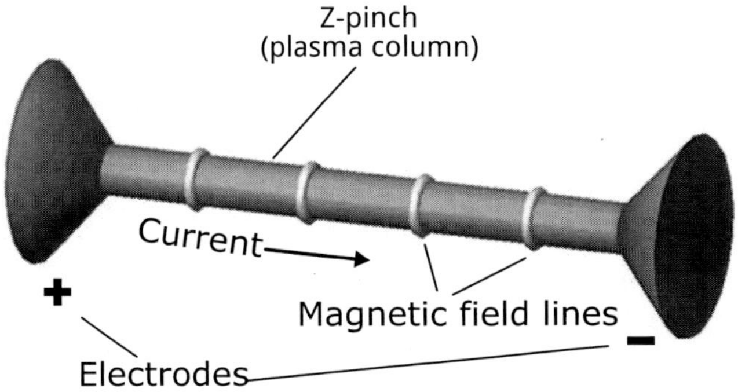

Figure 21: Magnetic field lines (four light grey rings) 'pinch' the Birkeland current into long filaments (central medium grey cylinder).

Now that we know how a single filamentary plasma current, or Birkeland current, behaves, let's see what happens to two nearby plasma filaments, as depicted in figure 22. At first, the magnetic fields generated by each filament attract each other and tend to merge. This electromagnetic interaction causes the filaments to draw closer together (top part). Then, the spinning magnetic fields cause the filaments to rotate around each other (bottom part). This is called a **plasma vortex**.

Notice that the two filaments are magnetically drawn towards each other at first, but once they are close enough, a force of repulsion is generated that holds them apart. What is happening is that the particles in the plasma are pulled toward each other by what is called the Lorentz force (see next chapter), thus the plasma contracts. But then, the contraction is counteracted by the increasing gas pressure of the plasma. The attractive and repulsive forces act together to create a **very stable configuration where the two filaments hold at a specific distance from each other**. They can't merge, but they can't separate either.

Keep these filamentary spiraling shapes and spinning movements in mind, because we will encounter numerous similar occurrences in nature in the coming chapters (e.g., whirlwinds, hurricanes, galaxy shapes, comet tails, solar winds, star and planet spins, etc.). For instance, Anthony Peratt[53] used the spiraling of plasma filaments observed under laboratory conditions to explain the formation of galaxies (figure 23).

of space and therefore the currents had to be generated by the Earth. However, in 1967 Birkeland's theory, referred to previously as 'fringe', was proved correct thanks to the data collected by U.S. Navy satellite 1963-38C. These magnetic field–aligned currents are now named Birkeland currents in his honor.

[53] Anthony L. Peratt is a leading plasma physicist. He's the author of a foundational book titled *Physics of the Plasma Universe*. Peratt is currently investigating archaeological evidence for major space plasma events in prehistory.

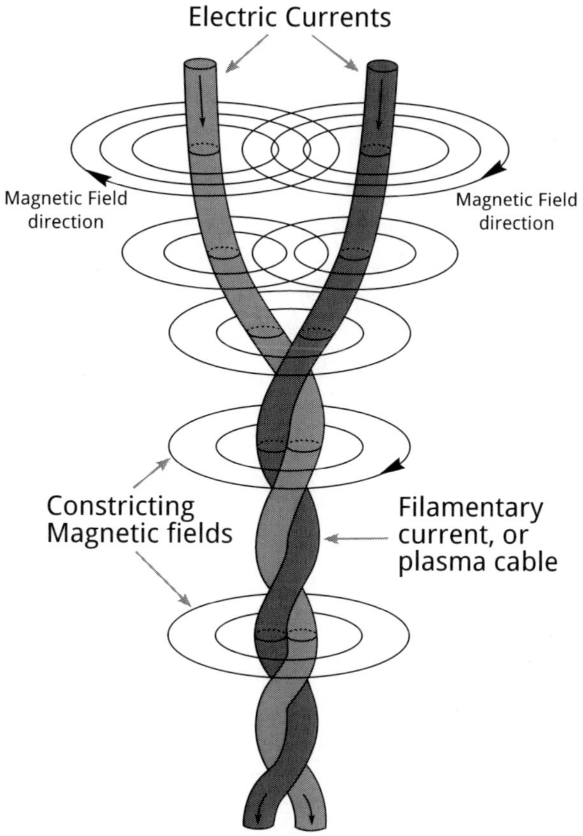

Figure 22: Electromagnetic interactions cause the two filaments to draw inward and rotate around each other to form a helical filament pair, also known as a 'plasma vortex'. (© Thunderbolts.info)

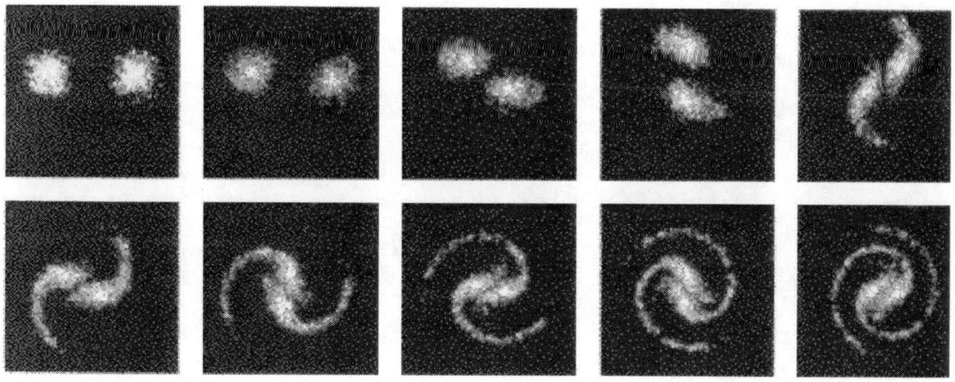

Figure 23: Supercomputer simulation of spiral galaxy formation by Anthony Peratt, based on charged particle interactions. (© Peratt)

Chapter 12.

Homopolar motors

Birkeland currents and the spiraling of double filaments are closely related to another concept: *homopolar motors.* Also known as Faraday motors, these are based on the force generated by the **interaction between an electric current and a magnetic field** (the Lorentz force or Laplace force). So, in nature, two kinds of invisible energy, namely magnetic fields and electric currents, can interact and generate a very tangible mechanical force, namely the Lorentz force.

The Lorentz force is *proportional to the electric current and the magnetic field.* The stronger the current and the electromagnetic field are, the stronger the subsequent Lorentz force is. This is the principle that drives homopolar motors, the simplest kind of motor. It is also the basic principle behind most other kinds of electric motors.

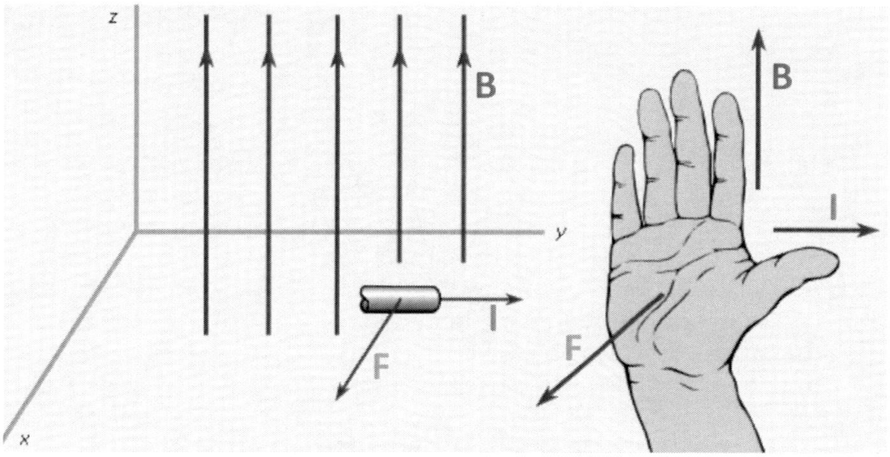

Figure 24: Direction of the Lorentz force relative to an electric current and corresponding magnetic field. (© Sott.net)

The Lorentz force is perpendicular to the plane formed by the electric current and the electromagnetic fields. If you hold up your right hand as shown in figure 24, the Lorentz force (F) would exert itself outwards from the palm of your hand if you imagined an electric current (I) running across your palm in the direction of your outstretched thumb, while the magnetic field (B) emerges upwards in the direction of your index finger.

Figure 25 shows a homopolar motor made of, from top to bottom, one battery, one screw and one magnet. An electric wire closes the circuit and connects the top of the battery to the magnet. The conjunction of the magnetic field (B – vertical ascending arrow) and the current intensity (I – three black arrows) generates the Lorentz force (F – diagonal descending arrow), i.e. the electromotive force that spins the magnet. The resulting spin is represented by the circular arrow.

Any conductor crossed by an electric current and an electromagnetic field is subjected to the same Lorentz force. *This also applies to celestial bodies.* Kristian Birkeland modeled the electric state of celestial bodies with his terrella experiment (see figure 26), where he plunged a charged,

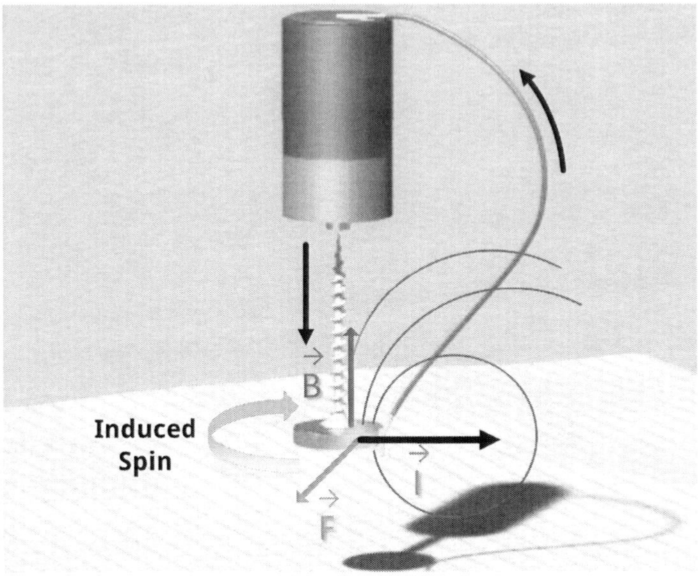

Figure 25: Representation of a homopolar motor with magnetic field, current intensity, subsequent Lorentz force and induced spin. (© Sott.net)

ball-shaped electrode into a vacuum. Here is an account of the first demonstration of the terrella:

> Once Birkeland was satisfied that electrons were streaming from the cathode, he flicked the switch beside the chamber and powered the electromagnet in the terrella. Within seconds, **a purple glow could be seen encircling Earth at the equator. As Birkeland increased the strength of the magnetic field around the Earth, the circle divided, and two circles began to move toward the poles.** The audience fell silent as the two spiral rings of glowing phosphorescent light hovered around the poles of the Earth, eerie and magical. After a few minutes Birkeland turned off the magnet and the cathode in the terrella; the glow disappeared, and the audience took a collective breath.[54]

Birkeland noticed that before the discharge circle divided, discharges were mostly localized in the equatorial and polar regions of the electrodes, as indicated in figure 26. This strongly suggests that most of the current injected into the electrode at the level of the polar regions escaped through the equatorial region. This is consistent with observations of the Sun, which shows a predominant glow and faster rate of rotation[55] around the equatorial region.

Reasoning by analogy and applying the principles of the homopolar motor to celestial bodies like stars and planets, we find that the 'internal' magnet of the celestial body plays the role of the cylindrical magnet of the motor. The external power source of the celestial body acts as the battery. The partial vacuum that constitutes outer space is equivalent to the partial vacuum generated in the lab. And the Birkeland current crossing the plasma that surrounds the celestial body plays the role of the electric wire which closes the circuit by connecting the battery to the magnet.

[54] Jago, L., *The Northern Lights*, Alfred A. Knopf, 2001

[55] According to the movements of the sunspots, the Sun rotates once every 27 days at the equator, but only once in 31 days at the poles.

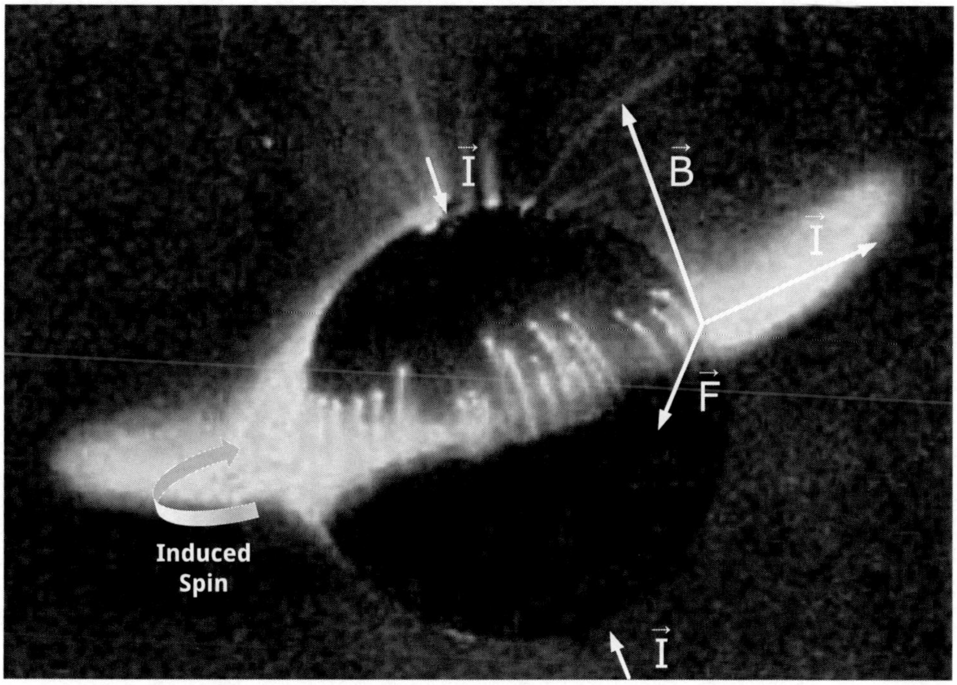

Figure 26: Birkeland's terrella experiment. (© Sott.net)

If a celestial body is a conductor crossed by an electric current and an electromagnetic field, it will also be subjected to the Lorentz force. In this sense, **stars and planets are giant homopolar motors, hence their spin**. Therefore, when the electric current and/or the magnetic field decreases in strength, the rate of rotation decreases as well.

Note that the Moon doesn't spin. As explained above, the Moon hasn't developed a double layer (DL) of its own. It has no plasmasphere because its electric potential is equal to that of its surrounding space. Electric potentials being equal, the Moon is not subjected to any electric current, so no Lorentz force can be generated, hence the absence of spinning.[56]

For plasma cosmologists the driving force of spinning stars is indeed electricity:

> ... within the visible universe we find magnetic fields linking galaxies, showing that the galaxies are 'threaded like beads on a string', along cosmic power lines. **The galaxies and stars within them are driven to rotate like the very simplest of electric motors, known as the 'homopolar' or Faraday motor.** The ubiquitous spiral arms of galaxies trace the current paths between the galactic nucleus and the periphery. ... From an electrical standpoint we make the simple observation that **increasing electric current input to stars results in increasing maximum rotational speeds.**[57]

In Part III, we will see how the Lorentz force (the result of the interaction between electric current

[56] But, as previously mentioned, the Moon has not always been devoid of a magnetosphere. The Moon's surface does bear remnant magnetism. The rocks returned to Earth by Apollo missions show evidence of this magnetism. See: Scott, D. E., *The Electric Sky*, p. 214

[57] *Ibid.*, p. 130

and magnetic field) plays a role in numerous natural phenomena on Earth.

So far, in Part I we have introduced some basic concepts of the Electric Universe theory and plasma cosmology: the primary role of electrically charged plasma, the way in which different electrical potentials form around celestial bodies and set up an electrical gradient through which current can flow, the relative charges of the bodies in our solar system, and the role of electricity (specifically the Lorentz force) in structuring galaxies and solar systems, making them spin. Now, in Part II, we'll take a closer look at the Nemesis theory, and how it might fit into the framework created by the above concepts.

Part II.

The Sun's companion and its accompanying cometary swarm

Chapter 13.

Solar discharges

Now that we know a bit more about plasma, we will focus on the electric properties and behavior of the Sun. As mentioned above, the Sun-heliopause couple can be likened to a giant capacitor. In addition to a permanent leakage current, the solar capacitor is subjected to episodic discharges that we know as 'solar activity'. These discharges are actually Birkeland currents piercing the photosphere (the bright and hot envelope surrounding our star), creating sunspots, which allow the colder internal dark solar matter to become visible. As shown in figure 27, the Sun's photosphere is granulated. These 'granules' are called 'anode arcs' or 'anode tufts'. Because all these tufts have the same polarity, the current filaments arrange themselves so that they avoid each other, hence their granulated appearance.

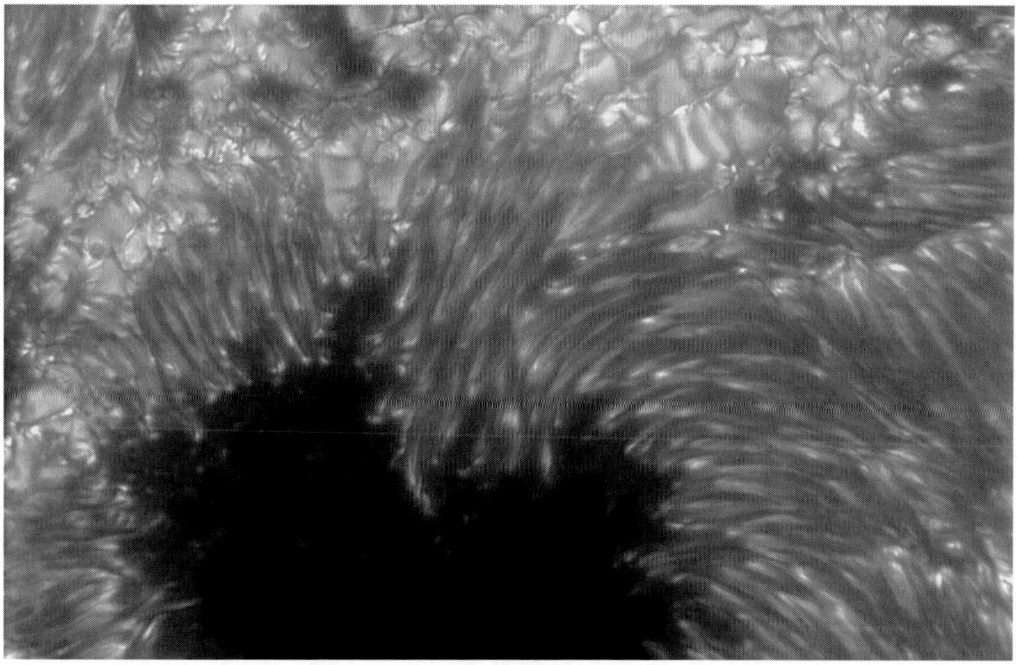

Figure 27: Contour of a sunspot. This image is one of the sharpest views of the Sun ever taken. The shot was made in 2002, from the Swedish Solar Telescope. (© SST)

Solar flares and coronal mass ejections (CMEs) are associated with sunspots.[58] Typically, a surge

[58] Hathaway, D. H., 'The Solar Cycle', *Solar Physics*.
See: solarphysics.livingreviews.org/open?pubNo=lrsp-2010-1&page=articlesu9.html

in solar activity results in a massive ejection of particles from inside the Sun (figure 28). These particles first pierce the photosphere (creating a sunspot) and then pursue their course outside the Sun, usually as flares,[59] or CMEs if the burst is powerful enough.

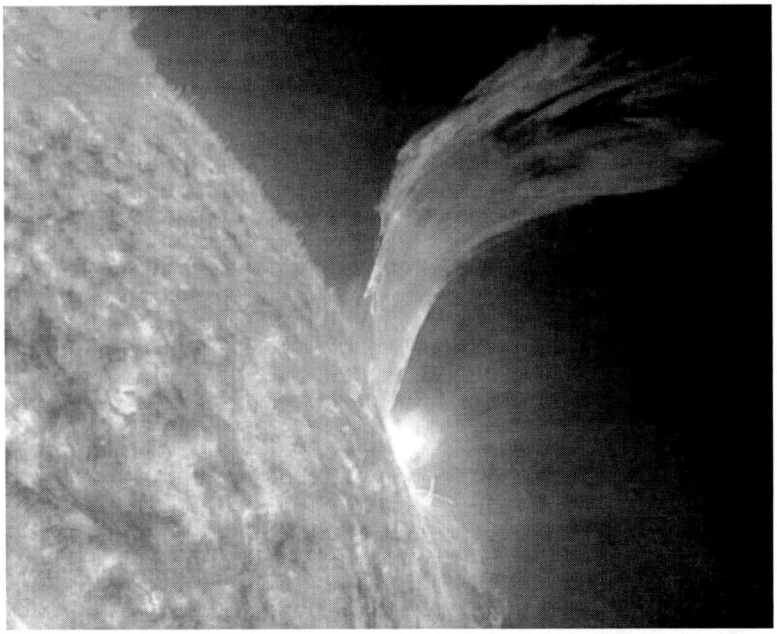

Figure 28: Solar flare captured by the Solar Dynamics Observatory on September 8th 2010. (© NASA)

What causes an increase in solar activity? What triggers solar discharges? The Sun's heliopause, the external boundary of the heliosphere, is almost 100 astronomical units (AU) (that is, 100 times the Sun-Earth distance)[60] from the Sun. Recall that, in electrical terms, the Sun-heliopause couple acts like a giant capacitor in which the Sun is the positive electrode (anode) – relatively speaking – and the furthest extent of the heliosphere, also known as the 'heliopause', is the negative electrode (cathode).

Bodies within the Sun's heliosphere, like comets and planets, can trigger solar discharges (solar flares, sunspots, coronal mass ejections) in the same way a mosquito flying between two wires of a bug zapper triggers an electric discharge, as previously mentioned. In both cases, the resistance between the two electrodes of the capacitor is lowered by the foreign body and enables an electric discharge.

The influence of a celestial body on solar discharges depends on several factors: its size, its electric charge, its trajectory and its location.

The more massive and negatively charged a body is, the more likely it is to trigger a discharge of the (relatively) positively charged Sun. From this perspective, Jupiter, Saturn and comets are the main solar dischargers. Jupiter and Saturn are major players because they are massive and highly charged planets. For example, Jupiter's magnetic field is about 10 times stronger than Earth's, reaching up to 14 gauss at the poles,[61] making it the strongest planetary magnetic field in the solar system.

In comparison to Saturn or Jupiter, comets are tiny. However, their high-density plasma tails can extend over millions of kilometers and display extremely high electric charges, partly due to the eccentricity of their orbits.[62] This is illustrated by the strong glow that comets display, and by

[59] One solar flare can contain as much energy as two and a half billion one-megaton hydrogen bombs. See: Felix, R. W., *Magnetic reversals and evolutionary leaps*, p. 77

[60] For comparison, Pluto, the farthest planet in the solar system, is on average only 40 AU from the Sun.

[61] 'Jupiter's Magnetic Field', *Hyperphysics*. See: hyperphysics.phy-astr.gsu.edu/hbase/solar/jupmag.html

[62] See Chapter 18: 'Comets or asteroids?'

their ability to hold a giant Langmuir sheath (DL) around them.[63]

Celestial bodies can also act in conjunction to modulate solar activity. Typically, planetary alignments will add up the electric discharge potential of each planet.[64] Of course, if electrically active planets like Jupiter and Saturn are involved in this alignment, the discharge will be even more massive. If the Earth happens to be part of this planetary alignment, or close enough, it can find itself located on the trajectory of strong solar emissions.[65] This leads to several earthly effects, one of which is the auroras at the poles, where a massive quantity of solar particles enters the Earth's atmosphere through its thinnest parts (the polar regions) and ionizes the sky, hence the glowing lights observed at high latitudes.

Geomagnetic storms are another direct consequence of surges in solar activity. The 1859 'solar superstorm' is one of the best-known examples:

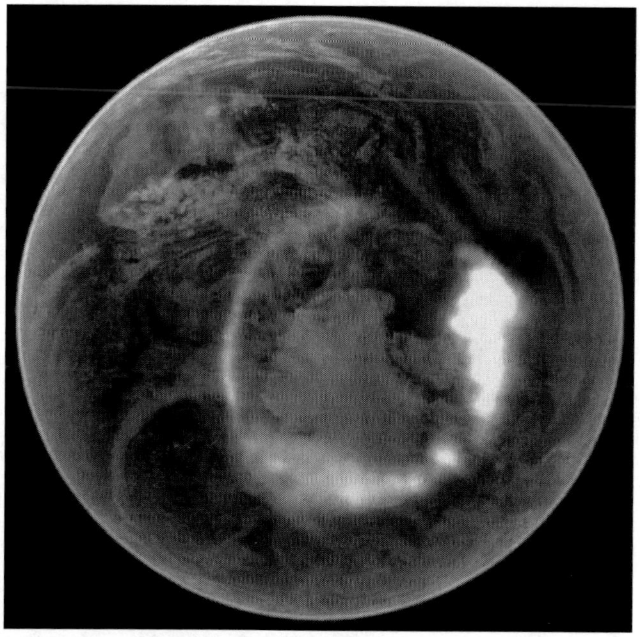
Figure 29: Aurora Australis (above Antarctica), captured by satellite in 2005. (© NASA)

> On September 1859, the largest recorded geomagnetic storm occurred. From August 28 until September 2, 1859, numerous sunspots and solar flares were observed on the Sun. This is referred to as the 1859 solar superstorm or the Carrington Event. It can be assumed that a **massive Coronal mass ejection (CME), associated with the flare, was launched from the Sun and reached the Earth** within eighteen hours – a trip that normally takes three to four days. Telegraph wires in both the United States and Europe experienced induced EMF, in some cases even shocking telegraph operators and causing fires. Aurorae were seen as far south as Hawaii, Mexico, Cuba, and Italy – phenomena that are usually only seen near the poles.[66]

The currents induced on Earth by the superstorm were so strong that they knocked out the whole power grid in the United States. Although telegraph lines lost power, they continued to function, powered solely by the current induced by the solar storm:

> This almost incredible feat was accomplished on the wires of the American Telegraph Company between Boston and Portland, upon the wires of the Old Colony and Fall River Railroad Company between South Braintree and Fall River, and upon other lines in various parts of the

[63] McCanney, J., *Planet X, Comets & Earth changes*, pp. 53–57

[64] Gribbin, J. R., *Beyond the Jupiter effect*, pp. 38-52

[65] *Ibid.*, pp. 47, 136

[66] 'Astronomer Q&A', *University of Arlington.*
See: www.uta.edu/planetarium/astronomy-101/ask-the-astronomer/qa.php?tag=43

> country. **Such was the state of the line on the September 2nd, 1859, when for more than one hour they held communication over the wires with the aid of celestial batteries alone.**[67]

Even a body as small as the Moon can exert very tangible effects on Earth, particularly in terms of weather. In fact, some weather-watchers[68] take the Sun-Moon-Earth positions into account when determining their weather forecasts, which often are strikingly accurate.

New moons are times of weather disruption. Just before the new moon appears, it is located directly between the Sun and the Earth, shielding Earth from solar activity. Then, during the few days following the appearance of the new moon, the Earth is no longer shielded from the incoming solar wind and suddenly receives a massive inflow of solar particles, leading to intensified weather events.

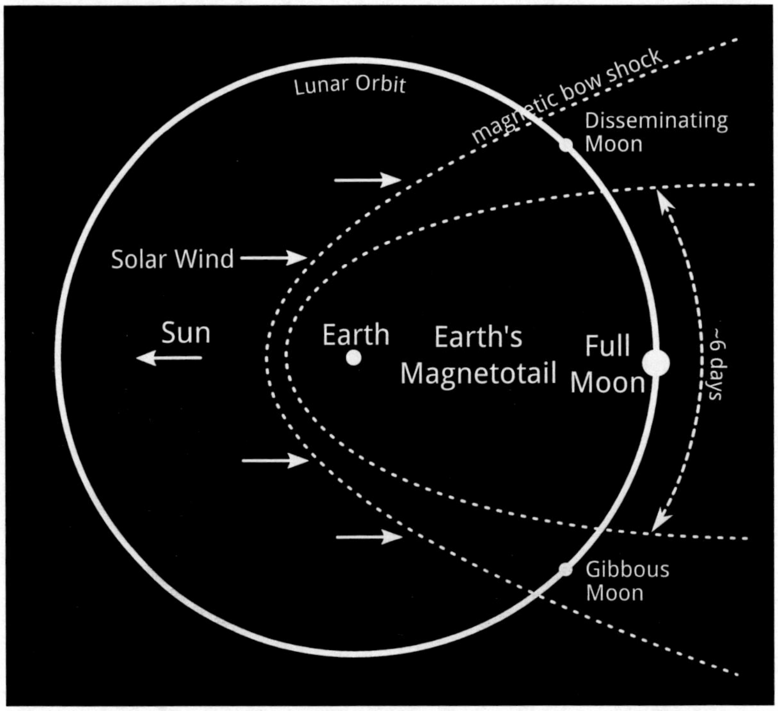

Figure 30: When the Moon is full it is located within the Earth's plasma tail, hence its increased influence. (© Sott.net adapted from SOHO / NASA)

Full moons are often associated with weather disruption and catastrophes because, as depicted in figure 30, during this phase the Moon enters the Earth's plasma tail (magnetotail) and disturbs its electromagnetic activity.[69] In such instances the Moon acts within the Earth's plasmasphere as previously mentioned comets act within the Sun's heliosphere. As such the full moon acts as a discharger of the Earth capacitor. In addition, during a full moon, the Moon and the Earth are

[67] Prescott, G. B., *History, Theory and Practice of the Electric Telegraph*, pp. 320–323

[68] To name a few: Tony Philips (spaceweather.com), Piers Corbyn (weatheraction.com), John L. Casey (spaceandscience.net), James McCanney (jmccanneyscience.com), Mitch Barros (earthchangesmedia.com).

[69] Scott, D. E., *The Electric Sky*, p. 106

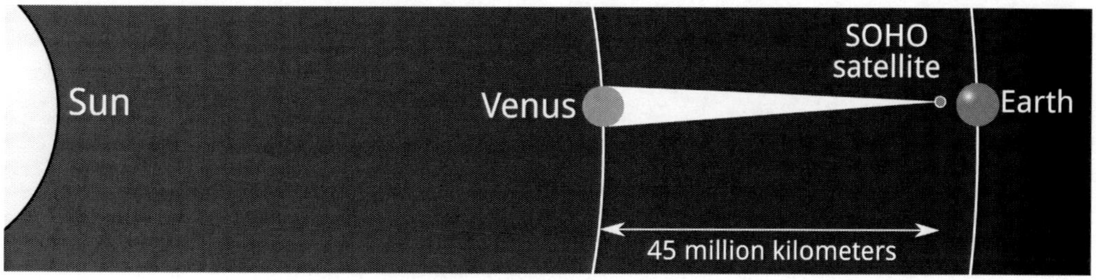

Figure 31: Venus plasma tail almost reaches Earth's orbit

aligned and add up their discharge capacity, triggering increased solar activity, hence the correlation between full moon phases and increased solar activity.[70]

In a similar manner, **Venus** can also exert a strong influence on Earth (see figure 31) because of its close proximity. Indeed Venus's plasma tail extends some 45 million kilometers in the direction of Earth's orbit.[71] This plasma tail almost reaches Earth when the two planets are at their closest, therefore it can strongly disrupt the Earth's electric state. This brushing constitutes an intermittent circuit for transferring charge between the two adjacent planets.[72] In addition, in such a case, Venus and Earth are aligned with the Sun and add up their discharge capacity.

[70] Radin, D. I., *Lunar Correlates Of Normal, Abnormal And Anomalous Human Behavior*, *Subtle Energies* (1994) 5(3): 220

[71] Van der Sluijs, R., 'Venus' Tail of the Unexpected', *Thunderbolts*. See: www.thunderbolts.info/tpod/2008/arch08/080 220venustail.htm

[72] Thornhill, W., 'Newton's Electric Clockwork Solar System', *Holoscience*. See: www.holoscience.com/wp/newtons-electric-clockwork-solar-system/

Chapter 14.

Variations in solar activity

As mentioned above, the Sun's activity is the result of solar discharges triggered by charged celestial bodies in the solar system and by planetary alignments. The three main discharge phenomena (flares, CMEs and sunspots) are intimately correlated and indicate an increase in solar activity. For now we'll focus on sunspots as the main indicators of solar activity.

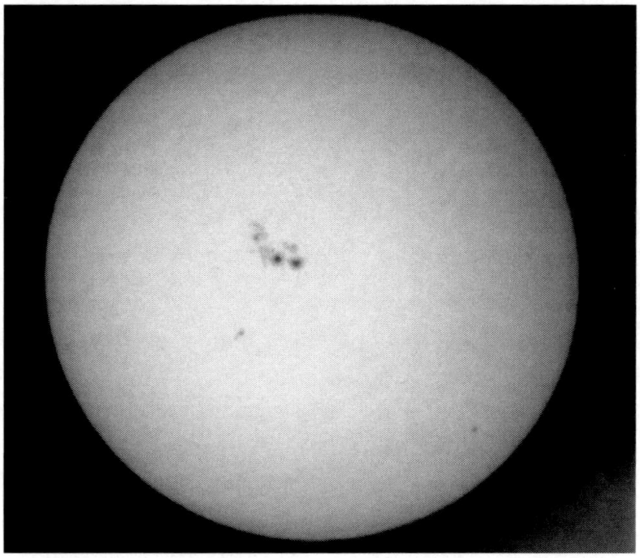

Figure 32: Sunspots on June 22nd, 2004. The sunspots are the dark spots located near the center of the image. (© Wikimedia Commons)

Sunspots have been observed, studied and recorded for centuries. Scholars in the East, starting with Chinese astronomer Gan De, began studying them as early as 364 BC.[73] The first clear mention of sunspots by a Western observer was made around 325 BC by Theophrastus,[74] a student of Plato and Aristotle.

From the middle of the 18th century, scientists began systematically counting the number of observable sunspots. This count provides us with a very useful record of solar activity spanning more than 250 years (from 1750 to the present). As depicted in figure 33, the **average number of sunspots during this time varied greatly, from 0 sunspots a month to more than 250.**

As of 2013, we are currently going through solar cycle 24 (SC24), which exhibits some unusual

[73] Rogers, L., 'The History of Trigonometry'. See: nrich.maths.org/6843

[74] 'A Primer on Spaceweather', *NOAA*. See: www.swpc.noaa.gov/primer/primer.html

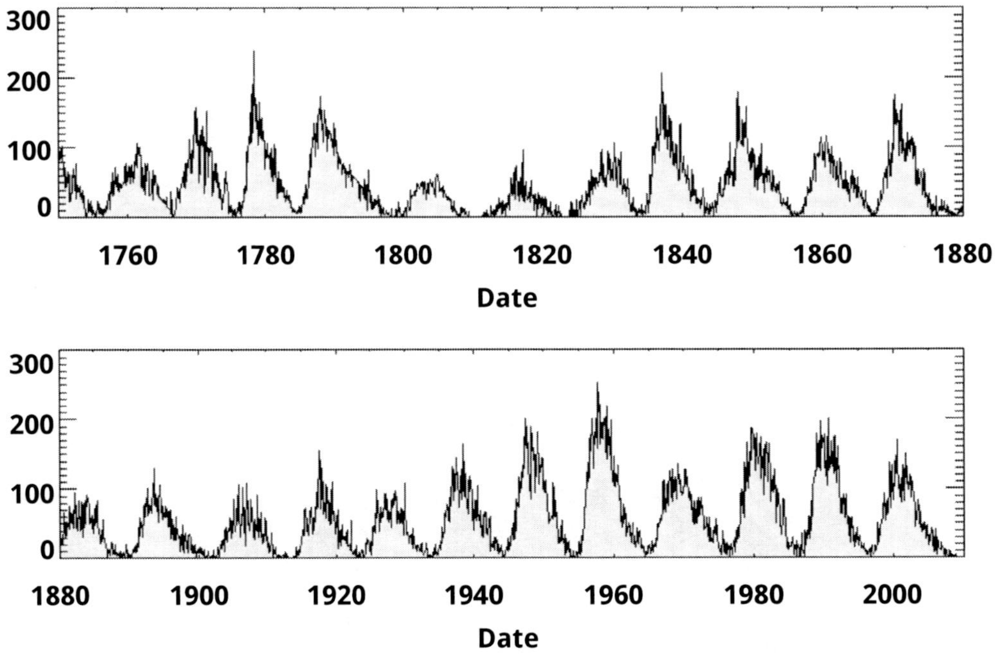

Figure 33: Average monthly number of sunspots (1750–present). (© Spaceweather.com)

and unexpected low activity. According to official predictions, SC24 was supposed to start in 2006, but solar activity was extremely low for the subsequent four years (2006–2010); the cycle only got off to a timid start in 2010.[75] In addition to this delayed and weak start, NASA, which initially anticipated a monthly maximum[76] of 180 sunspots for solar cycle 24, progressively revised this figure down to just 80 sunspots/month.

Figure 34 shows the weak solar cycle 24 (black curve) and the repeated downgrading of solar cycle predictions (four grey curves for the 2006, 2008, 2009 and 2010 predictions).[77]

As Joe Gurman, a project scientist for NASA's sun-observing mission STEREO,[78] recently stated about the current SC24:

> It's likely to be the **lowest solar maximum, as measured by sunspot 'number', in more than a century**.[79]

For comparison, cycle 23's maximum in 2000 exhibited 120 sunspots. As depicted in figure 35, solar cycle 23 (stars curve) was already weak compared to its predecessors. Solar cycles 22 (squares

[75] In 2008 and 2009, the average number of sunspots dropped to less than 5 per month. See: www.ips.gov.au/Solar/1/6

[76] Maximum of solar cycle 24 was due mid-2013.

[77] The observed number of sunspots in the graphic made by iceagenow.info stopped in August 2010. Average monthly sunspot counts extracted from the ISP database (Australian Bureau of Meteorology) have been added. See: www.ips.gov.au/Solar/1/6

[78] Solar Terrestrial Relations Observatory.

[79] Howell, E., 'Sun's 2013 Solar Storm Peak Expected to Hit Century Low', *Livescience*. See: www.livescience.com/26 005-sun-solar-space-weather-2013.html

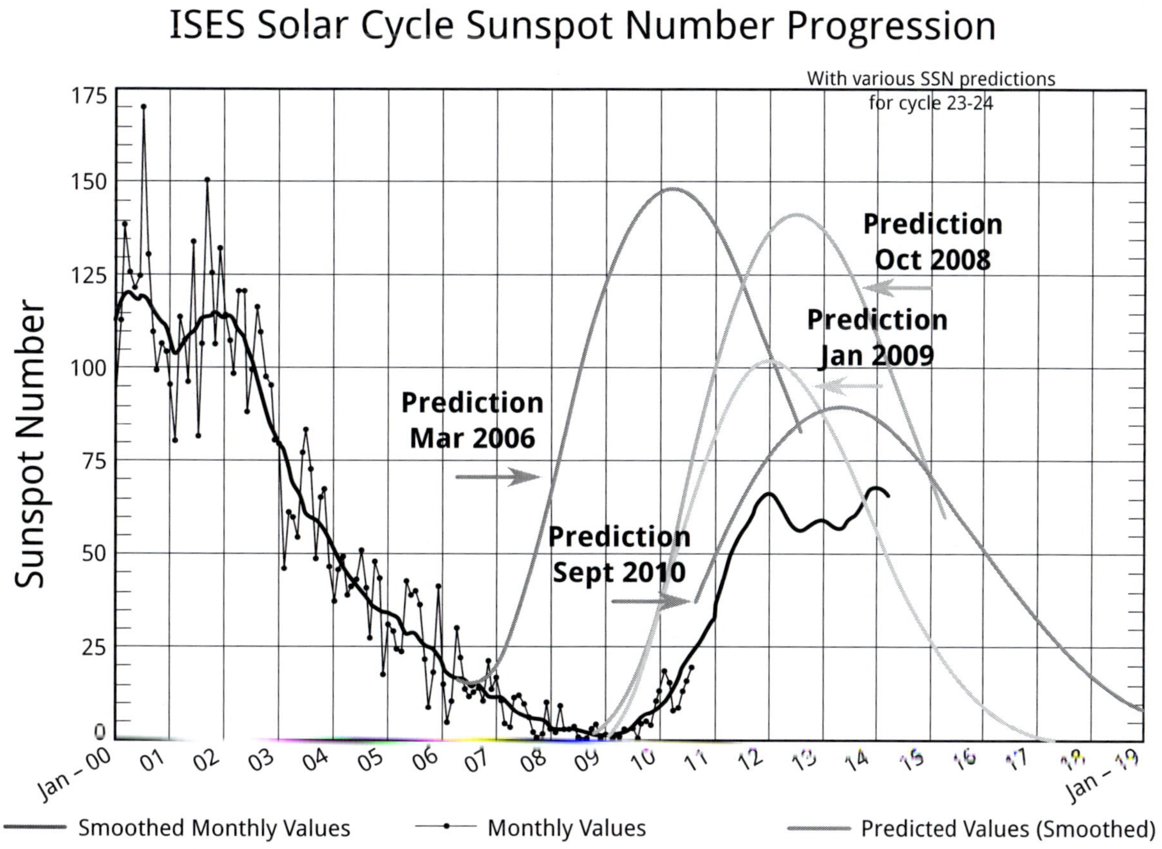

Figure 34: Solar cycle 24: observed data vs. NASA predictions. (© Sott.net, after iceagenow.info)

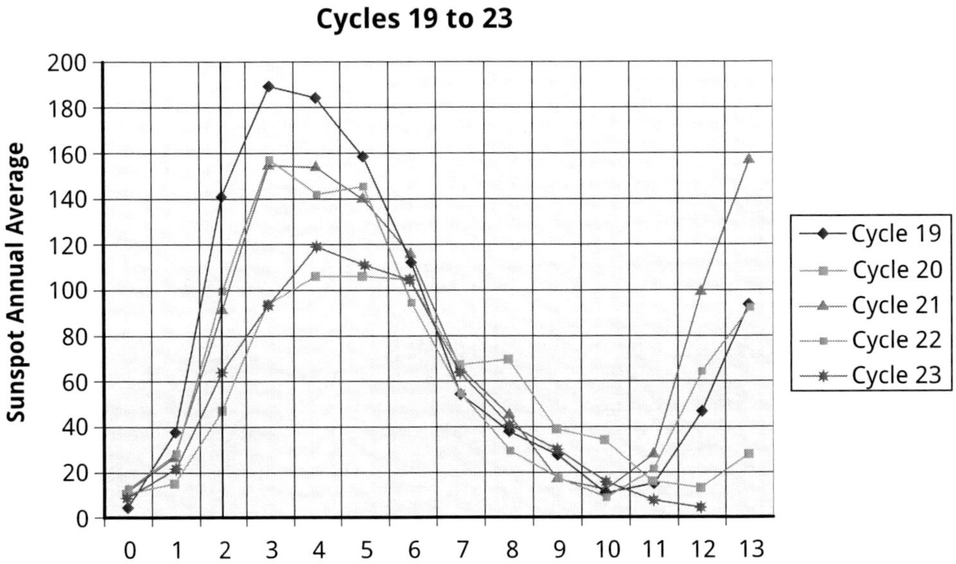

Figure 35: Comparison of solar cycles 19, 20, 21, 22 and 23 (© Wattsupwiththat.com)

curve) and 21 (triangles curve) reached a maximum of almost 160 sunspots. That's 35% more than solar cycle 23 and 100% more than the initial estimate for solar cycle 24's maximum.

SC23 not only had a low maximum; it also had a weaker start than its predecessors, as illustrated by sunspot levels for SC21, SC22 and SC23 at year 2 (the vertical black line in figure 35). It seems therefore that **solar activity started to decrease soon after the beginning of solar cycle 23; that is to say, around 1998.**[80] Since then, the decrease in solar activity seems to have accelerated.

The numerous 'adjustments' to the SC24 forecast strongly suggest that NASA doesn't comprehend all the factors modulating solar activity. It also provokes a number of questions. On what basis did they make the above-described repeatedly over-optimistic predictions? How is it that the observed maximum for SC24 produced results that completely contradicted the predicted maximum? What other factors are modulating solar activity that perhaps not even NASA is aware of?

Something else that makes solar cycle 24 stand out is that its unusually low solar activity *coincides with an exceptionally high level of meteoric activity* in our skies. We'll discuss this in greater detail later on,[81] but for now let's assume that this increase in fireball observations is due to increased cometary activity in the inner solar system.

The chart in figure 36 is based on data collected by the American Meteor Society (AMS).[82] **Reported fireballs increased by 1931% over the 2005–2012 period.** Unfortunately the AMS doesn't have data prior to 2005, although observers apparently began taking notice of fireball

[80] 'Solar cycle 23', *Wikipedia*. See: www.en.wikipedia.org/wiki/Solar_cycle_23

[81] See Chapter 18: 'Comets or asteroids' and Chapter 21: 'Increase in cometary activity'

[82] American Meteor Society. See: www.amsmeteors.org/members/fireball/browse_events/?year=2012&state=&num_-report=2

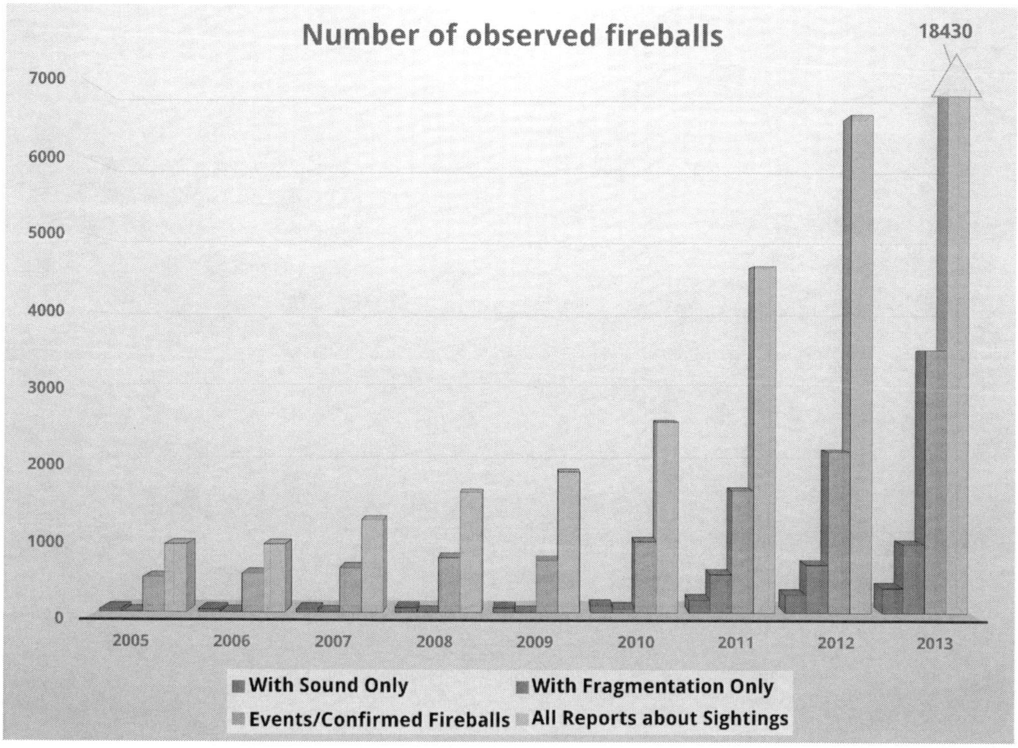

Figure 36: Acceleration of the increase in fireballs over the whole AMS observation period (2005–present). (© Sott.net)

and meteorite reports around 2002.[83]

So, on one hand we have an exceptionally quiet Sun and, on the other hand, we have exceptionally high cometary/meteor activity. But high cometary activity is supposed to trigger massive and frequent solar discharges. Instead, the Sun has not been this quiet for more than a century.

Did we miss something?

How can we reconcile high comet activity and reduced solar activity? What is causing the reduction in solar activity? What causes the increase in meteor activity? Could the same cause account for both?

[83] 'Watching the skies', *Fireballs and Meteorites*. See: fireballs-meteorites.blogspot.fr/2007/03/march-november-2002.html

Chapter 15.

Enter Nemesis

As everybody knows, our solar system is powered by a single star, the Sun. Well, it is *assumed* that ours is a single-star solar system because we see only one sun rise each morning. However, this is actually quite a peculiar configuration, since *most* stars astronomers have observed are part of multi-star systems (most often binary).

Based on data from NASA's Chandra X-ray observatory, it's estimated that over **80% of all stars may be in either binary or multiple-star systems.**[84] Grazia and Milton, who studied the 60 star systems nearest to our own reached a comparable conclusion:

> 61% of the 60 nearest stars are components of a double (binary) or triple star system.[85]

A twin-star model for our own solar system is a tantalizing prospect, not least because it could account for many 'anomalies' exhibited by the single-star hypothesis. As stated by the Binary Research Institute (BRI):

> ... elliptical orbit equations have been found to be a better predictor of precession rates than Newcomb's formula, showing far greater accuracy over the last hundred years. Moreover, a moving solar system model appears to solve a number of solar system formation theory problems including the sun's lack of angular momentum. **For these reasons, BRI has concluded our sun is most likely part of a long cycle binary system.**[86]

Bear in mind that the binary systems identified above are composed of stars bright enough to be detected with a telescope. This means that the percentage of binary systems may be even higher, since some systems can include 'unlit' stars, like so-called **'brown dwarfs'**, for example.

For plasma cosmologists, a binary system is the logical way for individual stars to cope with high electric stress, causing any given **star to go through a process of fission (i.e. splitting into two or more parts).**[87] When a sphere is divided into two equally-sized spheres, the total mass will remain the same (no matter disappears) but the total surface area of this pair will be about 26% larger than the area of the original single sphere.[88] This increases the total surface area

[84] Cruttenden, W., *Lost Star*, p. 111
[85] De Grazia, A. & Milton, E. R., *Solaria Binaria*, p. 17
[86] 'Introduction to Binary Companion Theory', *Binary Research Institute*.
See: www.binaryresearchinstitute.org/bri/research/introduction/theory.shtml
[87] Scott, D.E., *The Electric Sky*, pp. 157–159
[88] Scott, D.E., 'Electric cosmology – Stellar Evolution', *The Electric Sky*, online version.
See: electric-cosmos.org/hrdiagr.htm

exposed to the electric field and thus decreases the current density (amperes per square meter). Therefore, electrically-induced fission enables stars to reduce the electric stress they are subjected to by spreading it between two or more stars.

Because of the lower level of electric stress exerted on a binary system after fission, brown dwarfs (stars exposed to a weak electric field, hence their reduced brightness) should be quite common in binary systems:

> If the members of a resulting binary pair turn out to be unequal in size, the larger one will probably have the larger current density – but still lower than the original value. (This assumes that the total charge and total driving current to the original star distributes itself onto the new stars proportionally to their masses.) In this case, **the smaller member of the pair might have such a low value of current density as to drop it, abruptly, to 'brown dwarf'** or even 'giant gas planet' status.[89]

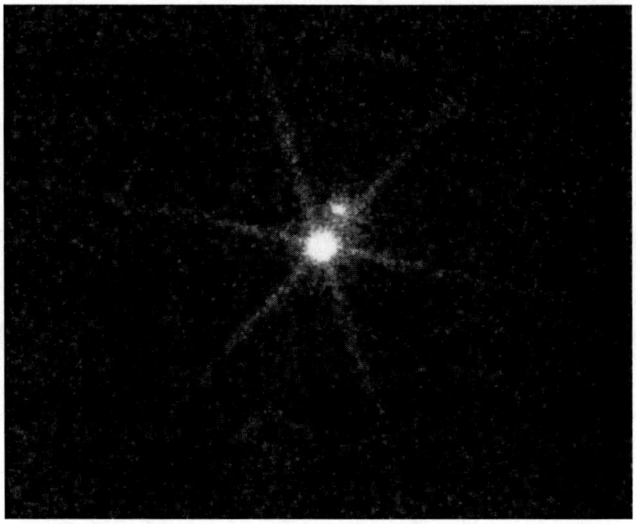

Figure 37: Sirius is a binary star. Sirius A is the brightest. Sirius B is dimmer and so close to Sirius A that it was not sighted until 1862. (© NASA)

It's clear that binary stars are very common, probably even more common than acknowledged in the scientific literature. So, is our Sun one more anomaly in the rather anomalous universe depicted by mainstream science? Is it really single?

A significant clue that our star may in fact be part of a binary system appeared in *Nature* on March 19th, 1982,[90] when the paleontologists David Raup and Jack Sepkoski unveiled a cyclical pattern of mass-extinction events in the fossil record.[91] Their research revealed that **over the last 250 million years, the Earth regularly experienced mass extinctions**, as depicted in figure 38.

Berkeley physicist Richard Muller discovered another significant clue in May 1986 when he compiled the iridium levels measured at the Cretaceous-Tertiary (C-T) boundary, which marked the mass extinction that occurred about 65 million years ago, in 66 different locations across the five inhabited continents. As shown on the map in figure 39, the results were unexpected: **in each of those 66 different locations, Muller found abnormal levels of iridium.**[92]

Then, Muller and his team checked level of iridium at the Eocene-Oligocene (E-O) boundary which marked the extinction of the dinosaurs and occurred between 39 and 35 million years ago.

[89] Scott, D. E., *The Electric Sky*, p. 158

[90] Raup, D. & Sepkoski, J., 'Mass extinctions in the marine fossil record', *Science* 215(4539): 1501–1503

[91] According to calculations made by Raup & Sepkoski, the probability of a 27-MY mass-extinction cycle being due to random chance is less than 1%.

[92] Muller, R., *Nemesis: The Death Star*, pp. 74–77

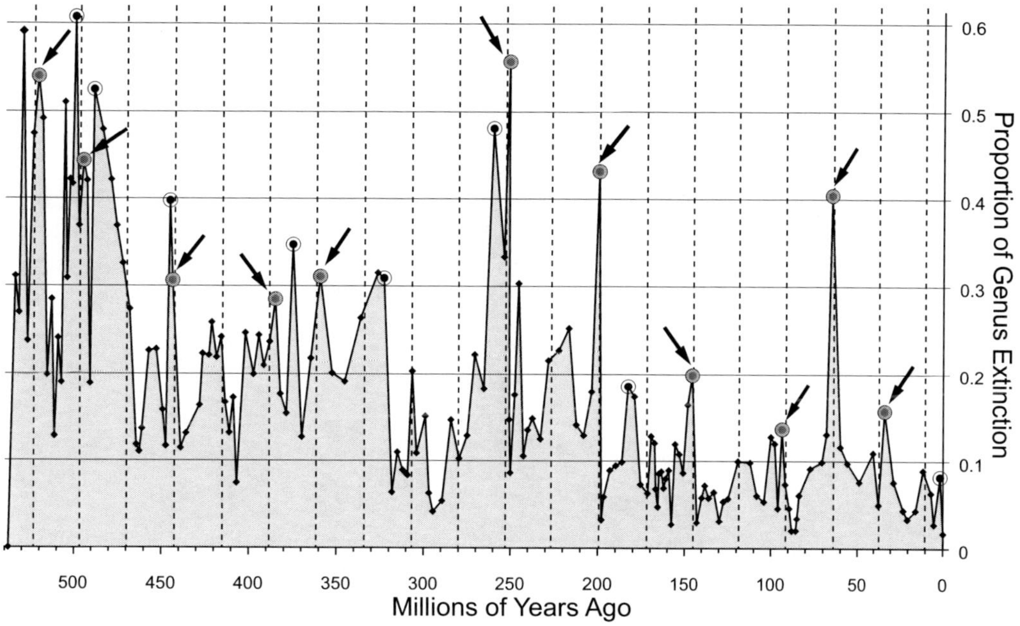

Figure 38: A plot of extinction intensity from the Sepkoski data. Out of 19 events (circled points), 11 (black arrows) lie on the vertical lines showing the 27-MY (million year) interval. (© Sott.net, adapted from Melott & Bambach)

They soon found similarly abnormal levels of iridium. Those results were independently confirmed by researcher R. Ganapathy.

What's the big deal with iridium, you might wonder. Together with gold, platinum, osmium and rhenium, iridium is one of the 'noble elements' which are 10,000 times as abundant in meteoric material than in the Earth's crust.[93] But it's not just meteors that deliver these elements to Earth; the massive radiations emitted by supernovas can also generate large quantities of heavy elements, including iridium.

So, while iridium anomalies at the E-O boundary signal that the mass extinction about 37 million years ago was due to a cosmic event, it remained to be seen whether the culprit was **a supernova or a meteor bombardment**.

Luis Alvarez was one of the proponents of the supernova theory.[94] To prove his case he reasoned as follows:

> Pu-244, one of the isotopes of plutonium, isn't naturally present in the Earth's crust, **nor in meteorites**. However, it is one of the heavy elements created by supernovas. So, if mass extinctions were due to supernova explosions, then we should find abnormally high levels of Pu-244 in the geological era boundary material because of its very long half-life of 80 million years. The analysis of Pu-244 concentrations in the clay sample from the E-O boundary led to the following conclusion:

[93] *Ibid.*, p. 44

[94] At the time, Luis Alvarez was the director of the Princeton research laboratories where Muller was conducting his research about Nemesis.

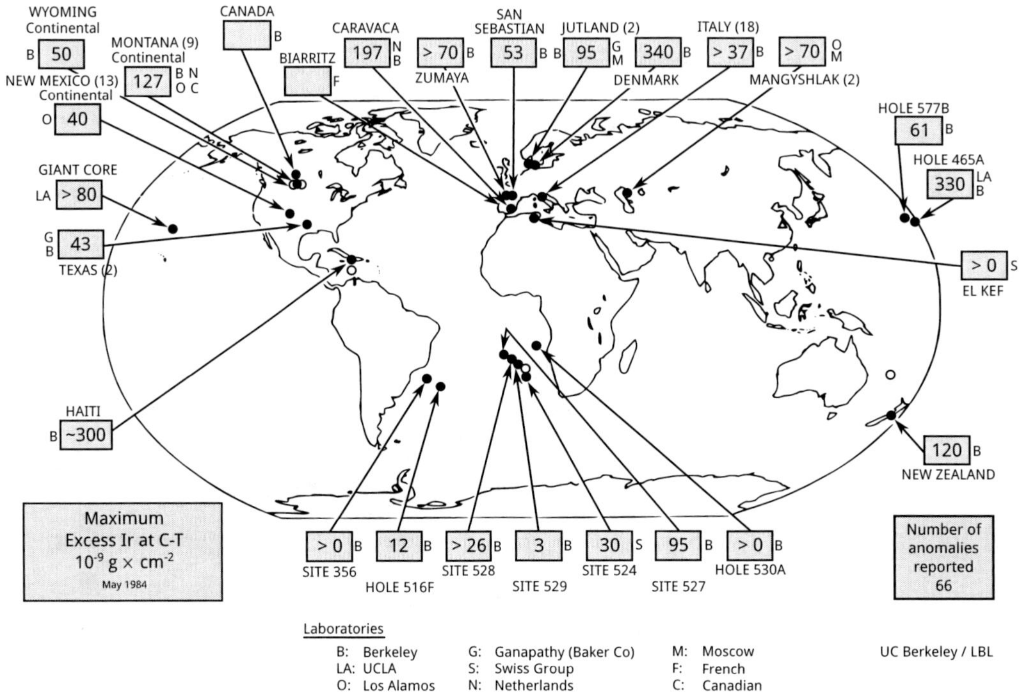

Figure 39: 66 iridium concentration anomalies measured at the C-T boundary. (© R. Muller)

> There was no plutonium. A supernova had not killed the dinosaurs. These results were later published in a paper titled 'Negative results for the Supernova hypotheses'. In this paper they [Alvarez et al.] described how the measurements showed **there was no plutonium-244 present, and how this result ruled out the supernova theory ...** [95]

If it was not a supernova, then the only possible cause was bombardment from asteroids or comets (we can use these terms interchangeably, as I'll explain in Chapter 18). Since then, several research teams, including University of Louisiana astronomer Dr. Daniel Whitmire and the Mellot and Bambach[96] team, have published papers confirming that meteor bombardments were indeed the cause of this mass extinction event on planet Earth.

Once it was established that a majority of cyclical mass extinctions were due to asteroids, another question arose: what is the main driver of these cyclical asteroid bombardments? The idea that a single asteroid or a swarm of asteroids can follow a stable 27-million-year orbit must be ruled out.

Small celestial objects cannot remain long in stable orbits. As we've seen repeatedly in recent years, when comets approach massive celestial objects like Saturn, Jupiter or the Sun, they are either destroyed, broken up, drastically weakened, 'sucked in' (hence the name 'sun-diving comets'), or the encounter alters their orbit as they're violently ejected back out of the inner solar system. So, it is very unlikely that an asteroid swarm could cross the solar system, maintaining a stable

[95] *Ibid.*, p. 59

[96] Melott, A. & Bambach, R., 'Nemesis Reconsidered', *Monthly Notices of the Royal Astronomical Society Letters*, p. 407

orbit throughout the journey, then return to the inner solar system on its next revolution and do it again and again for over 500 million years, as suggested by Raupp and Sepkovsky's research.[97]

In addition, such a long orbit implies that the hypothetical asteroid(s) would travel several light years away from the solar system and be subjected to the disruptive gravitational forces of other stars.[98]

If asteroid(s) can't follow such a stable 27-million-year orbit by themselves, then something else obviously does; something large enough to maintain a stable orbit; and something that regularly sends fresh comets our way. That's where Richard Muller came up with the idea that a solar companion – a twin sun – followed such an orbit. Every 27 million years, on its approach towards the solar system, such a companion would disturb asteroids or comets local to its orbit, pushing and pulling them along its trajectory. Muller named this hypothetical Sun's companion 'Nemesis',[99] after the Greek goddess who relentlessly persecutes the excessively rich, proud and powerful. Actually this is quite an apt choice of name, as we will see further on in our discussion of the possible role played by the 'rich, proud and powerful' in cosmic affairs.

Figure 40: Alfred Rethel, 'Nemesis', 1837. (Oil on canvas, Hermitage Museum)

After Muller's ground-breaking work, several research teams, including University of Arizona astronomer Frank Low[100] and Thomas Chester[101] from NASA's Jet Propulsion Laboratory (JPL), began searching the sky to find Nemesis. Their quest lasted for years but returned no positive results, so observers have yet to find Nemesis. But as we noted earlier, some stars are not bright at all, as is the case with brown dwarfs (Dan Whitmire's star category of choice for Nemesis[102]) and, to a lesser extent, extinct red dwarfs (Muller's choice[103]).

If the Sun's companion is a brown dwarf – and if it is also orbiting in the ecliptic plane – finding it would be literally like finding a dark, invisible object surrounded by millions of bright, shining ones, a feat even more difficult than searching for a needle in a haystack. This

[97] Raup & Sepkoski, *op. cit.*

[98] An asteroid of negligible mass (several tons) relative to the Sun would have an aphelion of roughly 200,000 AU, that is 3.16 light years. Such a remote aphelion (the farthest distance from the Sun) would subject the asteroid to the gravitational forces of nearby stars like *Proxima Centauri* (4.24 light years from the Sun) or the binary stars *Alpha Centauri A and B* (4.35 light years from the Sun), while the gravitational forces exerted by the Sun would be virtually null.

[99] Muller, *op. cit.*, p. 114

[100] 'Astronomer Frank J. Low, 1933–2009', *obit. UA News, University of Arizona.* See here: uanews.org/story/astronomer-frank-j-low-1933-2009

[101] Thomas Jay Chester's Website: tchester.org/znet/tchester/

[102] Whitmire, D. P. & Jackson, A. A., 'Are periodic mass extinctions driven by a distant solar companion?', *Nature* (19 April 1984) 308: 713–715

[103] Muller, *op. cit.*, p. 109

is quite an apt metaphor since up until now the technology hasn't existed to allow us to successfully and systematically detect dark celestial bodies like brown dwarfs, as Whitmire explains:

> Currently, I am searching the half billion point sources in the 2MASS database for evidence of this object. This survey covered 99% of the sky at near-infrared wavelengths of 1–2 microns. The optimum wavelength for our search is 5 microns but no such full sky survey exists, as yet.[104]

If Nemesis exists, the fact that it hasn't been found despite years of research strongly advocates for it being a dark star.

Figure 41: An artist's depiction of a brown dwarf. (© Wikimedia Commons)

As a side note, according to mainstream science, brown dwarfs are small stars nearing the end of their life. In standard cosmology, their 'internal Fermi reactions' are decreasing due to lack of fuel (hydrogen), making them progressively dimmer and dimmer. However, there are several problems with this model. For starters, brown dwarfs emit X-rays:

> The orbiting **x-ray telescope Chandra recently discovered a brown dwarf (spectral class M9) emitting an x-ray flare**. This poses an additional problem for the advocates of the stellar fusion model. A star this cool should not be capable of x-ray flare production. How 'gravitational collapse' can produce x-rays remains unexplained. 'We were shocked,' said Dr. Robert Rutledge of the California Institute of Technology in Pasadena, the lead author on the discovery paper to appear in the July 20 issue of *Astrophysical Journal Letters*. 'We didn't expect to see flaring from such a lightweight object. This is really the 'mouse that roared.'[105]

In standard astronomical models, brown dwarfs are 'supposed to be' too cool and too small to maintain fusion reactions in their cores. The minimum temperature 'should be' three million degrees Kelvin and the mass should be at least seven percent of the Sun's mass. However, some 'brown dwarfs' do not meet these criteria. Therefore, the absence of nuclear fusion cannot be explained by their size and/or temperature. Besides, they do emit X-ray radiation, which is the main marker of nuclear activity.[106]

But a brown dwarf presents no anomaly in electric models. It's simply a star that is not glowing because the local electric field is too weak. From this perspective, it's not the size (and therefore the limited gravitational field) that makes a star dark, but the electric stress. If the electric stress is too low, the star (whatever its size) doesn't glow. Thus, the size and temperature range determined by mainstream science to define brown dwarfs is irrelevant.

In figure 42, we refined Muller's simulation[107] for Nemesis' orbit. In this simulation, Nemesis is slightly bigger and its perihelion slightly closer to the Sun. The relative mass of Nemesis and the

[104] Bruce, A., *2012: Science or Superstition*, p. 67
[105] Scott, *op. cit.*, p. 127
[106] *Ibid.*, pp. 127–129
[107] Muller, *op. cit.*, p. 106

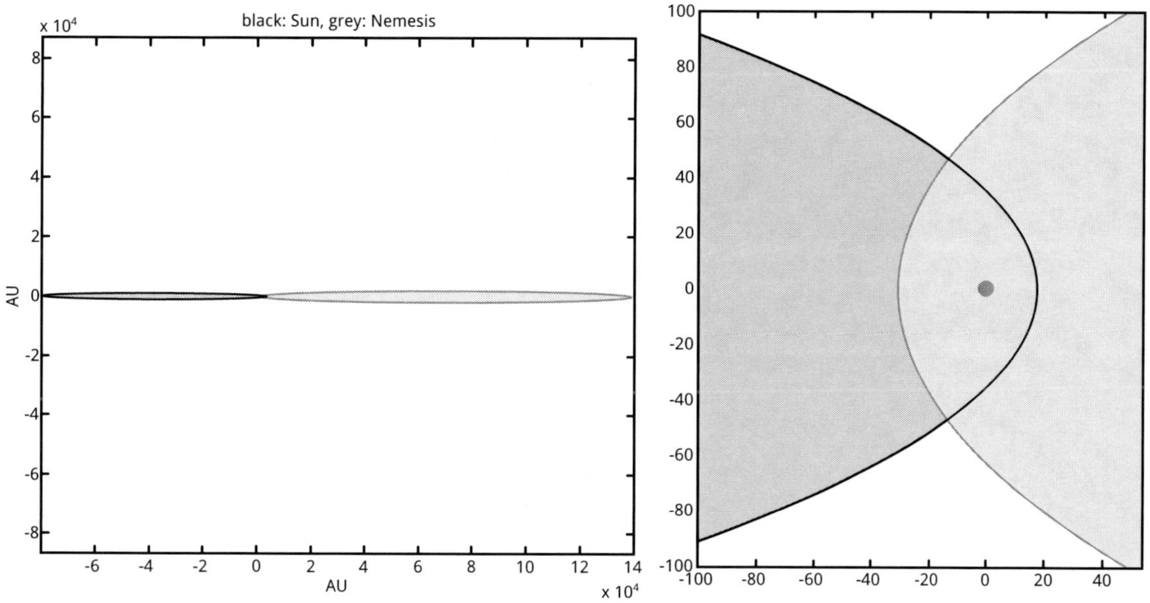

Figure 42: Simulation of Nemesis' orbit (mass = 0.56 of Sun mass – perihelion equals 49 AU). (© Sott.net)

Sun being quite similar, they turn around the same point known as 'center of mass', the grey dot in the right part of the drawing.

In this same figure, the Sun's orbit is the medium grey small ellipse on the left; Nemesis' orbit is the light grey large ellipse on the right. The graph on the right zooms in on the perihelion (the region in which the Sun and Nemesis make their closest approach). Notice that, because of the long orbital period of Nemesis, its orbit is very oblate (flat ellipse).

Figure 43 displays Nemesis' orbit and the solar system.

This simulation is based on the following hypothesis: Nemesis' mass = 56% of the Sun's mass. Nemesis' perihelion = 49 AU (roughly the distance between the Sun and Pluto). Duration of Nemesis' orbit = 26.9-million-year cycle as calculated by Mellot and Bambach.[108]

As a result the speed of Nemesis (relative to the Sun at the perihelion) would be 4.66 miles/s. The aphelion (the greatest distance between the Sun and Nemesis) would be 203,000 AU, i.e. 3.21 light years.[109] The eccentricity of Nemesis orbit would be 0.999, i.e. a very oblate ellipse shape.

However the simulation above doesn't give any information about the actual location of Nemesis relative to the Sun. According to Mellot and Bambach's graph showing cyclical mass extinctions, the last major extinction event, known as the Middle Miocene disruption,[110] occurred about 14 million years ago. Assuming that to be the case, the Sun's companion should currently be at or near aphelion along its 27-million-year orbit and should therefore pose no threat to life on Earth.

But did the Middle Miocene disruption really occur 14 million years ago? To create their timelines,

[108] Mellot and Bambach refined Muller's calculation of a 27-million-year period. See: Melott, A. & Bambach, R., 'Nemesis Reconsidered', *Monthly Notices of the Royal Astronomical Society Letters* 407

[109] For comparison, the nearest star is *Proxima Centauris* at a distance of 4.24 light years.

[110] 'Middle Miocene disruption', *Wikipedia*. See: www.en.wikipedia.org/wiki/Middle_Miocene_disruption

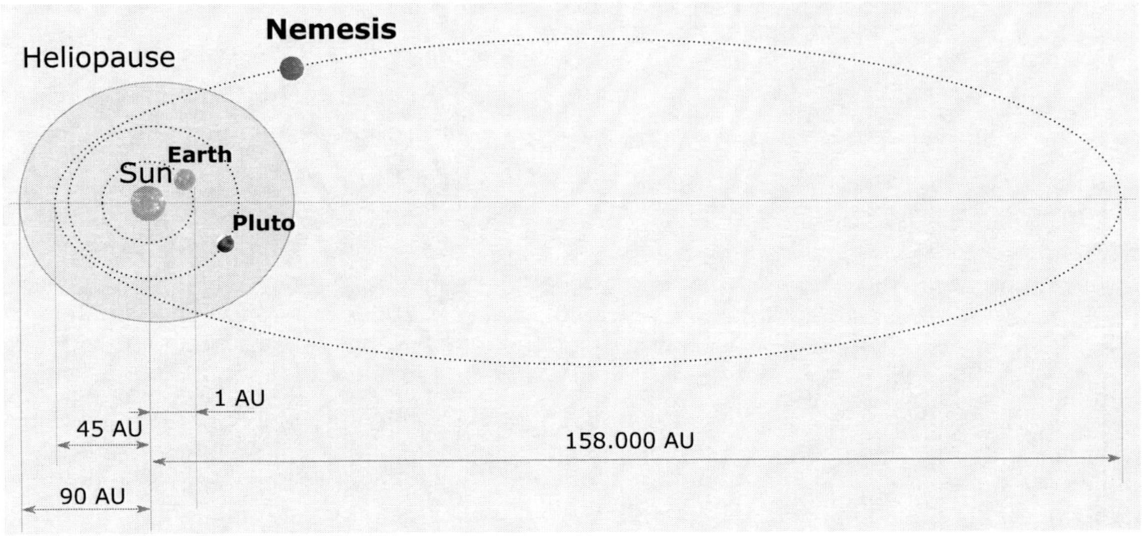

Figure 43: Nemesis' hypothesized orbit – the size of the celestial bodies has been exaggerated. (© Sott.net)

Muller, Raup and others tested samples in relevant geological strata; in particular, they tested some of the numerous fossilized animals and plants killed during the mass extinction, relying heavily on radiocarbon dating.

Chapter 16.

Radiocarbon dating 'anomalies'

Let's take a look at how radiocarbon dating works. Carbon-12 (12C) is 'normal' carbon (6 neutrons and 6 protons). When 2 neutrons are added to it, it becomes carbon-14 (14C). The transformation of carbon-12 into carbon-14 is induced by cosmic radiation. Specifically, cosmic rays (emitted by stars) impact the atmosphere to produce 2.4 atoms of carbon-14 per cm^2 every second. This means that, **on average**, for each atom of carbon-14 there are 10^{12} atoms of carbon-12.[111]

Carbon (both 12C and 14C) combines with oxygen to produce CO2, which is absorbed by plants and then by animals. So, when those living creatures die, they contain one atom of carbon-14 for every 10^{12} atoms of carbon-12. But carbon-14 has a unique characteristic: it decays. Every 5,568 years, half of the carbon-14 disappears (transformed into nitrogen-14). So when a fossil is discovered, scientists measure the carbon-12 over carbon-14 ratio; the higher this figure is, the more carbon-14 decay has occurred and, therefore, the older the sample is.

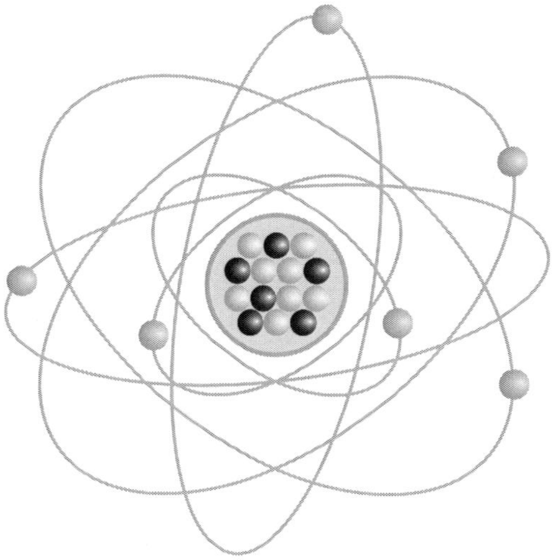

Figure 44: Depiction of a carbon-14 atom made of 6 protons (black), 8 neutrons (grey) and 6 orbiting electrons.

This straightforward method is, understandably, very seductive. Note, however, that the whole process is based on one fundamental assumption: the carbon-12 over carbon-14 ratio is constant. The problem is that when compared to other dating methods (electroluminescence, dendrochronology, archeology, geology, ice core analysis), radiocarbon dating repeatedly presents 'anomalies'. In fact, radiocarbon dating frequently attributes an age to samples that is actually much younger.

These 'anomalies' are due to the fact that the **12C/14C ratio is actually not constant**. This variability is due to several factors. The two first factors are linked to human activities.

> A clue to the possible source of artificially elevated 14C content of Pleistocene remains may be found in the well-documented 'atom bomb effect'. By the mid-1960s, **thermonuclear tests, with their enormous flux of thermal neutrons, had nearly doubled the volume of 14C in the atmosphere and – more important – nearly doubled the 14C activity of buried carbon-bearing materials**. (Taylor, 1987)

[111] Firestone, B. & Topping, W., 'Carbon and radiocarbon dating: A primer', *Mammoth Trumpet* (March 2001): 7–16

In other words, thermonuclear explosions artificially increased the level of 14C, making materials appear younger by radiocarbon dating than they actually were.[112]

The second factor linked to human activity is the burning of fossil fuel that releases carbon-12 (free of any 14C) and therefore messes up the 12C/14C ratio.[113]

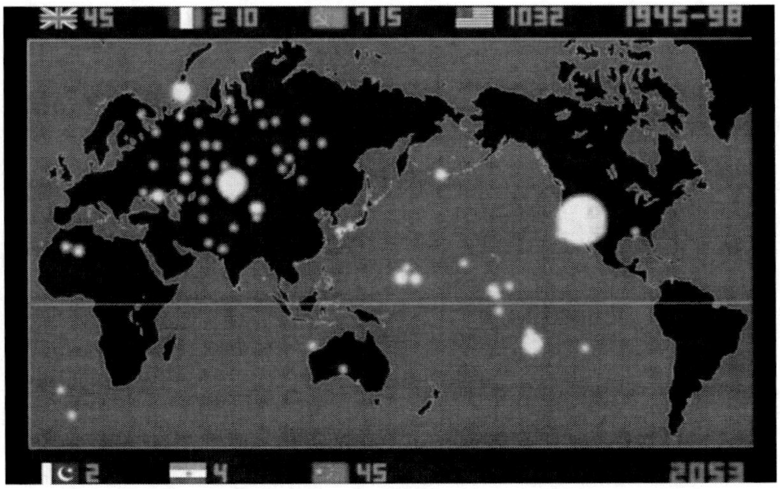

Figure 45: 2,054 nuclear explosions occurred between 1945 and 1998. The total number of explosions per country is displayed at the top of the chart. (© Isao Hashimoto)

The third factor is independent from human activity. It was identified by Lawrence Berkeley National Laboratory researchers Richard Firestone and William Topping, who discovered abnormally low levels of Pu-235 combined with abnormally high levels of Pu-239 in prehistoric samples. Pu-239 is enriched plutonium. This enrichment can only be due to massive neutron bombardment because Pu-239 is Pu-235 plus four neutrons. Since this occurred millions of years ago, before the emergence of civilization, such massive neutron inflows can only be explained by natural causes like supernovas, cometary bombardments or overhead cometary explosions. All three types of cosmic events produce massive neutron bombardments that alter the radiocarbon dating clock and make dated artifacts appear younger than they actually are.[114]

So, ironically, the cause of cyclical mass extinctions might also be the very reason why their dating is off. Cometary events cause mass extinctions while 'resetting' the clock of the fossilized organisms they kill. Like a good murder mystery, the killer manipulated the evidence in such a way as to fool the detective about the 'time of the crime'.

Figure 46 reveals the chasm between the actual 14C levels measured in each sediment strata (black triangle curve) and the theoretical 14C levels in a dating model that assumes *constant* cosmic ray/neutron inflow (grey smooth curve). Notice how chaotic the black curve with triangles is. Clearly the quantity of carbon-14 doesn't drop in a smooth, linear fashion. It exhibits several spikes (c. 32,000 BC, for example), corresponding to massive inflow of 14C – due to cosmic events, most likely cometary bombardment – followed by slow 14C decreases that fit approximately the theoretical model.

In order to be more precise, the theoretical model should take into account these 'resets'. For

[112] *Ibid.*

[113] Blöss, C. & Niemitz, H. U., 'The Self-Deception of C-14 and Dendrochronology', *Zeitensprünge* (1996) 8: 361–389

[114] For example, materials from the Gainey Paleoindian site in Michigan, radiocarbon dated at 2,880 Before Present, give an age by TL [Thermoluminescence] dating of 12,400 yr. B.P. Archaeologists Robson Bonnichsen and Richard Will report in *Ice Age Peoples* (1999) that out of 13 Paleoindian sites in northeastern North America, more than half yielded radiocarbon dates within the Holocene period (the current geological epoch), dates that site investigators consider to be too young. See: Firestone & Topping, *op. cit.*, p. 2

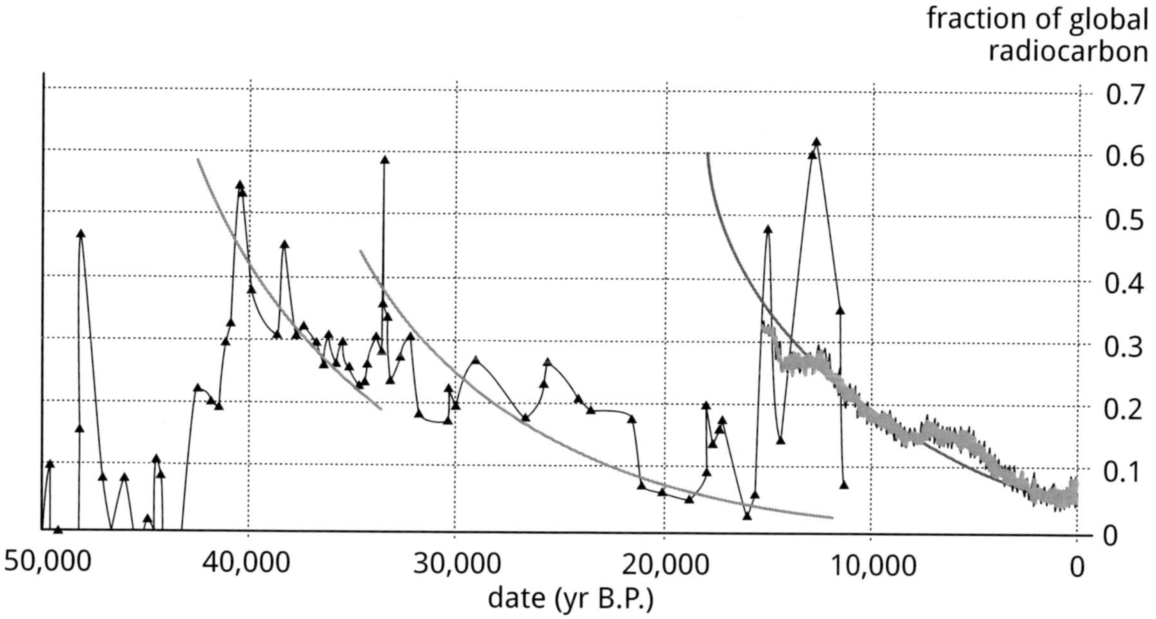

Figure 46: Grey jagged curve – measured 14C age calibration data (Voelker et al.). Grey smooth curve – expected decay rate (without 'reset'). Two descending light grey curves – decay rate including 'resets'. Black curve (triangles) – Icelandic marine 14C data. (© Firestone et al.)

this reason, Firestone has proposed an improved dating model made of three disjointed descending curves (the two descending light grey curves and the grey one). Each curve disjunction reflects a surge in neutron inflow.

Now, let's go back to Muller's 27-million-year (MY) cycle and factor in the above-mentioned dating resets. If the neutron inflows carried by each cometary event is relatively similar, then the 27-MY cycle might still be valid. It might also explain why 8 out of 19 mass extinctions don't exactly match with the 27 MY vertical lines: those 8 events might have triggered atypical neutron inflows

Because of its recent appearance on the global stage, human pollution (nuclear tests and burning of fossil fuel) doesn't alter the periodicity *per se*. It only changes the current 14C/12C ratio and therefore 'slides' all the events in the same direction along the timeline. If this is the case, the last Nemesis-induced extinction might have occurred much earlier or much later than 14 million years ago; **thus the Sun's companion might not currently be at its furthest distance from the Sun as suggested by traditional dating; on the contrary, it may be approaching perihelion.**

The hypothesized proximity of Nemesis is reinforced by the lunar gravitational 'anomaly' described in the following excerpt:

> A recent analysis of a Lunar Laser Ranging (LLR) data record spanning 38.7 yr revealed an anomalous increase of the eccentricity [e] of the lunar orbit ... The present-day models of the dissipative phenomena occurring in the interiors of both the Earth and the Moon are not able to explain it. We examine several dynamical effects, not modeled in the data analysis, in the

framework of long-range modified models of gravity and of the standard Newtonian/Einsteinian paradigm. It turns out that none of them can accommodate de/dt_meas. Many of them do not even induce long-term changes in e; other models do, instead, yield such an effect, but the resulting magnitudes are in disagreement with de/dt_meas.[115]

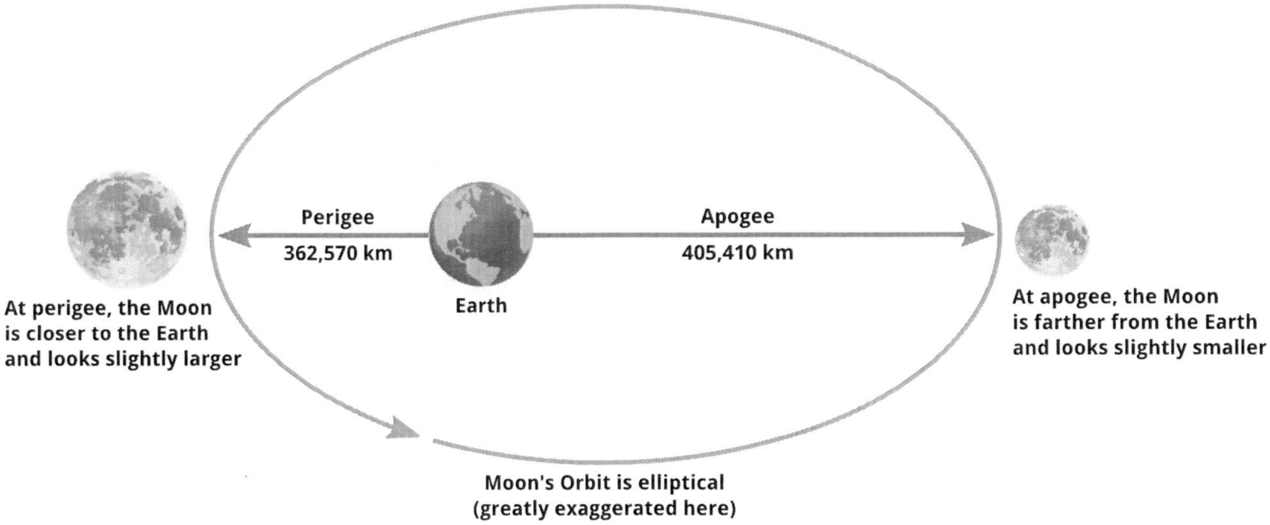

Figure 47: The elliptic (oval) orbit of the Moon. The flatter the ellipse, the higher the eccentricity. (© physics.usu.edu)

To put it simply, *the eccentricity of the Lunar orbit is increasing* (the orbit of the Moon is becoming more and more oblate) and such a lasting change can't be explained by any cause except one:

> A potentially viable Newtonian candidate would be a trans-Plutonian massive object.[116]

The lunar anomalous eccentricity can only be explained by the proximity of a massive unidentified celestial body located beyond Pluto. Over the past years several researchers have reached such a conclusion, naming the unidentified object 'Nemesis',[117] 'Planet X'[118] or 'Tyche',[119] depending on the sources.

The calculations relative to the above-mentioned lunar 'anomaly' point to *a specific mass and distance* from the Sun for the unidentified object. Because of their limited mass, Planet X (about half the mass of our planet) and Tyche (about 4 times the mass of Jupiter) should be close to or within our solar system,[120] and therefore observable.

[115] Iorio, L., 'On the anomalous secular increase of the eccentricity of the orbit of the Moon', *Mon. Not. R. Astron. Soc.* (2011)

[116] *Ibid.*

[117] Melott, A. L. & Bambach, R. K., 'Nemesis Reconsidered', *Monthly Notices of the Royal Astronomical Society Letters* (2010)

[118] Lykawka, P. S. & Mukai, T., 'An Outer Planet Beyond Pluto and the Origin of the Trans-Neptunian Belt Architecture', *Astronomical Journal* (2008)

[119] Matese, J. J. & Whitmire, D. P., 'Persistent evidence of a jovian mass solar companion in the Oort cloud', *Icarus* (2011)

[120] 30 AU for Planet X and 200 AU for Tyche. See: Iorio, *op. cit.*, p. 5

For such lunar perturbations to occur, Nemesis, however, (about 0.56 of the Sun's mass) should be about 4500 AU[121] from the Sun, close enough to interact with our solar system but too far away to be directly observed (especially if Nemesis is a dark body like a brown dwarf).

[121] 100 times the Sun–Pluto distance.

Chapter 17.

The grounding of the Sun

Okay, let's recap a little here. The Electric Universe theory posits that comets, planets and stars are giant capacitors undergoing constant electrical exchanges between one another and occasionally discharging powerfully. In the case of the Sun, discharges occur between its surface and its double layer or 'Langmuir sheath', located beyond Pluto's orbit. These discharges take the form of violent solar explosions generating a powerful solar wind made of ionized particles, which travel towards the boundaries of the solar system. The net charge of this solar wind is slightly positive. Concomitant with the solar wind are electron drifts in the opposite direction, from the heliosheath to the Sun (see figure 48).

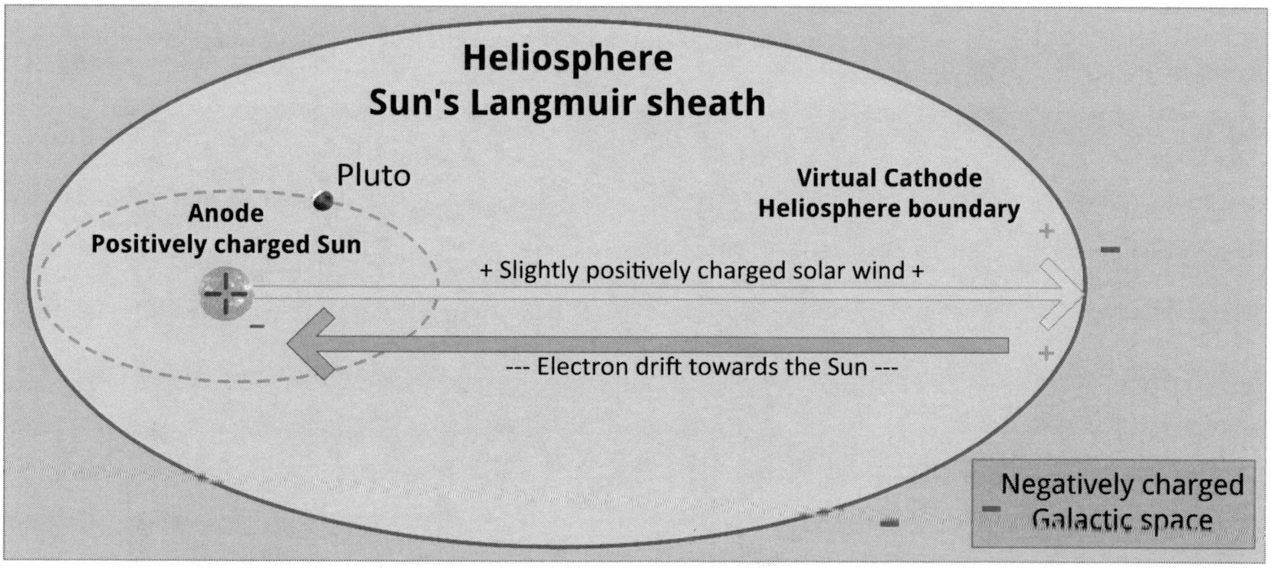

Figure 48: The Sun emits a permanent flow of ions and radiation. Where does this energy come from? (© Sott.net)

Solar activity produces spots on the surface of the Sun that result from intense magnetic activity. An increase in the number of sunspots is accompanied by an increase in the solar wind. With a velocity of about 250 miles/sec (400 km/sec), it takes the solar wind about 4.5 days to reach Earth.

Discharges in a capacitor reduce the electric charges of the anode and cathode until they eventually reach the same electric potential and the discharges stop. But if the electrodes maintain different potentials – if they are plugged into an external source of energy that will continually recharge the capacitor – then it can discharge again and again. In the same way, the Sun, despite countless discharges over billions of years, has not stopped shining. It seems that the Sun taps into an *external* source of energy (as proposed in Chapter 9). However, mainstream science states that the Sun is *internally* powered:

> In 1920 the eminent authority Sir Arthur Eddington asked whether the Sun got its energy from an external or an internal mechanism. **He could think of no external mechanism, but the new science of atomic physics provided a possible internal one:** the fusion of hydrogen into helium.[122]

The official explanation is the usual nuclear fusion theory we learn about in classrooms. According to this theory, the fusion of hydrogen atoms triggered by massive pressure and temperature leads to the creation of helium atoms (two hydrogen atoms fused together) and a massive quantity of energy. The problem with this theory is that it exhibits some major anomalies, the most striking of which relate to neutrinos, electrically neutral subatomic particles.

Neutrinos are the **main evidence that nuclear fusion occurs in the Sun's core**. But according to the measured energetic output generated by the Sun, the corresponding nuclear fusion should produce three times more neutrinos than what is observed. Mainstream scientists have attempted to brush aside this inconsistency by elaborating *ad hoc* theories where neutrinos have three different flavors which change en route from the core of the Sun to its surface.[123] But as is the case with 'dark matter', there's no direct evidence for this; the theory was created after the fact to explain the 'anomalous' data.

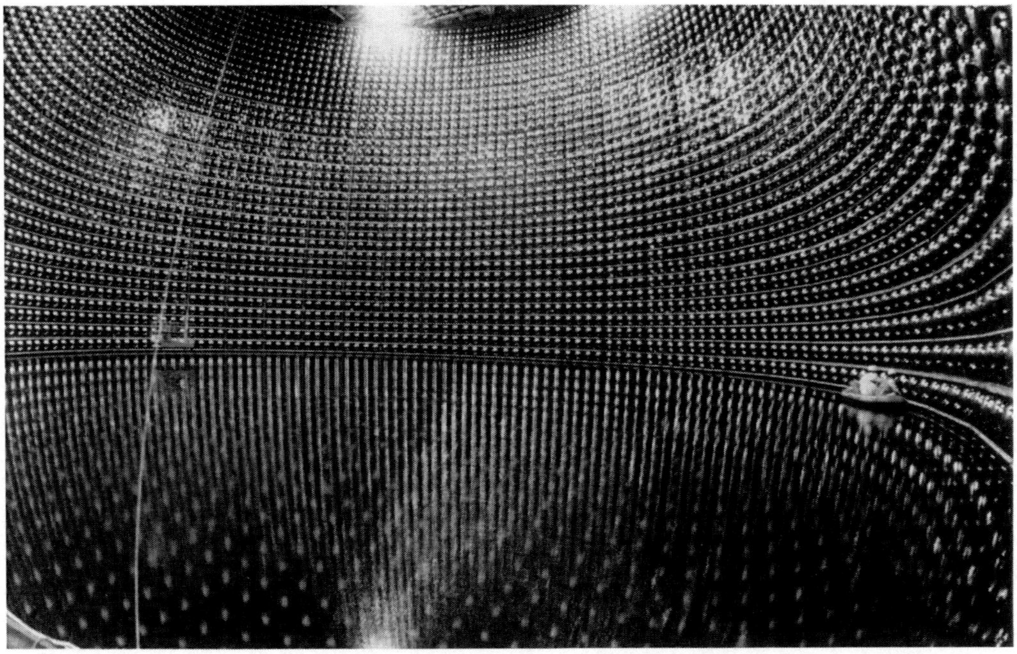

Figure 49: The Super Kamiokande neutrino observatory is built under Mount Kamioka near the city of Hida, Gifu Prefecture, Japan. See rubber dinghy on the right for scale. (© Cartwright)

[122] Scott, D. E., *The Electric Sky*, p. 84

[123] The three 'flavors' of neutrinos are 'tau-neutrino', 'mu-neutrino' and 'electron-neutrino'.

The second problem is something called 'neutrino variability'. Observations show that the Sun's neutrino output varies inversely with the sunspot cycle. If neutrinos were truly produced in the 'nuclear furnace' in the Sun's core, this relationship would be inconceivable, since solar physicists calculate that it takes about 200,000 years for the energy of internal fusion to affect the Sun's surface. In other words, neutrinos and sunspots show a direct correlation in time, not one separated by a couple of hundred thousand years. The *external* source theory, in contrast, doesn't struggle to accommodate these major 'anomalies'.

Despite the current high cometary activity in our solar system, the solar capacitor is not discharging at the level it should be, and the current solar cycle is both weak and 'late'. It is a paradoxical situation: cometary activity hasn't been this high for decades – perhaps centuries – while solar activity hasn't been this low for at least 100 years.[124] Obviously, the electric charge of the Sun has somehow dropped. Why?

What effect might a solar companion have on the Sun? Does it energetically connect to the Sun? Does it 'drain' the Sun? Does it 'ground' the Sun? Does it get its power from the same 'plug' or 'socket' as the Sun? At this point, we are, of course, entering speculative territory. Nevertheless, if we follow the observations previously made about the external power source of celestial bodies, it seems that stars are powered by galactic arms, hence their spiral alignment along galactic arms.[125] If Nemesis is approaching the Sun, it conceivably receives energy from the same source as the Sun – the local region of the Orion arm where the Sun is located – thus depleting the energetic source that normally feeds the Sun.[126]

In the same way you notice a drop in electric current in your house (lights getting dimmer) when you use powerful electric equipment, the local plugging in of Nemesis may reduce the electricity transmitted from this local region of the Orion arm to the Sun.

Alternatively, it could be that a direct connection is developing between the Sun and Nemesis. If that's the case, why isn't there any visible transfer of energy between the two stars? An example would be recurring solar flares and solar winds directed towards Nemesis' location. Is it possible for two celestial bodies to exchange energy in this way, invisibly and unmeasurable by the usual detectors? Mathematicians and quantum physicists consider this theoretically possible.

In 1921, German mathematician Hermann Weyl developed the 'wormhole' theory.[127] Wormholes are hypothetical 'shortcuts' through space-time, like a tunnel connecting two distant points in space-time. Although researchers have no observational evidence for wormholes, it is a robust theoretical concept that fits with the equations of the theory of general relativity.

Initially, wormholes were predicted to only exist at a microscopic level (less than 10^{-33} centimeters) and only for a very short period of time.[128] However, research conducted by Russian physicist Sergei

---

[124] 'Strange doings on the sun', *Wall Street Journal* (10 November 2013). See: online.wsj.com/news/articles/SB10001424052702304672404579183940409194498

[125] Scott, *op. cit.*, pp. 85–86

[126] Recall what was said above about binary stars: When a sphere is divided ... the total mass will remain the same ... but the total surface area of this pair will be about 26% larger than the area of the original single sphere. This increases the total surface area exposed to the electric field and thus decreases the current density (amperes per square meter). Thus, electrically-induced fission enables stars to reduce the electric stress they are subjected to by spreading it between two or more stars.

[127] 'Herman Weyl', *Stanford Encyclopedia of Philosophy*. See: plato.stanford.edu/entries/weyl/

[128] Redd, N. T., 'What is a Wormhole', *Space.com*. See: www.space.com/20881-wormholes.html

Krasnikov[129] suggests that wormholes could remain stable for long periods of time.

Figure 50: Artistic depiction of a wormhole. (©Igartist 79)

In addition, the possibility of macro-wormholes has been established by American theoretical physicist John Archibald Wheeler:

> The 'wormholes' predicted by quantum geometrodynamics are a property of all space, are submicroscopic, and they and the fluxes through them arise spontaneously, through quantum fluctuations. Nothing prevents one from considering also a single wormhole, of macroscopic dimensions, created *ab initio*, with a prescribed flux threading through it, and evolving deterministically in time in accordance with the classical field equations. However, this classical electric charge has not the slightest direct connection with the charges of the real world of quantum physics and requires no consideration here.[130]

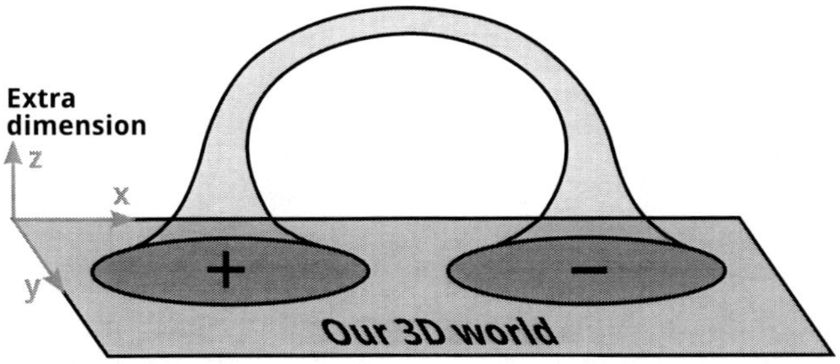

Figure 51: Classical geometrodynamical concept of charge as 'electric lines of force trapped in the topology of space.' (© Sott.net, adapted from J.A. Wheeler)

In Figure 51, the geometry of space is depicted in three dimensions. But there are actually four dimensions (three for our own space and one for the wormhole). Thus, for simplicity, our space is reduced to two dimensions (the light grey rectangle along the x-axis and the y-axis) while the third dimension (z-axis) is the invisible wormhole dimension. The wormhole (represented by the light grey connecting tube) connects two regions of our own three-dimensional space. An observer endowed with an instrument of inadequate resolving power sees one wormhole 'portal' as a positive charge, the other as a negative charge (dark grey ellipses).[131] Imagine that the distance between the two portals of the wormhole, as measured in our own space (represented by the flat plane) is 10 km. Einstein's field equations would permit such a distance, as measured through the wormhole, to be radically different, say 10 meters – or even less. Scale this up to the galactic level and wormholes can theoretically connect very distant points of the universe via a wormhole that is much, much shorter.

In summary, the energetic interaction between the Sun and Nemesis may be nonlocal in nature. Despite great distances and no measurable evidence (with current instruments) of interaction, the Sun and Nemesis might nonetheless interact in a way that triggers some very tangible environmental changes observable on the Sun and here on Earth.

129 Krasnikov, S., 'Traversable wormhole', *Physical Review D* (2000) 62(8)
130 Wheeler, J. A., *Superspace and Quantum Geometrodynamics*, pp. 265–267
131 *Ibid.*, p. 266

Chapter 18.

Comets or asteroids?

As depicted in figure 52, and routinely argued by mainstream science, comets are 'chunks of ice and rock', a.k.a. 'dirty snowballs'. This *belief*, however, is incompatible with the actual data. For instance, in 2011 Comet Lovejoy plunged into the Sun's atmosphere and emerged on the other side after an hour-long journey through the sun's corona. Its size and brightness didn't seem to have diminished.[132] Here are some (pretty typical) comments from observers of this event:

> This morning, an armada of spacecraft witnessed something that many experts thought impossible. **Comet Lovejoy flew through the hot atmosphere of the sun and emerged intact. 'It's absolutely astounding,'** says Karl Battams of the Naval Research Lab in Washington DC. 'I did not think the comet's *icy core* was big enough to survive plunging through the *several million degree* solar corona for close to an hour, but Comet Lovejoy is still with us.'[133]

But if the temperature of the Sun's corona is several million degrees,[134] and if Comet Lovejoy is no more than a chunk of ice estimated to be just a few hundred meters in diameter,[135] how was it possible that it wasn't vaporized?

In the same vein, in 2004 the space probe *Stardust* flew about 300 km from Comet Wild 2 and shot detailed pictures of it.[136] According to the textbooks, those pictures of the comet should have revealed a mass of 'dirty snow'. Here's what the *Stardust* program director said when the images came back:

> 'We thought Comet Wild 2 would be like a dirty, black, fluffy snowball,' said *Stardust* Principal Investigator Dr. Donald Brownlee of the University of Washington, Seattle. 'Instead, it was mind-boggling to see the diverse landscape in the first pictures from *Stardust*, including spires, pits and craters, which must be supported by a cohesive surface.'[137]

---

[132] If you open the following link to an official NASA page, you'll see footage of Comet Lovejoy plunging into the Sun's atmosphere and, one hour later, re-emerging on the other side: www.nasa.gov/mission_pages/sunearth/news/comet-lovejoy.html

[133] 'Comet Lovejoy Plunges into the Sun and Survives', *NASA* (16 December 2011). See: science.nasa.gov/science-news/science-at-nasa/2011/16dec_cometlovejoy/

[134] The corona is the plasma atmosphere that surrounds the Sun. Its temperature is estimated between one and five million degrees centigrade. See: North, G., *Astronomy in Depth*, p. 125

[135] 'Comet Lovejoy Plunges into the Sun and Survives', *NASA* (16 December 2011). See: science.nasa.gov/science-news/science-at-nasa/2011/16dec_cometlovejoy/

[136] 'Comet 81P/Wild2 and the Stardust Mission', *Dept. of Physics and Astronomy, University of Leicester*. See: le.ac.uk/departments/physics/research/src/res/planetary-science/comet-wild2-and-the-stradust-mission

[137] 'NASA Spacecraft Reveals Surprising Anatomy of a Comet', *JPL News, NASA*.
See: www.jpl.nasa.gov/news/news.php?release=2004-154

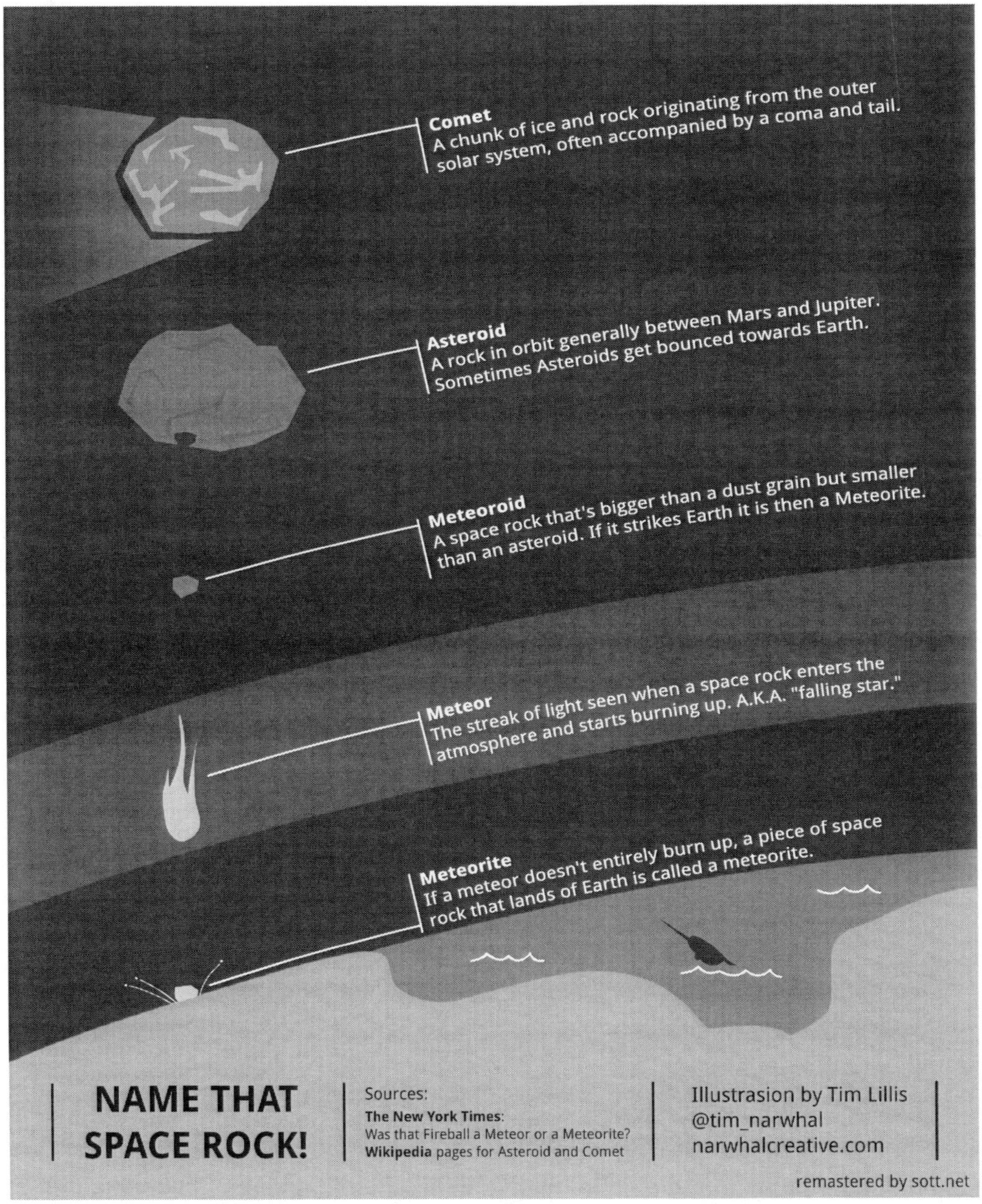

Figure 52: Comet, asteroid, meteor and meteorite classification, according to mainstream science. (© Narwhall)

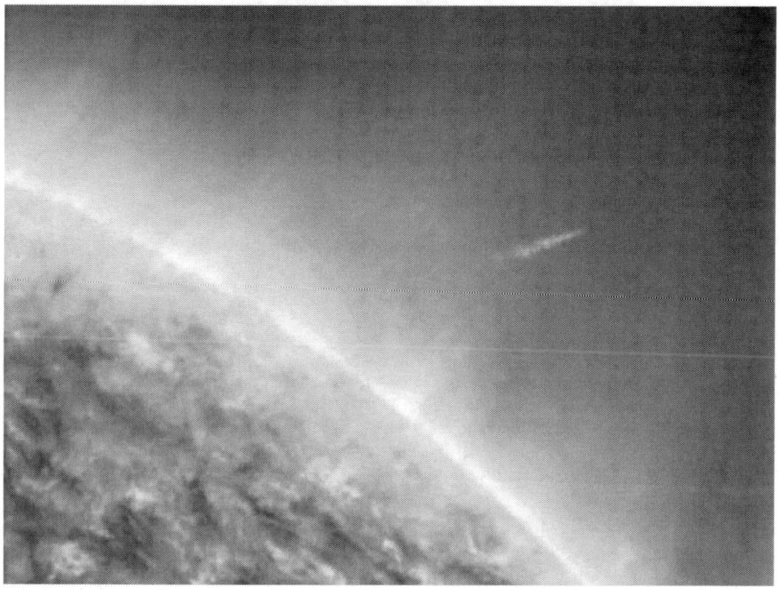

Figure 53: Comet Lovejoy re-emerging from behind the Sun on December 15th, 2011. (© NASA/SDO)

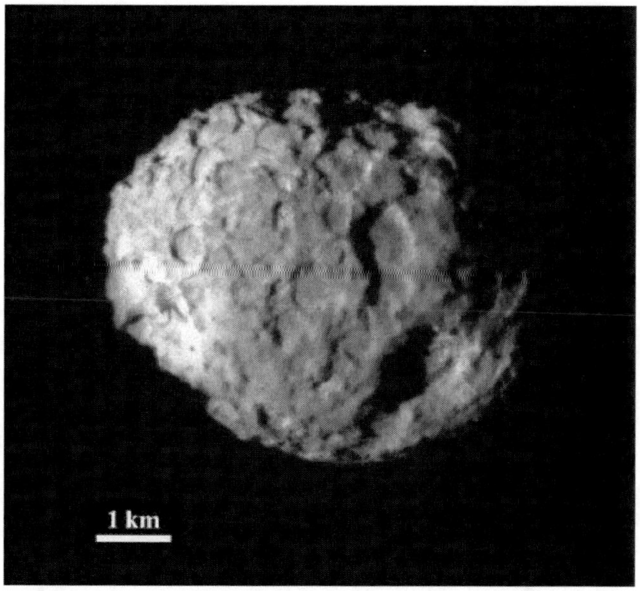

Figure 54: Comet Wild 2 looks like an asteroid. No ice, but a rocky surface scarred by craters. (© JPL/NASA)

Figure 55: 'Active' asteroid P/2013 P5. (© NASA, ESA, D. Jewitt/UCLA)

So, comets don't seem to be dirty snowballs after all. From the data presented above, they are glowing chunks of rock. On the other side, asteroids don't seem to be the non-glowing chunks of rock posited by mainstream science. For example, asteroid P/2013 P5 recently puzzled the whole scientific community when it started exhibiting a million-mile-long glowing tail. To rationalize this oddity, official science claimed the asteroid was spinning so fast that it was ejecting tons of dust, while finally acknowledging that the difference between 'comets' and 'asteroids' might not be so clear-cut.[138]

The fundamental difference between asteroids and comets is not their chemical composition, i.e. dirty, fluffy, icy comets vs. rocky asteroids. Rather, as has long been put forward by plasma theorists, what differentiates 'comets' from 'asteroids' is their electric activity.

When the electric potential difference between an asteroid and the surrounding plasma is not too high, the asteroid exhibits a dark discharge mode[139] or no discharge at all. But when the potential difference is high enough, the asteroid switches to a glowing discharge mode.[140] At this point the asteroid is a comet. From this perspective, a comet is simply a glowing asteroid and an asteroid is a non-glowing comet. Thus the very same body can, successively, be a comet, then an asteroid, then a comet, etc., depending on variations in the ambient electric field it is subjected to.[141]

Note that a comet can also exhibit the third plasma discharge mode, namely lightning or 'arc discharge mode', which is probably what happened when Comet Shoemaker-Levy entered the vicinity of Jupiter in July 1994:

---

[138] 'NASA's Hubble Sees Asteroid Spouting Six Comet-Like Tails', *NASA* (7 November 2013). See: nasa.gov/press/2013/november/nasas-hubble-sees-asteroid-spouting-six-comet-like-tails/#.Uq9i4bTDviF

[139] See Chapter 5: 'Discharge modes'

[140] An intense circulation of ions and electrons occurs between the asteroid and the surrounding space. The energy provided by this intense transfer 'excites' electrons which generate photons, hence the glow of the asteroid. See: Meichsner, J., *Nonthermal Plasma Chemistry and Physics*, p. 117

[141] Thornhill, W. & Talbott, D., *The Electric Universe*, pp. 95–99

Figure 56: 'Impact' of fragment 'G' of comet Shoemaker-Levy on Jupiter. (© Siding Spring telescope)

> Astronomers expected the encounter to be a trivial event. 'You won't see anything. The comet crash will probably amount to nothing more than a bunch of pebbles falling into an ocean 500 million miles from Earth.' Then came the encounter and an about face. As reported by *Sky & Telescope*, "When Fragment 'A' hit the giant planet, it threw up a fireball so unexpectedly bright that it seemed to knock the world's astronomical community off its feet..."
>
> The Hubble Space Telescope (HST) detected *a flare-up* of fragment 'G' of Shoemaker-Levy *long before impact at a distance of 2.3 million miles* from Jupiter. For the electrical theorists this flash is the expected electrical discharge as the fragment crossed Jupiter's plasma sheath, or magnetosphere boundary.[142]

Mainstream cosmology, on the other hand, has no explanation for such an event.

The last two categories in the above-mentioned classification[143] are 'meteors' and 'meteorites'. Meteors are simply asteroids or asteroid fragments that reach the Earth's atmosphere, and meteorites are the chunks of meteors that manage to reach the Earth's surface. So throughout its life, an asteroid can be a comet (when it is in glowing discharge mode), a meteor (when it enters the atmosphere) and finally a meteorite (if it reaches the Earth's surface).

In late November 2013, Comet ISON provided us with real-time observations that, in terms of mainstream expectations of how comets *ought* to behave, revealed a number of inconsistencies:

1. Expected brightness versus observed brightness

[142] 'Deep Impact and Shoemaker-Levy 9', *Thunderbolts.info* (18 July 2005). See: www.thunderbolts.info/tpod/2005/arch05/050 718deepimpact.htm

[143] See figure 52 at the very beginning of this chapter.

Figure 57: November 12th versus November 14th: Images of Comet ISON showed a sudden increase in brightness (© Juanjo Gonzalez)

ISON was supposed to be the 'Comet of the Century' because of its predicted brightness. It was going to be even brighter than the full moon, according to some popular sources.[144] However, ISON didn't reach any such level of brightness. The brightness of the full moon[145] is magnitude -13 and ISON, during its journey towards the Sun, exhibited[146] a brightness between 19, when it became observable by amateur telescopes, and -2, its brightest magnitude[147] on November 28th, when it passed Delta Scorpii.[148]

Perhaps the belief that comets are 'dirty snowballs' and the universe is fundamentally non-electric contributed to overestimated magnitude forecasts.

An understanding of the electrical nature of celestial bodies and their interactions tells us that the brightness of comets should be *proportional to the electrical field they cross.* Weak solar activity weakens the heliospheric electric field, which then reduces the electric potential difference between the comet and its surrounding space. That could explain why ISON was dimmer than expected.

2. Expected outburst versus observed outburst

En route towards the Sun, ISON experienced several unexpected outbursts when its brightness suddenly increased dramatically. For example, between November 12th and November 14th, ISON's brightness jumped from magnitude 8 to 4.[149] This means that in just 72 hours, it increased nearly 16-fold in brightness.[150]

---

[144] Grego, P., 'New comet might blaze brighter than the full Moon', *Astronomy Now* (25 September 2012). See: astronomynow.com/news/n1209/25comet/

[145] The higher the magnitude number, the lower the brightness. Typically the Sun's brightness is about -27; the full moon is about -13; the brightest modern comet, Ikeya-Seki (1965), was -10; the brightness of Venus is -5. Objects with a magnitude up to 6.5 are visible to the naked eye.

[146] Bakich, M., 'Comet ISON will light up the sky', *Astronomy Magazine* (25 September 2012). See: www.astronomy.com/news/2012/09/new-comet-will-light-up-the-sky

[147] Battams, K., 'Very quick update', *NASA Comet ISON Observing Campaign* (27 November 2013). See: www.isoncampaign.org/karl/quick-update

[148] Star located in the constellation Scorpius.

[149] 'Comet ISON Visible To Naked Eye After Outburst Of Activity, Observers Say', *Huffington Post* (15 November 2013). See: www.huffingtonpost.com/2013/11/15/comet-ISON-visible-naked-eye_n_4280 731.html

[150] The magnitude scale is logarithmic. For example, increasing magnitude from -9 to -6 (+3) means doubling the brightness.

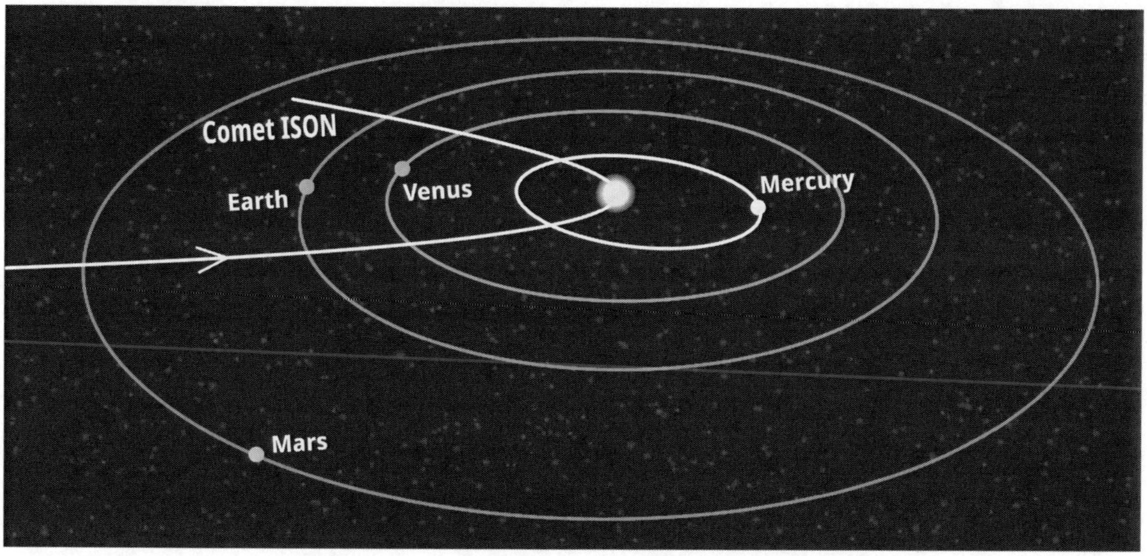

Figure 58: The eccentric orbit of Comet ISON. (© Unknown)

NASA *et al.* didn't predict these outbursts and had great difficulty explaining them, invoking outlandish factors such as water production,[151] jets of gas and steam,[152] and spinning.[153]

If we take into account the electrical interaction of celestial bodies, ISON's outburst makes sense because a few days before this event, two class-X solar flares occurred: an X1.1 flare on November 8th and an X1.1 flare on November 10th.[154]

Solar flares are massive discharges of solar particles that have an overall positive charge. When these powerful solar winds reached ISON a few days after their ejection, they subjected this negatively charged body (because it comes from the negatively charged edges of the solar system) to a positively charged environment. This electric difference led to:

a.) A massive discharge between the comet and its surrounding space, resulting in the comet glowing more intensely.

b.) An increase in the comet's electric potential, bringing it closer to the electric potential of the surrounding space. That's one of the two factors explaining ISON's survival to make the rendezvous for a solar fly-by (see inconsistency number 3 below).

Comet ISON's orbit (see figure 58) was similar to that of most comets and follows a highly elliptical orbit.[155] Usually aphelion lies beyond the region of Jupiter, while perihelion can be less

[151] Ferrin, I., 'Comet ISON is in Outburst', *The Comets Page of Ignacio Ferrin.* See: astronomia.udea.edu.co/cometspage/OUTBURST.html

[152] Gary, B. L., 'Comet ISON Observations by Three Non-Professional Observers', *Bruce Gary personal website.* See: brucegary.net/ISON/

[153] 'Will Comet ISON Survive Its Brush with the Sun?', *Space Ref.* (9 October 2013). See: spaceref.com/news/viewpr.html?pid=41 750

[154] 'My Solar Alerts'. See: mysolaralerts.blogspot.fr/p/solar-flare-list.html

[155] 'Eccentricity of an Orbit', *Windows to the Universe.* See: windows2universe.org/physical_science/physics/mechanics/orbit/eccentricity.html

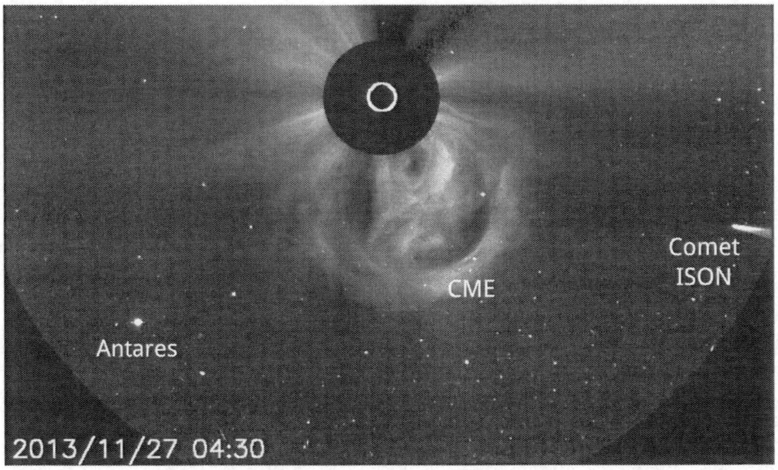

Figure 59: Glowing Comet ISON approaching the Sun during a massive electric discharge in the form of a 'coronal mass ejection' (CME). (© SOHO/NASA)

than one astronomical unit (i.e. one Sun-Earth distance) from the Sun.[156] ISON's aphelion is unknown. It can't be exactly determined since ISON experienced its first identified passing in 2013, therefore its orbital period (if any) can't be calculated. However, its very elongated trajectory suggested a remote aphelion near or beyond the heliopause (the outer boundary of the DL described in Chapter 7).

Because of their highly eccentric orbits, the trajectory followed by most comets is almost perpendicular to the Sun's electric field. This means that the surrounding electric potential rapidly changes during the comet's journey across the solar system.[157] This subjects the comet to increasing electric stress brought on by increasing electric potential difference between the comet and its surrounding space. This imbalance in electric potential triggers massive solar discharges and comet outbursts, as illustrated in figure 59.

By comparison, the non-electric comets we've come to know as 'asteroids' usually follow more circular orbital paths, characterized by lower electric potential variations (because they maintain roughly the same distance from the Sun throughout their orbits). This would explain why usually asteroids are not in glowing mode, as is the case with asteroids in stationary clouds, asteroids following a circular orbit around the Sun, and asteroids located in the belt between Mars and Jupiter.[158]

3. Predicted perihelion events versus observed perihelion events

Perihelion is generally considered a dangerous time for comets. This was especially the case for Comet ISON because it was supposed to be a one-mile diameter 'dirty snowball' passing less than

[156] In the case of ISON, the perihelion was about one million miles from the center of the Sun, i.e. 0.012 AU. 'MPEC 2013-W16: COMET C/2012 S1 (ISON)', *International Astronomical Union – Minor Planet Center* (26 November 2013). See: www.minorplanetcenter.net/mpec/K13/K13W16.html

[157] Thornhill & Talbott, *op. cit.*, pp. 90–95

[158] Also known as the 'main asteroid belt'. Its total mass is supposed to be 4% of our Moon. See: Krasinsky, G. A., Pitjeva, E. V., Vasilyev, M. V., Yagudina, E. I., 'Hidden Mass in the Asteroid Belt', *Icarus* (July 2002) 158(1): 98–105

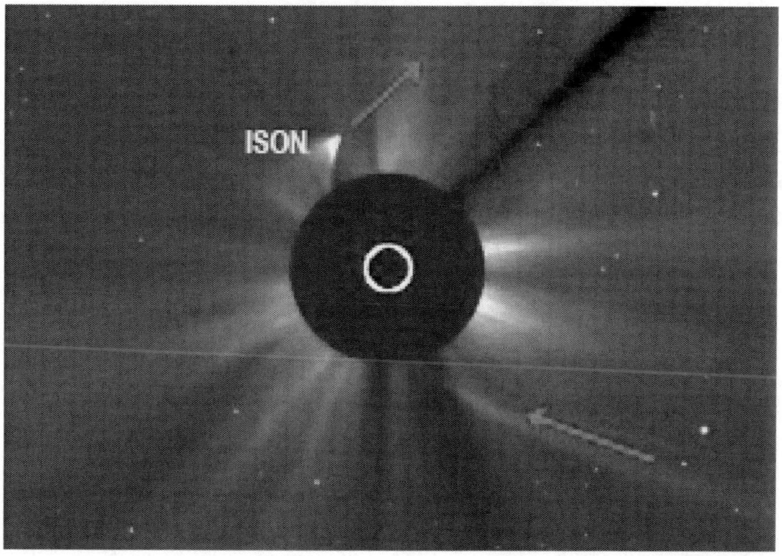

Figure 60: ISON survived its fly-by around the Sun and appeared after a period of invisibility. (© SOHO, NASA)

one million miles away from the Sun's center where temperatures reach about 2,700° Celsius,[159] more than twice the temperature necessary to melt iron.[160]

Probably recalling Comet Lovejoy's 'mission impossible' when it passed through the Sun's millions °C corona without sustaining much damage, NASA scientists were more cautious as ISON's solar rendezvous approached and wondered if it might survive the encounter.[161]

When ISON entered the solar vicinity and disappeared from observation for several minutes, commentators announced its death.[162] It would have been a relief for them, interpreted as confirmation of their dubious 'dirty snowball' theory.

But Thanksgiving Day 2013 had a surprise in store: after a period of unbearable suspense, ISON reappeared on the other side of the Sun (see figure 60).

The ISON fly-by revealed two interesting pieces of data:

a.) It became *dimmer* as it approached the Sun;[163]

b.) It didn't disintegrate.

These two observations run contrary to the dirty-snowball model. When an 'icy comet' gets closer

[159] 'Can You Spot Comet ISON at Perihelion?' *Sky and Telescope* (22 November 2013). See: www.skyandtelescope.com/observing/highlights/Spot-Comet-ISON-at-Perihelion-233011 581.html

[160] That's 4,900° Fahrenheit. For reference, iron's melting point is 1,200° Celsius, i.e. 2,100° Fahrenheit. See: www.engineeringtoolbox.com/melting-temperature-metals-d_860.html

[161] 'Comet Ison's sun flyby survival odds weighed by NASA', *CBC News* (27 November 2013). See: www.cbc.ca/news/technology/comet-ISON-s-sun-flyby-survival-odds-weighed-by-nasa-1.2442 205

[162] 'Is Comet ISON Dead? Astronomers Say It's Likely After Icarus Sun-Grazing Stunt', *Universe Today* (28 November 2013). See: universetoday.com/106 813/is-comet-ISON-dead-astronomers-say-its-likely-after-icarus-sun-grazing-stunt/

[163] Pesnell, D., 'You would expect it to get brighter and brighter but unfortunately it got dimmer and dimmer as it got close to the sun', *The Globe And Mail.* See: theglobeandmail.com/news/world/comet-ISON-pulls-a-disappearing-act-after-close-approach-to-the-sun/article15654 037/

to the Sun, it should become brighter (more evaporation, more gas jets) and it should eventually melt away. Neither happened. In fact, the opposite happened, which makes sense from the point of view of an electric universe.

During its solar fly-by, a comet is subjected to a rather constant electric field. In figure 61, we can see that ISON passed through different electric field lines (illustrated by the concentric circles +1, +2 ...) during its approach. As described previously, this electric potential difference between the comet and its surrounding space triggered solar flares, intense glowing, and raised ISON's electric potential.

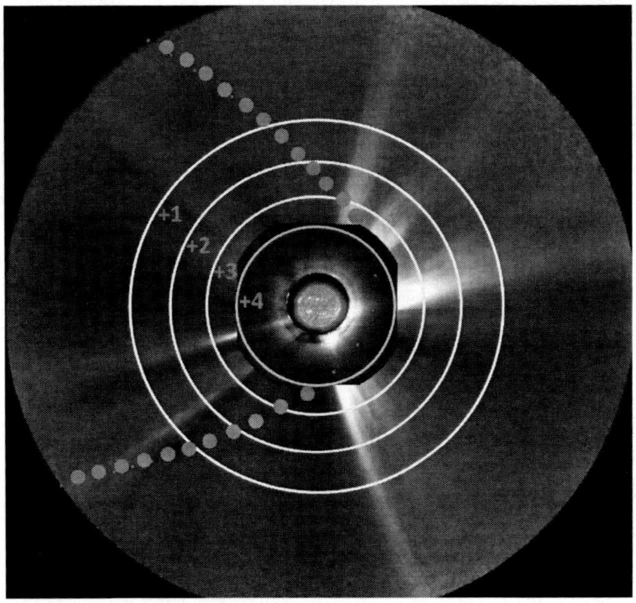

Figure 61: ISON's fly-by: electric fields and glowing intensity. (© Sott.net)

By the time ISON entered the solar vicinity (demarcated by the smallest concentric circle, the +4 region in the drawing), its potential would already have become quite positive. In addition, at this point it entered a region of rather constant electric potential.

Unlike its crossing of the electric field lines[164] +1, +2 and +3, which it crossed almost perpendicularly, by the time it reached the +4 field line, its trajectory was almost parallel to it, meaning that the surrounding electric potential was almost constant.

As a result, ISON 'had a breather' and was subjected to reduced electric stress. Less electric difference between the comet and the surrounding space therefore produced less discharge, less glowing, and it also reduced the chances that the comet would disintegrate.

Karl Battams, the U.S. Naval Research Laboratory astrophysicist who operates NASA's SOHO and STEREO observatories' Sungrazing Comets Project, had announced the death of ISON, making the following statement:

> 'ISON is **just weird**. It has **behaved unpredictably at times**. When it's done something **strange**, we spent some time scratching our heads, figuring out what is going on and we think we know what it's doing ... **it then goes and does something different**.'[165]

When observations keep providing data that are opposite to what the conventional model predicts, the logical conclusion should be that the model is flawed. Of course, it could be that the scientific agenda here is not to find truth, but rather to maintain the dominant dogma, in which case one

[164] An electric field line defines locations where the electric potential is the same. It's similar to the altitude lines on a geographic map where every point of the line is at the same altitude.

[165] 'Is Comet ISON Dead? Astronomers Say It's Likely After Icarus Sun-Grazing Stunt', *Universe Today*, 28 November 2013. See: universetoday.com/106 813/is-comet-ISON-dead-astronomers-say-its-likely-after-icarus-sun-grazing-stunt/#ixzz2m2YUyNIU

would naturally blame the observed phenomena (instead of the official explanation) by calling them 'weird', 'strange' and 'unpredictable'.

In the end, ISON didn't 'melt away' as expected. The solar fly-by only fragmented it (which is understandable when rocky material is exposed to very high temperatures) and in January 2014, it provided a final show for planet Earth in the form of several fireballs that were reported by several observers.[166]

[166] Lunsford, R., 'Meteor activity from Comet ISON', *American Meteor Society.* See: www.amsmeteors.org/2014/01/meteor-activity-from-comet-ison/

Chapter 19.

Comets: Cycles and origins

Muller's description of a companion star pushing and pulling asteroids or comets along its trajectory, and the cometary bombardments induced by Nemesis' journey through the solar system, account for just one cycle of cometary activity (albeit a highly destructive one) among many others. Comets exhibit all kinds of orbital periods, from a few years to centuries in duration. Usually the longer the period, the more eccentric the orbit and, therefore, the brighter the comet.

Short-period comets have orbits that average between 3.3 and 20 years.[167] These comets usually circle around the Sun and Jupiter and exhibit limited brightness.[168] Comet Encke is one of the 400+ known Jupiter-family comets.[169]

Figure 62: Comet Encke was one of the comets visible with a naked eye at the end of 2013. The image above was shot on October 30th, 2013. (© D. Peach)

Medium-period comets generally follow larger and more eccentric orbits, lasting between 20 and 64 years. They usually circle around the Sun and outer planets[170] of the solar system. Comet Crommelin,[171] with its 28-year orbital period circling the Sun and Uranus, is fairly typical in this category.

Long-period comets complete their orbits in 64 to 164 years.[172] Their aphelia are usually outside the solar system. Most long-period comets have not been observed more than once or twice. One exception is the famous Halley's Comet, with an orbital period of

[167] 'Comet Encke', *Encyclopaedia Britannica.* See: global.britannica.com/EBchecked/topic/186 530/Enckes-Comet

[168] Moore, P., *Guide to the Comets*, p. 32

[169] *Encyclopedia of Astronomy*, Swinburne University of Technology. See: astronomy.swin.edu.au/cosmos/J/Jupiter-family+comets

[170] The 4 outer planets are Jupiter, Saturn, Uranus, and Neptune. Cessna, A., 'The Outer Planets', *Universe Today.* See: www.universetoday.com/33 292/the-outer-planets/

[171] 'Countdown to 500 comets', *Earth Institute.* See: www.earthriseinstitute.org/coms48.html

[172] Comet Grigg-Mellish, for example. See: Matheus, P. & Jenniskens, M., *Meteor Showers and Their Parent Comets*, p. 315

Figure 63: The famous Halley's Comet. This image was taken during its last passage in 1986. (©NASA/ESA/Max-Planck-Institute for Solar System Research)

about 75 years. Each of its seven come-backs has been documented since its discovery in 1531.[173]

Forty or so comets are orbiting the Sun with periods of 100 to 1,000 years, while the majority take much longer. Some have been estimated to exhibit orbits greater than 40,000 years.[174] This last category of comet is labeled 'non-periodic'. This is a misleading term because most 'non-periodic' comets follow an elliptical orbit,[175] but since they reach perihelion only once every several centuries (at best), it's difficult to determine whether they are periodic and whether their trajectories are parabolic, hyperbolic or elliptic (see Comet ISON for example).

Some of these comets are understood to have an interstellar origin; they enter the solar system because of the movement of our galactic arm and then exit it for the same reason. Some of them, during their journey through the solar system, fragment into 'daughter comets' which are captured by the Sun and its planets and then orbit within the solar system, becoming short- or medium-period comets. Other comets partly disintegrated in the distant past, leaving 'stationary' cometary clouds (Taurids, Aquariids, Orionids) that the Earth crosses every year on specific dates.[176] Others travel with their own cometary swarm. That's the case of Comet Giacobini-Zinner, which is very dim but surrounded by a massive swarm.[177]

---

173 'Halley's Comet: Facts About the Most Famous Comet', *Space* (20 February 2013). See: www.space.com/19 878-halleys-comet.html

174 Brand, J. C., 'Comets', *Scient Am* (1981), pp. 29–36

175 Moore, *op. cit.*, p. 32

176 Beta Taurids (24 June – 6 July) and Taurids (15 September – 15 December) are both associated with Encke's Comet. Eta Aquariids (21 April to 12 May) and the Orionids (18 to 26 October) are associated with Halley's Comet. *Ibid.*, p. 88

177 In 1933, Earth made contact with the meteor swarm moving in the comet's orbit, resulting in a strong rain of meteors that lasted for an hour or two. Over some parts of Europe, more than 15,000 meteors were estimated to have been recorded within sixty minutes. *Ibid.*, p. 88

Figure 64: The draconid meteor shower is caused by debris following Comet Giacobini-Zinner. It crosses Earth orbit every 6.6 years. During its last passage on October 8th, 2011, it is estimated to have left a ton of meteoric material on Earth. (© NASA)

There also appear to be cycles of cometary swarms much greater than the 27 MY Nemesis one. The galactic-year cycle is one of them. Every 186 million years, our galactic arm crosses the same region of high cometary activity:

> ... the four largest probable impact events: the K/T, P/Tr, O/S, Stu/V boundaries are evenly spaced by ~186 million year intervals. The Galactic Orbit model implies that these four events occurred when our solar system passed through the most dangerous zone, or Zone-1. The impact/extinction data also suggests two minor Zones, also causing extinctions with a **186 million year period** (just like Zone-1).

An analogue to this model would be the annual meteor showers seen from Earth. The comet debris trails that cause the showers are basically fixed with respect to Earth's orbit. We hit them the same time every year like clockwork as we orbit the sun. **On the galactic scale, the time required for our solar system to make a lap around the galactic nucleus has been estimated to be in the range of 200–250 million years.** The Galactic Orbit model suggests that 186 million years might be the effective period for our solar system. Perhaps we encounter stationary zones or features (relative to our galactic orbit) where gravitational flux or possibly shockwaves cause Oort cloud comets to venture into the inner solar system in large numbers.[178]

So apart from the 27-million-year Nemesis cycle, our solar system is exposed to various cycles of cometary activity, from short one-year cycles to 186-million-year cycles.

According to Muller, Nemesis disrupts comets in the Oort cloud en route towards our solar system. No one has actually seen the Oort cloud, which was first hypothesized in 1950 by Dutch astronomer Jan Oort.[179] It is supposedly a gigantic, spherical cloud located at about 1000 times the distance

178 Janke, P. R., *A Correlated History of the Universe*, Pan Terra Inc, 2002

179 'Oort Cloud', *Universe Today* (15 June 2009). See: www.universetoday.com/32522/oort-cloud/

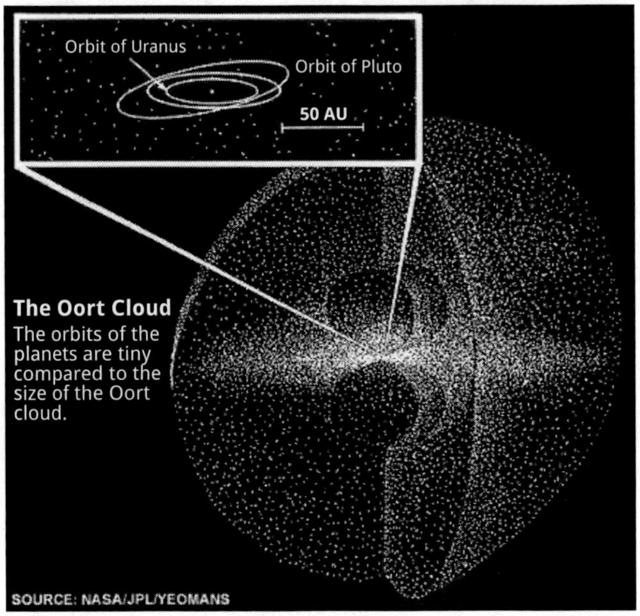

Figure 65: Depiction of the hypothesized Oort cloud. (© NASA)

of Pluto from the Sun[180] that surrounds the solar system and contains something like 10^{11} comets.

But according to astronomer Tom Van Flandern, the Oort cloud theory is impossible simply because its requirements (quantity of produced comets) far exceeds its capabilities (its size and the matter it contains).[181]

Victor Clube also disagrees with the Oort cloud model. He posits that most comets and asteroids located in our solar system come from fragmented giant comets.[182] Those giant comets originate outside our solar system in the Kuiper Belt, which is located along the galactic plane. The solar system periodically bobs up and down through the galactic plane of the galaxy as it orbits the galactic center (see figure 66). Each passage dislodges giant comets from the Kuiper Belt and diverts them closer to the Sun.

In any case, these theories about the origin of comets and asteroids (giant comet, Kuiper belt, Oort cloud) don't explain how comets form in the first place.

Comets have been carefully observed for millennia. Ancient astronomers in Mesopotamia made detailed observations and wrote accounts of what they saw in the skies. Ancient Greece also developed its own astronomical theories.[183] From the Renaissance to the present day, scientists are still proposing possible explanations for the origin of comets. In the end, all kinds of theories have been proposed: ejecta from volcanic eruption occurring on other planets or from star explosions, chunks of planets, chunks of the Sun, accretion of cosmic dust, accretion from the protosolar disk,

[180] About 40,000 AU. See: Covey, J., 'Short-Period Comets and the Age of the Solar System', *Creation in the Crossfire*. See: www.creationinthecrossfire.org/Articles/ShortPeriodComets.html

[181] Tuttle, R. J., *The Fourth Source: Effects of Natural Nuclear Reactors*, p. 196

[182] Bailey, Clube, & Napier, *The Origin of Comets*, pp. 220–270

[183] *Ibid.*, pp. 7–87

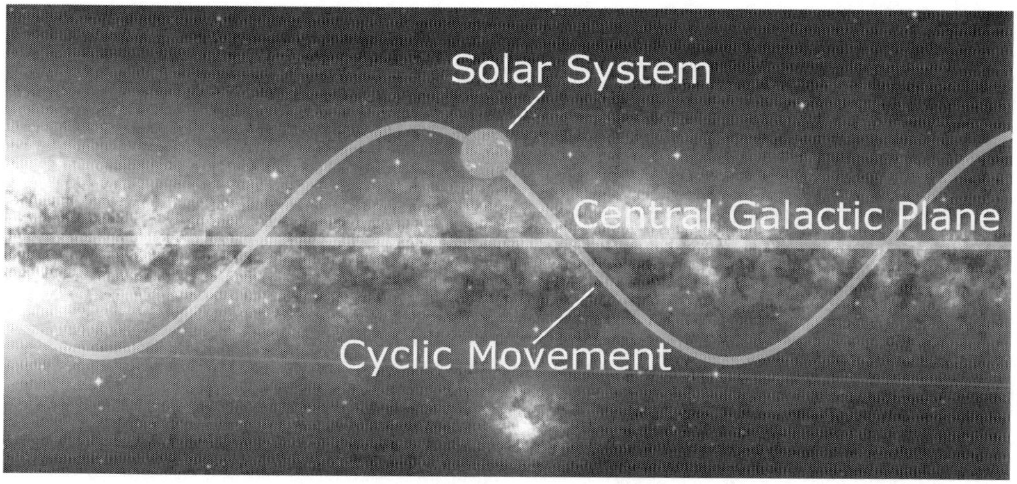

Figure 66: The oscillation of the solar system through the galactic plane. (© Sott.net)

condensation, etc.[184]

When scientists propose so many different theories to explain the very same phenomenon, one can suspect that they have yet to discover the true explanation.

This being said, all those official theories postulate a *non-electric* universe. If we also take into account the fact that space is filled with plasma, then the formation of comets becomes clearer. A charged body is much more likely to undergo an accretion process (i.e. the attraction of oppositely-charged particles) than an electrically neutral body. This could explain the accretion of matter into larger objects like comets and planets. In addition, the previously described[185] fission of celestial bodies due to excessive electric charge can explain the high quantity of chunks of matter in space. Also, because of the similar rocky nature of most planets, moons, comets and asteroids, the same formation process might apply to those similarly composed bodies whose only difference is the size.

To sum up so far, the Nemesis hypothesis may be able to account for both the increase in cometary activity and the decrease in solar activity experienced in recent years. Nemesis may be dragging comets into the vicinity of the solar system while grounding the Sun, thus depleting its electric charge. This results in an overall decrease in solar discharges. Due to problems with radiocarbon dating techniques, it's hard to say when exactly our last encounter with Nemesis took place, but recent solar and cosmic developments suggest that something big is afoot. In Part III, we will focus on the consequences these phenomena are having here on planet Earth.

[184] *Ibid.*, pp. 105–144

[185] See Chapter 15: 'Enter Nemesis'

Part III.

Earthly effects of the approaching Nemesis and its cometary swarm

Chapter 20.

Impact of the Sun's reduced activity on planet Earth

Now that we know more about the electric nature of the Sun, and have taken note of its reduced activity, it's time to take a look at how some of the ideas we've discussed could be correlated with, or even be causing, unusual natural phenomena taking place here on Earth.

First we need to explore another key linkage in the Electric Universe model: how *the Earth itself is analogous to a capacitor*. We know that the Earth is powered by the Sun via the solar wind, which sweeps into and around the Earth and its ionosphere, charging them electrically. This is why, despite numerous discharges occurring between the Earth's surface and its ionosphere (in the form of lightning, for example), the electric field between the Earth and the ionosphere doesn't disappear: it is being regularly recharged by the Sun.

Together, the upper and lower parts of the Earth's atmosphere might be thought of as its insulating bubble, its double layer (DL). The ionosphere extends from a height of about 50 km to over 500 km.[186] This region is highly ionized compared to the lower atmosphere because of its direct interaction with solar radiation. Solar particles ionize the molecules of the ionosphere, which are mostly gaseous in nature. In fact, it's because of these ions that the upper part of the atmosphere is called the ionosphere.

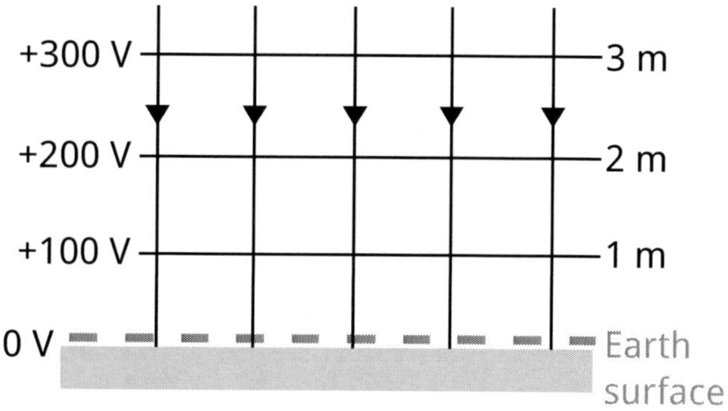

Figure 67: The vertical atmospheric electric field. (© Sott.net)

The electric charge of the ionosphere is positive.[187] Since the Earth's electric charge is negative,[188] a vertical electric field exists within the atmosphere between the ionosphere and the Earth's surface. As depicted in figure 67, the atmospheric electric field is, on average, equal to 100 volts per meter,[189] although it's stronger at the equator and decreases with latitude.

[186] Chavalier, G., *The Earth's electrical surface potential*, p. 60

[187] Bennett, A. J. & Harrison, R. G., 'Variability in surface atmospheric electric field measurements', *Journal of Physics: Conference Series* (2008) 142

[188] G. A. Erman was the first scientist to mention the negative overall electric charge of the Earth in his research paper titled 'Kritische Beitrage zur atmospharischen Electromie', published in 1803, *Annln. Phys.* 15.

[189] Bennett, A. J. & Harrison, R. G., 'Variability in surface atmospheric electric field measurements', *Journal of Physics: Conference Series* (2008) 142

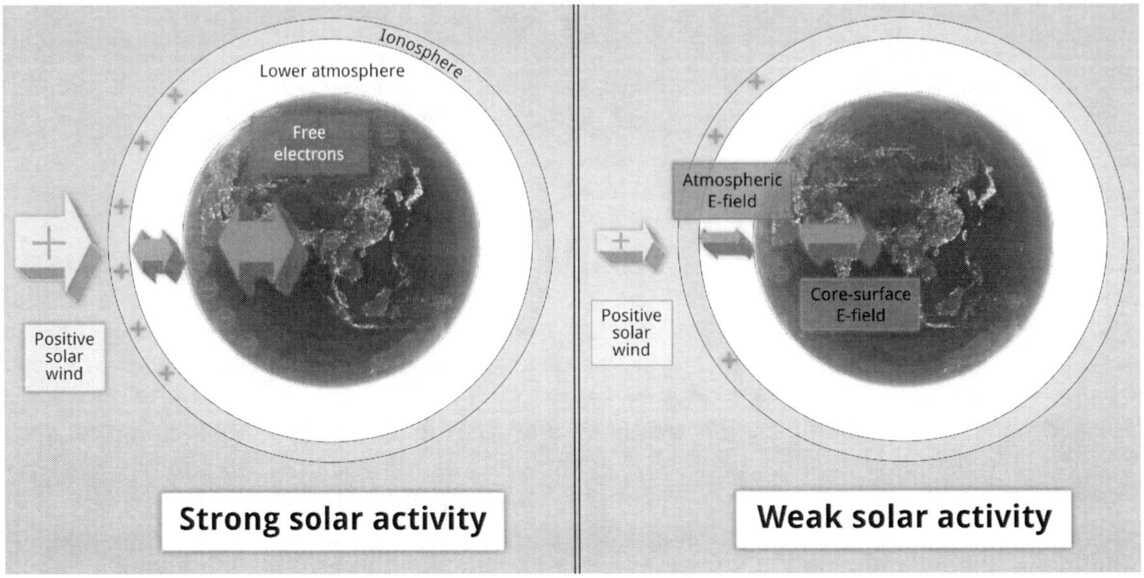

Figure 68: Earth's electric fields and potentials according to solar activity. (© Sott.net)

Why, you may be wondering, if such a vertical electric field exists in the atmosphere, don't we get electrocuted? The air around our heads, after all, has an electric potential that is about 180 volts higher than our feet. One reason is that despite the strong difference in electric potential, the *current density* in air is extremely small, about 10^{-12} amperes per square meter.[190] So the current flow resulting from this electric field is, most of the time, almost imperceptible.

Figure 68 depicts the influence of solar activity on the Earth's atmospheric E-field and its core–surface E-field.

On the right of figure 68, solar activity is weak; therefore the Earth receives less (positively charged) solar winds (small light grey arrow). As a consequence, the electric potential of the ionosphere is less positive and it tends to attract fewer free electrons from inside the Earth to its surface, making the Earth's surface less negatively charged. As a result, the E-field between the ionosphere and the Earth's surface (atmospheric E-field) is reduced (small dark grey arrow on the right picture).

With fewer free electrons attracted from the inside of the Earth to its surface, the electric field between the Earth's surface and its core is also lowered (small medium grey arrow on the right picture).[191]

Let's keep in mind this idea that weak solar activity reduces the atmospheric E-field and the surface–core E-field, because it will be an important explanatory factor in many of the unusual natural phenomena we'll be looking at in the coming chapters.

---

190 Chavalier, *op. cit.*, p. 4

191 Remember that a relative potential difference is enough for an electric field to exist. Therefore, the two regions between which the E-field appears don't have to exhibit opposite polarities. In the case of the Earth, the electric potential of the surface and that of the core can both be negative, but their relative difference makes it possible for an E-field to exist.

Chapter 21.

Increase in cometary activity

In a previous chapter[192] we briefly addressed the contemporary increase in cometary activity. In this chapter we will deal more extensively with this phenomenon. To do so we will focus on the following subtopics that are directly correlated with cometary activity: increase in fireballs (fireballs being cometary material entering our atmosphere), global dimming (related to the increase in atmospheric cometary dust), increase in newly discovered comets, moons and asteroids (the three of them being cometary material of various sizes). Then we will deal with the way this increased activity is covered up by the mainstream media and, finally, we will cover some of the unexpected earthly effects induced by cometary activity.

Increase in fireballs

As we've already seen in Chapter 14, data collected by the AMS[193] indicates that the number of confirmed meteor fireballs in our atmosphere has increased by almost 700% over the period 2005–2013. More than 2,000 confirmed meteor fireballs were observed over the USA in 2012; nearly 3,500 were observed in 2013.[194] That amounts to a 64% increase.

Notice that the sharp increase in reported fireballs (+1931% over the 2005–2013 period) is very *unlikely to be solely due to an increase in the number of observers* and/or an increase in popularity of the AMS database. Indeed, if it was the case, the number of observations should keep increasing year after year, but this is not the case. For example, between 2008 and 2009 the number of confirmed fireballs dropped by 4%.

The AMS database focuses on the U.S., which constitutes less than 2% of the Earth's surface. Can we deduce from this that the total number of observed fireballs across the planet in 2013 was 50 times the number observed over the U.S.? If so, the world total for 2013 would then amount to almost one million[195] sightings.

An increase in fireball sightings was not only measured in the U.S. The same phenomenon has also been noted in Japan. Figure 70 compiles the number of observed fireballs as collected by the SonotaCo Network Japan.[196]

---

[192] Chapter 14: 'Variations in solar activity'

[193] American Meteor Society.

[194] 'Fireball Logs', *American Meteor Society.*
See: www.amsmeteors.org/members/fireball/browse_events/?year=2013&state=&num_report=1

[195] 50 (2%) X 18,459 (number of sightings in the U.S.) = 922,950

[196] SonotaCo is an observation network covering the Japanese territory and including numerous observation cameras. The increase in observed fireballs was based on the very same cameras that collected data over the period 2007–2013.

Year	Sound Only	*Yearly variation*	Fragmentaion Only	*Yearly variation*	Confirmed Fireballs	*Yearly variation*	All Reports about Sightings	*Yearly variation*
2005	49	/	2	/	467	/	909	/
2006	41	-16%	1	-50%	517	11%	923	2%
2007	50	22%	2	100%	591	14%	1231	33%
2008	69	38%	0	-100%	730	24%	1604	30%
2009	71	3%	2	/	700	-4%	1885	18%
2010	104	46%	53	2550%	954	36%	2548	35%
2011	183	76%	517	875%	1637	72%	4625	82%
2012	248	36%	643	24%	2151	31%	6648	44%
2013	337	36%	929	44%	3525	64%	18459	178%
2005-2013 variation	*588%*		*46350%*		*655%*		*1931%*	

Figure 69: Observed fireballs over the U.S. over the period 2009–2013. (© Sott.net – data extracted from the AMS database)

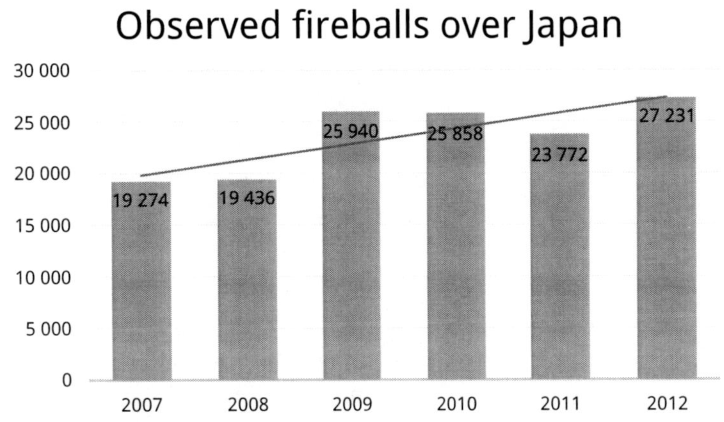

Figure 70: Number of observed fireballs over Japan collected by SonotaCo and linear regression (dark grey line). (© Sott.net)

Concurrent with this increase in observed fireballs, there is also a growing number of reports of 'sonic booms', i.e. exploding sounds that observers generally report as having come from overhead, and which are considered unexplained because the source usually identified with sonic booms – jet aircraft breaking the sound barrier – was ruled out. Hard data on this is lacking, but together with my colleagues at SOTT.net, we have collected hundreds of global media reports of unexplained sonic booms over the past ten years. By all accounts, it appears to have gone from being a sporadic, perhaps once-monthly occurrence, to a daily and familiar experience.[197] Even Wikipedia, a resource known to err on the side of officially-sanctioned (and often corrupt) science, lists 39 recent reports of mysterious sonic booms reported by the mainstream media.[198] The first report is from 2006,[199] and

Therefore, the fireball increase can't be due to a growing number of cameras. See: sonotaco.jp/

[197] Visit: www.sott.net/#searching, then enter 'intext: +sonic +boom' as a search query.

[198] 'Unexplained boom', *Wikipedia*. See: en.wikipedia.org/wiki/Unexplained_boom

[199] Roth, A., 'What's behind mysterious booms? Phenomena produce theories, but no answers' *San Diego Union-Tribune* (23 April 2006). See: legacy.utsandiego.com/news/metro/20060 423-9999-1n23bigboom.html

Figure 71: One of the flashes generated by the Chelyabinsk meteor. Screen-capture from a dashboard video camera.

over half of them were published in 2012 and 2013. These were the only cases where, apparently, the official number of witnesses mandated an official response.

Not only have fireballs increased in number, but they have also increased in mass. On February 15th, 2013, an asteroid with an estimated mass of 10,000 tons entered the Earth's atmosphere over the Ural Mountains in Russia and exploded at a height of about 23 km (15 miles) over the city of Chelyabinsk.

The energy of the bolide was about 500 kilotons, 20–30 times more energy than was released by the atomic bomb dropped by the U.S. government on the civilians of Hiroshima. The largest known object to 'hit' the planet since the Tunguska event in 1908, the Chelyabinsk fireball shone *30 times brighter than the Sun* and set off a powerful shockwave that left several thousand people injured from flying glass as the windows in almost 8,000 city buildings were blown in,[200] while radiation from the exploding space rock burned people's eyes and skin.[201]

The material left behind in the atmosphere formed a thin but cohesive and persistent stratospheric dust belt that was tracked by a NASA satellite. By February 19th, four days after the explosion, the plume had been carried by the jet stream entirely around the northern hemisphere and arrived back at Chelyabinsk. Three months later, a detectable belt of bolide dust persisted around the planet.[202]

While the Chelyabinsk event stole the limelight in 2013, and deservedly so, it was just one of many spectacular overhead meteor explosions. To give you an idea of the extent and regularity of

[200] 'Chelyabinsk meteor', *Wikipedia.* See: en.wikipedia.org/wiki/Chelyabinsk_meteor

[201] Sample, I., 'Scientists reveal the full power of the Chelyabinsk meteor explosion', *The Guardian* (7 November 2013). See: www.theguardian.com/science/2013/nov/06/chelyabinsk-meteor-russia

[202] 'Around the world in four days: NASA tracks Chelyabinsk meteor plume', *NASA* (14 August 2013). See: nasa.gov/content/goddard/around-the-world-in-4-days-nasa-tracks-chelyabinsk-meteor-plume/#.Urh6RLTDviG

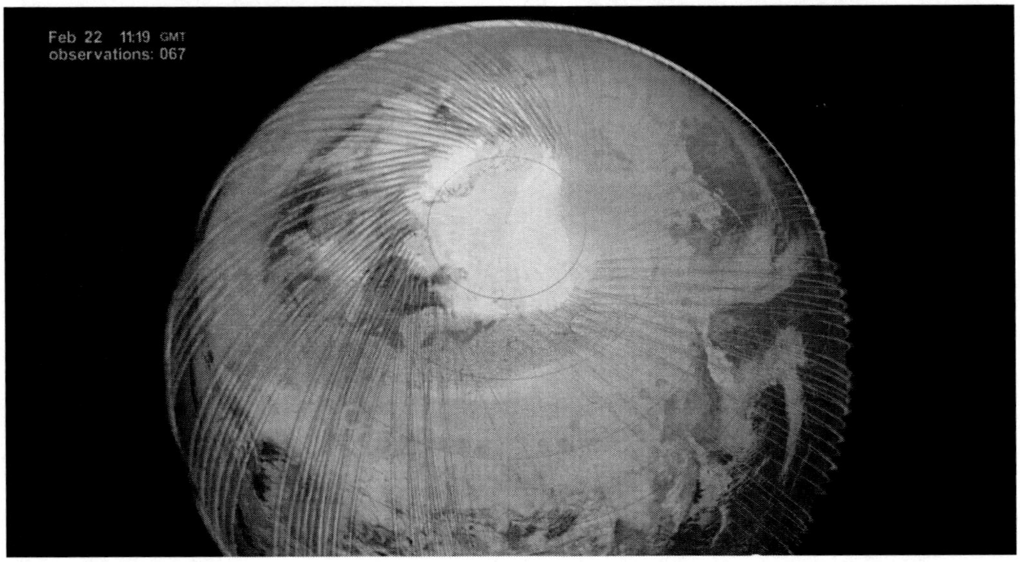

Figure 72: The Chelyabinsk atmospheric plume on February 22nd, 2013, i.e. one week after the overhead explosion. (© Discovery Magazine)

global coverage, consider that on the very same day as the Chelyabinsk event, astronomers were tracking an apparently unrelated asteroid (2012 DA14) which was making its closest approach to Earth. This other asteroid had been discovered a year beforehand and was regularly discussed in the media in anticipation of its fly-by, not least because it was roughly the same size as estimates for the Tunguska body, but also because it passed *inside* the orbit of NASA's geosynchronous satellites.[203]

Just two days previously, on February 13th, another separate fireball had exploded over the Caribbean, sending out a shockwave that shook building foundations in Cuba.[204] Two days before this event, on the morning of February 11th, another daytime fireball was videoed from the dashcam of a vehicle travelling on a highway just 250 km west of Chelyabinsk.[205]

Global dimming

Global dimming is the reduction in the amount of solar radiation received on the Earth's surface during fair weather. One of the main causes of global dimming is atmospheric dust.[206] Numerous scientists have demonstrated that *a global dimming trend has been in process for decades.*[207]

There's been some debate as to whether atmospheric dust induces a net warming effect on the

[203] Howard, J., 'Asteroid 2012 DA14 Will Miss Earth By Just 15 Minutes', *Huffington Post* (12 February 2013). See: huffingtonpost.com/2013/02/11/bill-nye-asteroid-2012-da14-february-15-space-rocks_n_2662 826.html

[204] 'Meteorite explodes in skies above Cuba Wednesday 13th Feb.', *Sott.net* (15 February 2013). See: www.sott.net/article/258 474-Meteorite-explodes-in-skies-above-Cuba-Wednesday-13th-Feb

[205] 'Russia Bolide 11FEB2013', *Lunar Meteorite Hunters blog* (19 February 2013). See: lunarmeteoritehunters.blogspot.fr/2013/02/russia-bolide-11feb2013-314.html

[206] Haywood, J., 'Global Dimming', *MetOffice UK*. See: www.metoffice.gov.uk/climate-change/guide/science/explained/dimming

[207] Depending on sources, global dimming has been between 2% and 5.3% per year over the period 1960–1990.

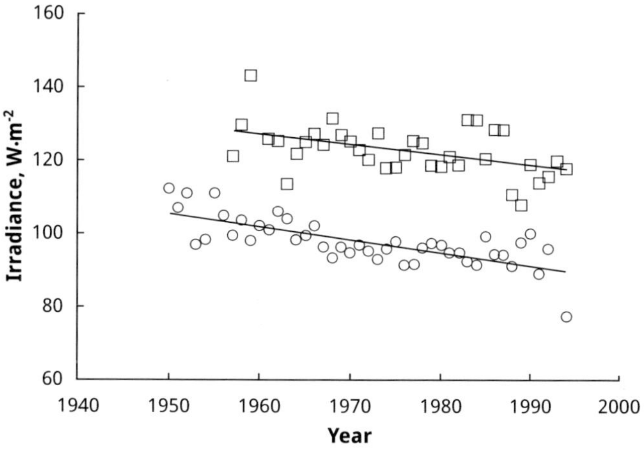

Figure 73: Reduction in solar irradiance over the period 1950–2000. (© Stanhill & Cohen)

planet (because it absorbs more radiation than it reflects), or whether it induces a net cooling effect (because it reflects more than it absorbs radiation). In 2008, atmospheric scientist Richard Hansell tested and measured the net effect of atmospheric dust particles on temperatures and concluded that, although atmospheric dust both absorbed and reflected solar radiations, it induced an overall cooling effect:

> The analysis showed that **over half of dust's cooling effect is compensated for by its warming effect**. The finding, published in the *Journal of Geophysical Research, Atmospheres*, could clarify scientists' understanding of how dust influences moisture fluctuations in the atmosphere and surface temperatures around the planet.[208]

As shown in figure 73, researchers from the Institute of Soil, Water and Environmental Sciences[209] found a significant reduction, globally averaged 2.7% per decade, in solar radiation reaching the Earth's surface over the last 50 years.

In the '90s an inversion occurred and our planet experienced a global brightening in some regions.[210] Then, after the year 2000, global dimming restarted in certain areas and became overall more chaotic with different continents experiencing opposite trends.[211]

Now, according to mainstream science, global dimming is man-made, caused by the accumulation of aerosol particles in the atmosphere due to industrialization. The trend inversion noticed in the '90s is attributed to the ban of several types of aerosols and other anti-pollution measures.[212]

[208] 'Dust: It Warms and it Cools', *Astrobiology Magazine* (3 November 2012). See: www.astrobio.net/pressrelease/5120/dust-it-warms-and-it-cools

[209] Stanhill, G. & Cohen, S., 'Global dimming: a review of the evidence for a widespread and significant reduction in global radiation with discussion of its probable causes and possible agricultural consequences', *Agricultural and Forest Meteorology* (19 April 2001) 107(4): 255–278

[210] Hegerl, G. C., Zwiers, F. W., Braconnot, P., *et al.*, 'Chapter 9, Understanding and Attributing Climate Change' in *Climate Change 2007: The Physical Science Basis*, IPCC (2007)

[211] Wild, M, *et al.*, 'Global dimming and brightening: An update beyond 2000', *Journal of Geophysical Research Atmospheres* (16 May 2009)

[212] Denman, K. L. *et al.*, 'Couplings between changes in Climate System and the Biogeochemistry', IPCC (2007)

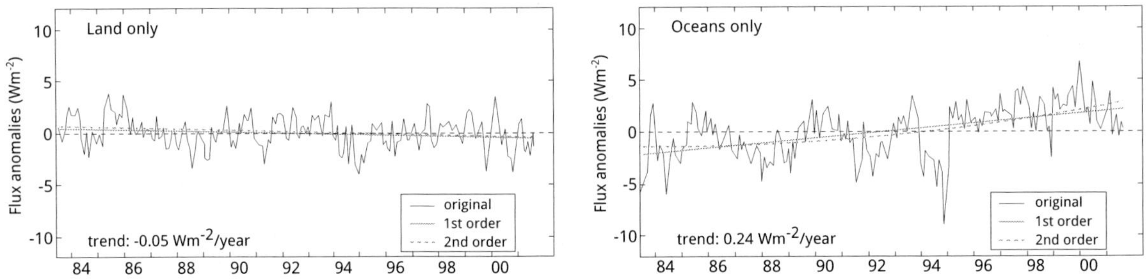

Figure 74: Irradiance over the period 1982–2002. Land measurements on the left (global dimming), ocean measurements on the right (global brightening). (© Pinker et al.)

However, a paper published in 2005 showed that over the period 1986–2000, although a slight dimming was occurring over land, a brightening occurred over the oceans.[213]

If human activity was indeed the cause of global dimming, and the reduction in human aerosol use the cause for the brightening observed in the 1990s, a brightening over land should have been observed and, possibly, a delayed brightening over the oceans (due to air circulation), since most industrial sources are located on continents. But the paper referenced here shows exactly the opposite.

Since 2000, dimming has been observed in numerous places, including China, India and the whole southern hemisphere, *despite the relatively lower presence of anthropogenic pollution in this less industrialized hemisphere.*[214] We can deduce from this that while human pollution might indeed affect the amount of sunlight reaching the surface, it's obviously not the only cause and its effects are probably negligible in the larger context.

Unlike human pollution, cometary activity could, at least partly, explain both the global dimming observed above the oceans during the 1990s and the dimming since 2000. Between 40 and 400 tons of extraterrestrial material is estimated to enter the Earth's atmosphere daily,[215] [216] most of it arriving in the form of cometary dust. These estimates calculated years ago *do not, however, take into account the recent surge* in cometary activity.

If we factor in the 655% increase in confirmed fireballs (see AMS statistics above) over the past eight years, the quantity of cometary dust entering Earth's atmosphere should be at least six times higher than that generally estimated; that is, daily incoming dust measuring between 260 and 2,600 tons.

Increase in asteroids, comets, and moons

Where is all this debris coming from? Well, there has been a significant increase in the discovery of asteroids in recent decades: In 1980, there were 9,000 known asteroids in the solar system. By 2000, there were 86,000. By 2007, there were 380,000.[217] As of 2013, the International Astronomical

[213] Pinker, R. T. *et al.,* 'Do Satellites Detect Trends in Surface Solar Radiation?', *Science* (2005) 380

[214] Wild, M., 'Enlightening Global Dimming and Brightening', *Bull. Amer. Meteor. Soc.* (2012) 93: 30

[215] Leinert C. *et al.,* in 'Interplanetary Dust', *Physics and Chemistry in Space*, pp. 204–275

[216] Alvarez, Luis, quoted in Muller, R., *Nemesis*, p. 44

[217] Stone, R., 'Preparing for Doomsday', *Science* (7 March 2008)

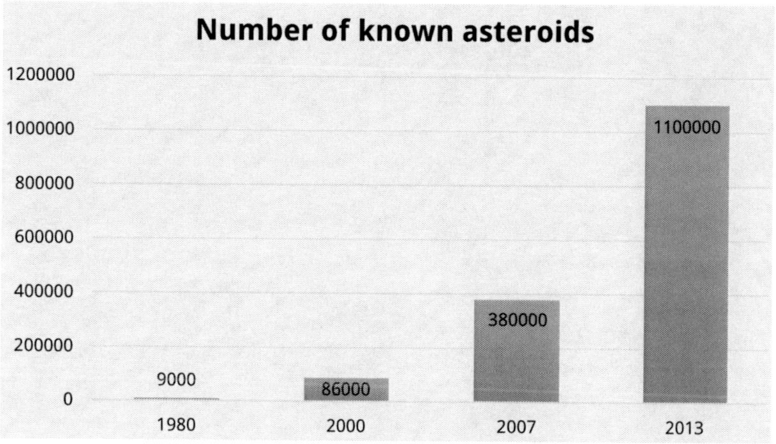

Figure 75: Number of known asteroids (1980–2013). (© Sott.net)

Union's Minor Planet Center had data on 1.1 million asteroids.[218]

The data in figure 75 are scattered and were compiled from various sources because no centralized systematic count of known asteroids exists for the period 1980–2013.

Of the 1.1 million asteroids known in 2013, NASA announced the discovery of the 10,000th NEO – Near Earth Objects on Earth-crossing orbits – in June 2013.[219] An alarming number, especially when we know that most NEOs are discovered just as they fly past Earth, and very often inside lunar distance.[220] Other estimates place between 100,000 and 1,000,000 more undiscovered NEOs on similar Earth-crossing orbits,[221] increasing the current estimated total range of asteroids in the solar system to 100 million.

The number of discovered comets has also increased in recent years. Figure 76 is based on official data released by the *British Astronomical Association* in November 2013.[222] By 2000 over 150 new comets were being discovered per year, a figure that increased to 265 by 2009.

Note that the figures between 2009 and 2013 are not yet finalized. Once updated, I fully expect the final count for those years to be higher than the current count.

Rather appropriately, 2013 was declared the *'Year of the Comets'*, with a high number of active comets in the Sun's vicinity. By November 21st, 17 comets[223] were visible with a small telescope,

[218] 'Provisional Designations', *Minor Planet Center*, IAU. See: www.minorplanetcenter.net/iau/lists/Desigs.html

[219] 'Ten Thousandth Near-Earth Object Unearthed in Space', *NASA* (24 June 2013). See: www.nasa.gov/mission_pages/asteroids/news/asteroid20130 624.html

[220] Atkinson, N., 'Space Shuttle-Sized Asteroid 2013 XY8 to Fly Past Earth on Dec. 11', *Universe Today* (10 December 2013). See: www.universetoday.com/107 091/space-shuttle-sized-asteroid-2013-xy8-to-fly-past-earth-on-dec-11/

[221] 'Current Map Of The Solar System' *Armagh Observatory* (2010). See: szyzyg.arm.ac.uk/~spm/

[222] 'Latest Discoveries - Comet Section', *British Astronomical Association & Society for Popular Astronomy.* See: https://people.ast.cam.ac.uk/~jds/. The medium grey bars show the total number of comets identified by NASA's SOHO project while the light grey bars show the total number of new comets. While newly discovered comets were virtually nil at the beginning of the '90s (for example, only one comet was discovered in 1995, namely 1995 O1 Hale-Bopp ('BAA Comet Section: 1995 comets', *British Astronomical Association & Society for Popular Astronomy* See: https://people.ast.cam.ac.uk/~jds/coms95.htm).

[223] 'Observable comets for November 21st 2013', *British Astronomical Association & Society for Popular Astronomy.* See here: https://people.ast.cam.ac.uk/~jds/.

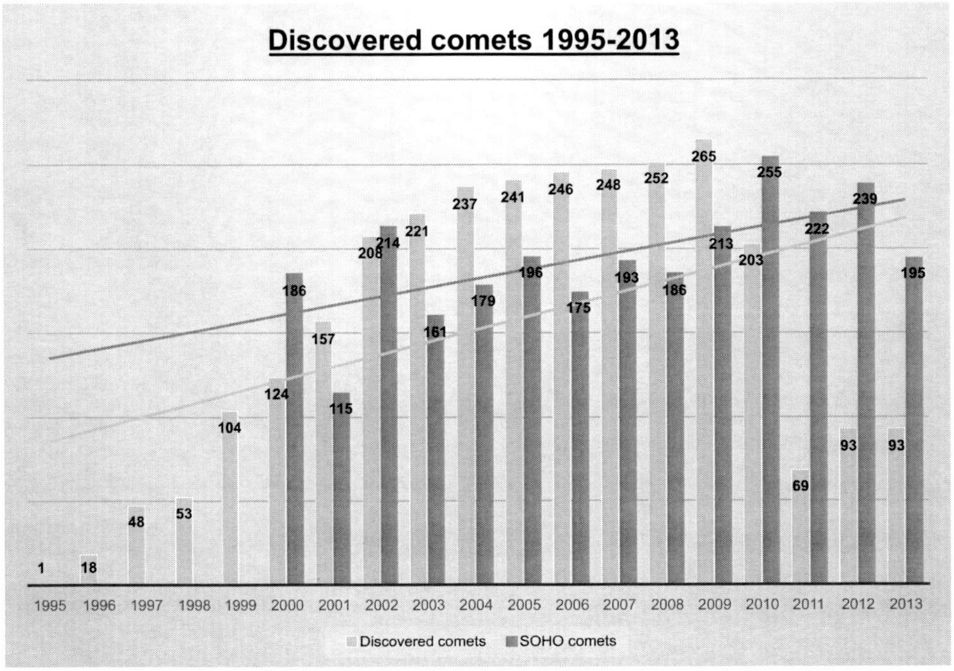

Figure 76: Discovered comets (1995–November 2013) according to data provided by the British Astronomical Association. (© Sott.net)

of which five[224] exhibited a magnitude between +4 and +7.5 (meaning they could be seen with binoculars[225]) and two of which (ISON and Encke) could be seen with the naked eye.

Remember, these are only the *identified* comets, bright enough to be observed at a long distance. Most 'comets' remain invisible (as 'asteroids') until they start glowing as a result of the heliosphere-induced electric stress they experience when they enter the solar system. If the exceptional fireball events in 2013 were anything to go by, 2014 may have even more celestial surprises in store for us.

Directly correlated with this increase in new comets, and another piece of evidence that strongly points to an increase in cometary activity, is the growing number of 'moons' orbiting the planets in our solar system. For example, in 1974 Jupiter had 9 moons.[226] Today it has 67 confirmed moons (see figure 77).[227]

The explanation usually given for this surge in the number of moons is that telescopes have improved. While this certainly may explain the discovery of *some* new moons, particularly the smaller ones, 'anomalies' persist. If the only factor in finding new moons was telescope efficiency, then year on year we should expect to find smaller and smaller moons around the same planet, but this is not the case.

One of the new moons around Saturn is called 'Calypso' and was discovered in 1980. Its diameter

---

[224] There were Comets ISON (2012 S1), Lovejoy (2013 R1), 2P/Encke, Linear (2012 X1), and P/Nevski (2013 V3).

[225] 'Comet Chasing in April', *Comet chasing*. See: cometchasing.skyhound.com/

[226] Abell, G. O., *Exploration of the Universe* (3rd edition), 1975

[227] Sheppard, S. S., 'The Jupiter Satellite and Moon Page', *Scott Sheppard's homepage*. See: www.dtm.ciw.edu/users/sheppard/satellites/

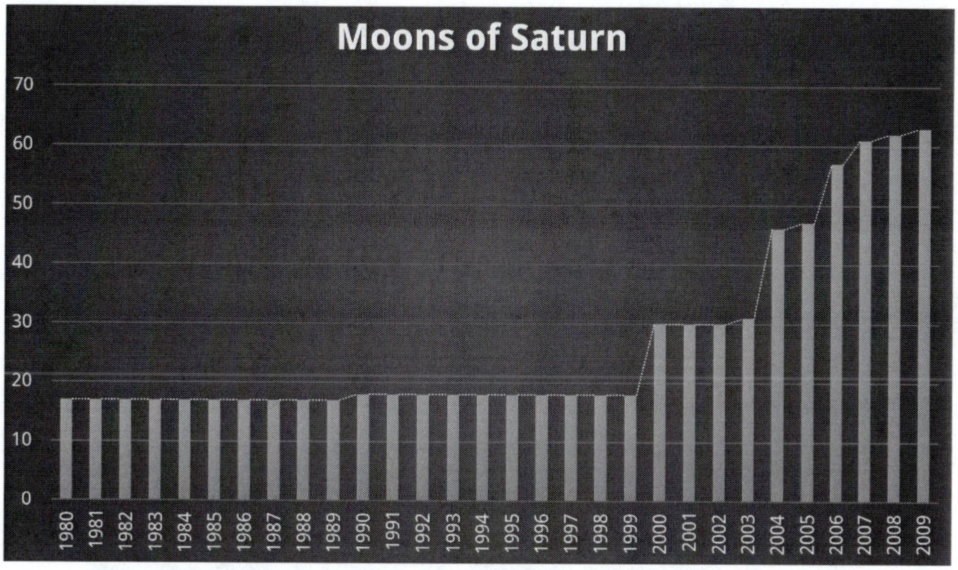

Figure 78: Number of known moons of Saturn (1980–2009). (© Sott.net)

is about 20 km (13 miles).[228] We could therefore assume that all the moons discovered after 1980 were smaller. But in 2000, a new moon of Saturn called 'Siarnaq' was discovered from a ground-based telescope.[229] Its diameter is about 40 km, meaning that twenty years later, a moon twice as big as Calypso was discovered, thanks to 'improved technology'.

Another flaw in the assumption that improvements in telescope technology are the unique driver of discovery is the 'discovery gap'. Sticking with Saturn as our case study, we note that in 1980, the planet had 17 identified moons.

Planet	1975	2013
Jupiter	9	67
Saturn	10	62
Uranus	5	27

Figure 77: Number of moons orbiting Jupiter, Saturn and Uranus (1975 and 2013). (© Sott.net)

Between 1981 and 2000, just one new moon was discovered. This moon was Pan, discovered in 1990 from Voyager's picture taken in 1980.[230] Then, between 2000 and 2009, 44 moons were discovered.[231]

Are we to assume that for two decades (1980–2000) technological progress ground to a halt? Did astronomers go on strike? The Voyager probe in 1980 arguably brought data that was well ahead of its time. Apparently it took 20 years to fill the technological gap and have pictures taken from the Earth that meet the accuracy of those taken from Voyager. And yet new moon databases show that in 1980 small moons were being discovered from ground-based observatories. For example Calypso, the small moon mentioned above, was discovered from ground-based observations by Dan Pascu.[232] As a matter of fact, in 1980, the year when Voyager

[228] 'Calypso: Overview', *NASA*. See: solarsystem.nasa.gov/planets/profile.cfm?Object=Sat_Calypso
[229] 'Siarnaq: Overview', *NASA*. See: solarsystem.nasa.gov/planets/profile.cfm?Object=Sat_Siarnaq
[230] 'Pan: Overview', *NASA*. See: solarsystem.nasa.gov/planets/profile.cfm?Object=Pan
[231] 'Saturn: Overview', *NASA*. See: solarsystem.nasa.gov/planets/profile.cfm?Object=Saturn
[232] 'Satellites of Saturn', *Central Bureau for Astronomical Telegrams*. See: www.cbat.eps.harvard.edu/iauc/03400/03496.html

was nearby Saturn, 6 new moons were discovered;[233] only half of them were discovered by Voyager. The other half were discovered by ground-based telescopes, so the observers were not lagging behind, technologically speaking.

A more logical explanation for this sudden discovery of new moons is that before their discovery they were simply not there; that the new moons were only recently acquired.

One other peculiar feature is the time sequence of moon acquisition, planet by planet. It appears from the data that, following a 20-year period where no new moons were discovered (1980s and 1990s), planets in the solar system all of a sudden started acquiring many new moons. Uranus[234] started acquiring most of its recent moons in 1999, Saturn[235] in 2000 and Jupiter[236] in 2001. This sequence is interesting because, in the solar system, Uranus is the 7th planet from the Sun, Saturn is the 6th and Jupiter is the 5th.

Figure 79: Outer planets Jupiter, Saturn and Uranus. (© Sott.net, adapted from wall321.com)

If the discovery of new moons was due to technological progress, then moons of the closest planet should have been discovered first. But the exact opposite occurred. The early acquisition of new moons by the farthest planet from the Sun suggests that these 'moons' originate from the same source, approaching the Sun from *outside* the solar system. So, as comets or asteroids passing the orbits of the outer planets first, they would only arrive at the inner planets later. This inward trajectory (from outside the solar system toward its center) is consistent with our description of Nemesis' orbit and the accompanying cometary swarm.[237] Once inside the solar system, we would expect some of the swarm to be attracted by massive celestial bodies in general, and the Sun in particular.

An objection made by mainstream science against this idea that new moons are recently captured asteroids or comets is that moons are conventionally thought of as being a totally different type of celestial body. But there actually doesn't seem to be any physical difference between them, as we

[233] Atlas, Prometheus, Pandora, Telesto, Calypso and Helene.

[234] 'Moons of Uranus', *Wikipedia.* See: en.wikipedia.org/wiki/Moons_of_Uranus

[235] 'Moons of Saturn', *Wikipedia.* See: en.wikipedia.org/wiki/Moons_of_Saturn

[236] *Ibid.*

[237] See Chapter 15: 'Enter Nemesis'

can see when we compare images of the two.

Figure 80: This composite image shows two asteroids among five moons of Saturn. If you had not been told which were 'moons' and which were 'asteroids' (or 'comets' for that matter), would you be able to tell them apart? (© NASA/JPL)

After years of denial, the penny appears to have dropped for mainstream astronomy, and it was finally conceded that moons might indeed be captured asteroids. An international team of astronomers modeled the ultraviolet light reflected from the surface of Phobos, one of the two small moons of Mars, and compared it to the asteroid 624 Hektor and the Tagish Lake meteorite found on Earth and concluded that they bore 'strong similarities'.[238]

Figure 81: Phobos, one of the two moons of Mars, along with Deimos. (© NASA)

One could think that the recent increase in new moons, comets, asteroids and NEOs might be due to a sudden surge in the NASA budget dedicated to detecting celestial bodies. As a matter of fact, just the opposite has been happening. NASA Astrophysics departments (the ones concerned with the detection program) had a $1.11-billion budget in 2007. By 2013, the budget dropped to $610 million. That's a 45% drop over six years.

[238] Pajola, M., *et al.*, 'Phobos as a d-type captured asteroid, spectral modeling from 0.25 to 4.0 μm', *The Astrophysical Journal* (2013) 777(2)

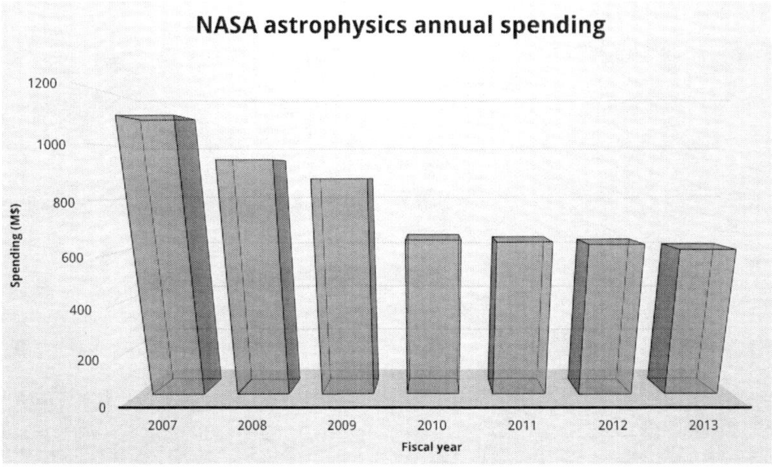

Figure 82: NASA astrophysics annual spending over the period 2007–2013. Source: U.S. House of Representative spending bills. (© Sott.net)

Disguising celestial intentions

The mainstream media's response to the increase in earthbound space rocks has been similar to its response to the increase in atmospheric dust: they downplay the magnitude of the phenomenon and blame it on human activity. But, sadly for the authorities, the proliferation of mobile phones equipped with cameras has provided extensive evidence that some fairly large-sized meteors are raining down on Earth, with fresh accounts of spectacular fireballs now being reported on a daily basis. With outright denial of the phenomenon no longer being possible for them, it has become imperative that the media find rational explanations to ease the inevitable apprehension people feel about what they are seeing.

So the authorities now regularly claim to be conducting bizarre, and previously unannounced, 'missile tests'[239] and dealing with equally outlandish 'accidents'[240] to account for 'sonic booms'[241] and 'explosive earthquakes'[242] that are often accompanied by eyewitness reports and video footage of fireballs[243] turning night into day over densely populated areas. Another explanation trotted out with some frequency is 'falling space junk'.

239 Bradley, N., 'Incoming! Meteor or Comet Fragment Explodes Above Southwestern US, Prompting US Army 'Missiles' Cover-up', *Sott.net* (16 September 2012). See: www.sott.net/article/251 268-Incoming-Meteor-or-Comet-Fragment-Explodes-Above-Southwestern-US-Prompting-US-Army-Missiles-Cover-up

240 Quinn, J., 'Meteorite Impacts Earth in Minden, Louisiana - Media and Government Cover It Up', *Sott.net* (12 October 2012). See: www.sott.net/article/252 515-Meteorite-Impacts-Earth-in-Minden-Louisiana-Media-and-Government-Cover-It-Up

241 'Another 'booming' earthquake - this time near Niagara Falls: Earthquake causes early morning 'boom'', *YNN.com* (26 October 2012). See: www.sott.net/article/252 802-Another-booming-earthquake-this-time-near-Niagara-Falls-Earthquake-causes-early-morning-boom

242 Werthmann, M., 'Thousands reported bang, rumble of Maine quake in Massachusetts', *Boston* (16 October 2012). See: www.boston.com/metrodesk/2012/10/17/witness-hollis-maine-quake-felt-like-boom-rumble-and-sinking-feeling/ZW40TFPUIKKbXXPu02fx8H/story.html

243 'California Fireball Meteor 17 October 2012', *Lunar Meteorite Hunters blog*. See: lunarmeteoritehunters.blogspot.fr/2012/10/breaking-news-ca-fireball-meteor.html?m=1

This media cover-up of the increase in cometary activity by conflating it with military missiles and/or space junk lends an almost prophetic tone to the following text written in 1996:

> ... the Christian, Islamic and Judaic cultures have all moved since the European Renaissance to *adopt an unreasoning anti-apocalyptic stance*, apparently unaware of the burgeoning science of catastrophes. History, it now seems, is repeating itself: it has taken the Space Age to revive the Platonist voice of reason but it emerges this time within a modern anti-fundamentalist, anti-apocalyptic tradition over which governments may, as before, be unable to exercise control. ... **Cynics (or modern sophists), in other words, would say that we do not need the celestial threat to disguise Cold War intentions; rather we need the Cold War to disguise celestial intentions!** [Emphasis in the original][244]

The problem is that, unlike conventional missiles, cometary bodies, because of their electric activity, have a strong electromagnetic signature that can trigger lightning bolts between incoming asteroids and the Earth's surface, which can fry electrical devices over a wide area. One naturally wonders if this is related to the media frequently reporting, since around 2008,[245] the development of electromagnetic pulse (EMP) weapons by the USA,[246] Israel,[247] China,[248] South Korea[249] and the convenient suspect, Iran.[250] The source mentioning the development of Iranian EMP weapons specifically states the following:

> An EMP is an **above-atmosphere level detonation of a nuclear device** that produces enough radiation to wreak havoc with electrical systems.[251]

Because of the protection provided by our atmosphere, an overhead cometary explosion is a far more probable event than a direct impact and could easily be mistaken for an 'above-atmosphere detonation'. And the EMP generated by such an explosion could, of course, be blamed on the 'Iranian EMP missiles'. In addition to 'disguising celestial intentions', a cover story about Iranian EMP weapons would provide a *casus belli* for the U.S. and Israel to attack Iran. At the very least, it keeps the population (of the U.S. at least) focused on the enemy 'over there' rather than the real enemy 'out there', and in that way, dependent on the authorities for protection. The point being, our 'authorities' cannot protect us from the celestial threat and, as we shall see, they may in fact play a central role in provoking it.

---

[244] Clube, V., 'The Hazard to Civilization from Fireballs and Comets' (1996)
[245] 'Iran Developing EMP Weapon', *Koinonia House* (12 August 2008). See: www.khouse.org/enews_article/2008/1347/
[246] 'Doomsday For Iran? U.S. Tests EMP Bomb', *Jewish Awareness Ministries* (2 January 2013). See: www.jewishawareness.org/doomsday-for-iran-u-s-tests-emp-bomb/
[247] Geertz, B., 'Report: Possible EMP Strike on Iran?' *Washington Free Beacon* (29 August 2012). See: freebeacon.com/analyst-sees-possibility-of-emp-strike-on-iran/
[248] Geertz, B., 'Report: China building electromagnetic pulse weapons for use against U.S. carriers', *Washington Times* (21 July 2011). See: www.washingtontimes.com/news/2011/jul/21/beijing-develops-radiation-weapons/
[249] Shah, D., 'South Korea develops Electromagnetic Pulse Bomb', *fareastgizoms*, (6 September 2011). See: fareastgizmos.com/other_stuff/south_korea_develops_electromagnetic_pulse_bomb.php
[250] Shamah, D., 'An electromagnetic pulse attack — the 'other' Iranian nuclear threat', *Times of Israel* (14 October 2012). See: www.timesofisrael.com/an-electro-magnetic-pulse-attack-the-other-iranian-nuclear-threat/
[251] *Ibid.*

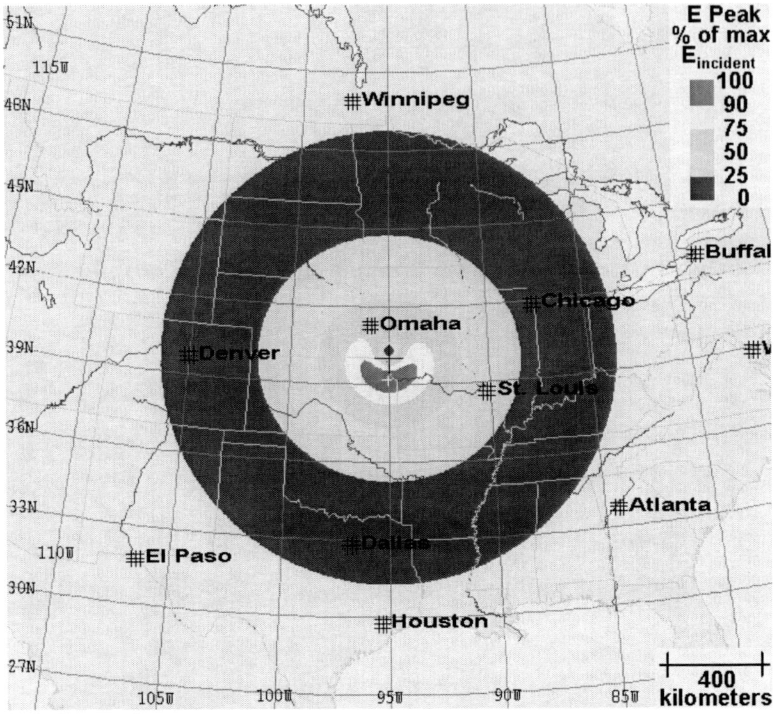

Figure 83: Theoretical peak electric field induced by a high-altitude electromagnetic pulse. Burst height is 75 km. (© Metatech corp.)

The recent series of 'disclosures' about new EMP weapons is all the more suggestive because the prime generators of electromagnetic pulses are nuclear weapons. This effect was already suspected by the time the Manhattan Project's Trinity Test in the desert of New Mexico on July 16th, 1945, detonated the first atomic bomb. Their electronic equipment had been shielded[252] because project scientist Enrico Fermi had predicted that an EMP would accompany the explosion. By the 1960s, the EMPs generated by nuclear explosions and their disruptive effects on electronic and electric equipment was being extensively measured and documented.[253] Put simply, nuclear weapons are EMP weapons, an established fact for over half a century. Why then does the media depict EMP bombs as a 'newly discovered' weapon? Why all this fuss about half a dozen countries suddenly developing EMP bombs at the same time that such a dramatic increase in comets and fireballs is being recorded?

Another feature of the electric model of the universe we've covered is how a Nemesis-induced grounding of the Sun weakens the E-field between the Earth's surface and its ionosphere (the 'atmospheric E-field'). This E-field, through the electric constraint it exerts on incoming meteors, acts as a protective shield:

> **If a body having a different electrical potential (voltage) penetrates the double layer boundary and moves into the plasmasphere surrounding a planet, charges will flow to try to cancel the voltage difference. Electrical discharges will occur.** Thus, if any other body, such as a large meteor, asteroid, or comet should come close enough to Earth to penetrate our plasma sheath, violent electric discharges would occur between the two bodies. It would, of course, be unfortunate to be standing on Earth's surface at the point of origin (or reception) of such a mega-lightning discharge. But the massive discharge itself might either deflect the intruding body or break it up and thus protect Earth from a collision. Such large-amplitude arcs would undoubtedly cause scarring on the surface of both bodies. A very small intruder would be broken up by the discharge. It is likely that this is what happened to comet

[252] Bainbridge, K. T., *Report LA-6300-H*, p. 53

[253] Glasstone, S. & Dolan, P., *The Effects of Nuclear Weapons*, United States Department of Defense and the Energy Research and Development Administration, p. 514

> Shoemaker-Levy 9 as it entered Jupiter's giant plasmasphere a few years ago.[254]

Note that the same phenomenon occurs on the scale of the solar system. Because of reduced solar activity, comets entering the solar system are subjected to a lesser E-field and less electric stress, therefore being less likely to fragment or disintegrate.

If Nemesis' comet swarm was approaching, the probability of encountering more and bigger asteroids should increase and, because of the weakened heliospheric and atmospheric E-field, the probability increases that more of them would come closer to the Earth's surface and trigger overhead explosions or even ground impacts.

In this electric model, we thus see Nemesis and its cometary swarm subjecting our planet to the threat of cometary bombardment while simultaneously neutralizing the very defense mechanism our planet and our solar system possess against such a threat.

We should keep in mind here that we're not looking at a scenario where a single large object hits a particular location, as depicted in some popular Hollywood productions. We are probably talking about passing through a vast cloud – composed of cometary dust and fragments of varying sizes (up to and including whole, miles-wide comets) – in which we could expect to encounter countless numbers of objects over an extended period of time. Such cometary bombardment could indeed last centuries if not millennia. Geologist David Jolley discovered that the extinction of the dinosaurs (Cretaceous-Paleogene boundary about 65 MY ago) was due to cometary bombardments that lasted several millennia:

> The very short period of time, as little as 2000–5000 years, between two large asteroid impacts on Earth close to the K-Pg boundary constrains the likely impactor delivery mechanism since it necessitates a high probability of **delivering several large bodies into the inner solar system within a few thousand years.**[255]

Notice that a lasting cometary bombardment involving several objects also explains why high levels of iridium in the K-Pg boundary were consistently found all over the planet[256] (one single impact would have raised iridium levels only locally).

Action at a distance: discharge events

As long as these space rocks continue discharging, fragmenting and exploding up there, they present no danger to us down here, right? Long assumed to be true, that may no longer be the case.

There were multiple reports of meteor fireballs in the skies above Colorado when devastating wildfires broke out there in June, 2012, with one local fire chief stating that meteorites had actually hit the ground.[257] Officially, meteorites are not supposed to be red-hot when they make impact, but many recent reports contradict the assumption that rocks must be icy cold by the time they reach

[254] Thornhill, W. & Talbott, D., *The Electric Universe*, pp. 44–45

[255] Jolley D., *et al.*, 'Two large meteorite impacts at the Cretaceous-Paleogene boundary', *Geology* (April 2010)

[256] See Chapter 15: 'Enter Nemesis'

[257] Glionna, J., 'Colorado firefighters hampered by winds, heat – and meteors', *L.A. Times* (21 June 2012). See: articles.latimes.com/2012/jun/21/nation/la-na-nn-meteors-colorado-firefighters-20120621

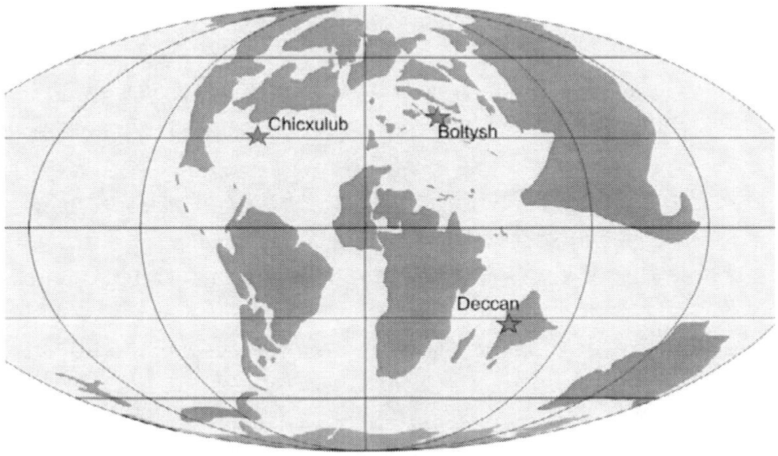

Figure 84: Location map of three major impact sites: Chicxulub, Boltysh and Deccan at time of Cretaceous-Paleogene (K-Pg) events. (© David Jolley)

the ground.[258] When we look back at available data for wildfires in the U.S. for the period 2001–2012, a marked upward trend reveals itself,[259] though this doesn't necessarily mean that increased meteor fireballs translate directly into increased wildfires. Other factors are probably involved, not least of which is increased electrical discharge activity in the form of increased lightning activity,[260] the traditional trigger for many wildfires.

Another phenomenon is an apparent spike in unexplained surface-level explosions, one sub-set of which is explosions or fires at industrial plants. There were nearly 40 such hazardous incidents, including five at fertilizer plants, between April and August 2013:

Date	Place	Dead/wounded	Activity	Notes
04/18[261]	West, Texas	15/160 hundreds	Fertilizer	Lightning
06/14[262]	Donaldsville, Louisiana	1/several	Fertilizer	
06/24[263]	Union Mills Indiana	1/2	Fertilizer	
07/29[264]	Shandong, China	Unknown	Explosives	
08/06[265]	Horlivka, Ukraine	5/3	Fertilizer	

[258] 'Narrow escape for British family after red-hot sulphuric meteorites rain down on their Shrewsbury garden', *Yahoo! News* (26 June 2013). See: uk.news.yahoo.com/meteor-strike-shrewsbury–sarah-marston-jones-asteroid-fragments-shropshire-141706 345.html

[259] Nelson, J., 'Major Fires Since 2001', *UX blog* (13 July 2012). See: uxblog.idvsolutions.com/2012/07/major-fires-since-2001.html

[260] Hough, A., 'Record 110,000 lightning bolts strike during UK 'superstorms'', *The Telegraph*, 29 June 2012. See: telegraph.co.uk/topics/weather/9365 699/Weather-record-110 000-lightning-bolts-strike-during-superstorms

[261] Levs, J. & Mungin, L., 'As many as 35 killed in Texas plant explosion', *CNN*. See: www.cnn.com/2013/04/18/us/texas-explosion

[262] Scherer, C., 'Deadly fertilizer plant explosion in Louisiana', *agprofessional*. See: www.agprofessional.com/news/Deadly-fertilizer-plant-explosion-in-Louisiana-211797 051.html

[263] 'Indiana fertilizer plant explosion kills one person', *CBS News*. See www.cbsnews.com/8301-201_162-57590 777/indiana-fertilizer-plant-explosion-kills-one-person/

[264] 'At least 12 killed and others buried during blast at Chinese explosives factory', *New York Daily News*. See: www.nydailynews.com/news/world/12-killed-blast-chinese-factory-article-1.1348 849

[265] 'Ukraine's largest nitrogen fertilizer plant explosion killed at least two people', *Xinhua*.

08/04[266]	Bartow County, Georgia	0/2	Power plant	
04/17[267]	Mount Isa, Australia	0/0	Sulphuric acid	
04/22[268]	Dexter, Michigan	1/1	Wastewater	
04/26[269]	Shakopee, Minnesota	0/0	Biomass	
05/01[270]	Mishazi, China	119/54	Poultry	
05/01[271]	Newmarket, Canada	0/6	Doors	
05/07[272]	Eufaula, Oklahoma	0/2	Sandblasting	Double boom
05/07[273]	Cartersville, Georgia	0/0	Power plant	Sonic boom
05/13[274]	Putnam, West Virginia Virginia	0/2	Acetylene storage	
05/15[275]	Cheyenne, Wyoming	0/0	Power plant	
05/23[276]	Ramat Hovav, Israel	1/31	Pharmaceutical	
06/02[277]	Seward, Illinois	0/1	Hexachlorodisilane	
06/17[278]	Postville, Iowa	0/3	Plastic laminates	
06/19[279]	Montreal, Canada	2/0	Fireworks	
06/20[280]	Foley, Alabama	0/2	Aircraft engines	
06/23[281]	Pudong, China	Unknown	Acrylic acid	

See: www.english.peopledaily.com.cn/90 777/8355 056.html

266 'Explosion rocks Plant Bowen in Bartow County; 2 injured', *myfoxatlanta.* See: www.myfoxatlanta.com/story/21883 961/incident-investigated-at-georgia-power-plant-in-bartow-county

267 'Explosion at Mt Isa acid plant', *Australian Mining.* See: www.miningaustralia.com.au/news/explosion-at-mt-isa-acid-plant

268 'Dexter wastewater treatment plant explosion kills one, injures another', *Detroit Free Press.* See: www.freep.com/article/20130 422/NEWS05/304220 171/explosion-Washtenaw-County-death

269 'Shakopee plant explosion under control but monitored through the night', *kare11.* See: www.kare11.com/news/article/1023 217/391/Shakopee-plant-explosion-under-control-but-monitored-through-the-night

270 'Explosion, fire reportedly kill 119 at poultry plant in China', *Fox News.* See: www.foxnews.com/world/2013/06/03/as-many-as-43-reportedly-killed-in-fire-at-poultry-farm-in-china/

271 'Explosion at Ontario plant sends 6 to hospital', *CBC News.* See: www.cbc.ca/news/canada/toronto/explosion-at-ontario-plant-sends-6-to-hospital-1.1303 768

272 'Eufaula propane explosion at Ford Sandblasting Plant forces neighborhood evacuations, closes highway', *kjrh.* See: www.kjrh.com/dpp/news/local_news/eufaula-propane-explosion-at-ford-sandblasting-plant-forces-neighborhood-evacuations-closes-highway

273 'Power Plant Explosion 'Sounded Like a Sonic Boom, Earthquake'', *Cartersville Patch.* See: www.cartersville.patch.com/groups/politics-and-elections/p/power-plant-explosion-sounded-like-a-sonic-boom

274 'Two injured in explosion at Putnam County plant', *Charleston Gazette.* See: www.wvgazette.com/News/201305130 068

275 'Explosion, fire at Laramie River Power Plant near Wheatland; no injuries', *KNEB.* See: www.kneb.com/news/index.php?more=kxt3gr83

276 'Explosion, fire at Teva plant kills one, injures 31', *pharmatechnologist.com.* See: in-pharmatechnologist.com/Ingredients/Manufacturing-Restarts-at-Teva-API-Plant-After-Deadly-Explosion

277 'Explosion fire reported at Winnebago County chemical plant', *rrstar.* See: www.rrstar.com/news/x776199 936/Explosion-fire-reported-at-Winnebago-County-chemical-plant

278 'Three injured in Postville plant explosion', *Des Moines Register.* See: blogs.desmoinesregister.com/dmr/index.php/2013/06/17/three-injured-in-postville-plant-explosion/article

279 'Victims of Quebec fireworks plant explosion named', *metronews.* See: www.metronews.ca/news/canada/715 309/women-killed-in-quebec-explosion-are-named/

280 'Explosion in Foley: Two UTC Aerospace employees injured, one seriously', *AL.* See: www.blog.al.com/live/2013/06/explosion_in_foley_two_utc_aer.html

281 'Pudong chemical plant suffers explosion', *Global Times.* See: www.globaltimes.cn/content/790 974.shtml#.UgtlAG3wHGo

06/28[282]	Mt Bethel Pennsylvania	0/0	Silicon	
07/05[283]	Takoradi, Ghana	0/1	Power plant	
07/11[284]	Suizhou, China	0/2	Biochemicals	
07/16[285]	Auburn, Tennessee	0/1	Thermite	
07/17[286]	Auckland, New Zealand	0/0	Steel recycling	
07/21[287]	Sibley, Iowa	0/5	Drainage tiles	
07/22[288]	Nuevo Leon, Mexico	2/12	Steel	
07/23[289]	Statesville, NC	0/1	Composites	
07/24[290]	Derry, Penn.	0/0	Nickel alloys	
07/25[291]	Lowell, Arkansas	0/0	Recycling	
07/30[292]	Zhangzhou, China	0/0	Paraxylene	
07/31[293]	Tavares, Florida	0/8	Propane	
08/01[294]	Makurdi, Nigeria	1/0	Cement	
08/06[295]	Luhju, Taiwan	Unknown	Nitrocellulose	
08/06[296]	Byram, Mississippi	0/0	Metal painting	
08/11[297]	Puerto la Cruz, Venezuela	0/0	Refinery	Lightning

Before investigations had even begun, official explanations in all of these cases invoked similar

[282] 'Explosion, fire at Voltaix started in chemical tank' *The Morning Call.* See: www.articles.mcall.com/2013-06-28/news/mc-f-upper-mount-bethel-explosion-20130628_1_chemical-tank-water-and-foam-fire-crews

[283] 'Explosion at Takoradi Thermal Plant', *Ghanaweb.com.* See: www.ghanaweb.com/GhanaHomePage/NewsArchive/artikel.php?ID=278718

[284] 'Two killed in China chemical plant explosion', *Z News.* See: www.zeenews.india.com/news/world/two-killed-in-china-chemical-plant-explosion_861441.html

[285] ''Chemical explosion' reported at Auburn plant', *oanow.* See: www.oanow.com/news/auburn/article_80209d9a-ee54-11e2-a94b-001a4bcf6878.html

[286] 'Explosion at Auckland recycling plant', *Otago Daily Times.* See: www.odt.co.nz/news/national/265166/explosion-auckland-recycling-plant

[287] 'Plant explosion injures five in Sibley, Iowa', *Sioux City Journal.* See: siouxcityjournal.com/news/local/plant-explosion-injures-five-in-sibley-iowa/article_f40f0877-2cc9-56a9-b10b-1545048b3ad7.html

[288] 'Explosion at Ternium Mexico steel plant kills two, wounds 12', *Reuters.* See: reuters.com/article/2013/07/23/mexico-steelmaker-explosion-idUSL1N0FT00V20130723

[289] '1 injured in Statesville plant explosion', *wcnc.com.* See: www.wcnc.com/news/local/1-hurt-in-Statesville-plant-explosion-216700961.html

[290] 'Explosion rocks ATI plant; no one injured', *triblive.com.* See: www.triblive.com/news/westmoreland/4418315-74/ati-plant-explosion#axzz2bw6qYrgh

[291] 'Crews battle fire, explosion at AERT recycling plant', *4029tv.com.* See:4029tv.com/news/arkansas/northwest/silo-explosion-at-aert-recycling-plant/-/8897460/21159286/-/cv6c65/-/index

[292] 'Zhangzhou chemical plant blast highlights safety fears', *South China Morning Post.* See: www.scmp.com/news/china/article/1293070/paraxylene-plant-expodes-zhangzhou

[293] 'Sabotage not suspected in Florida gas plant explosion', *Fox News.* See: www.foxnews.com/us/2013/07/31/sabotage-not-suspected-in-florida-gas-plant-explosion/

[294] 'One dies as explosion rocks Dangote cement plant', *vanguard.* See: www.vanguardngr.com/2013/08/one-dies-as-explosion-rocks-dangote-cement-plant/

[295] 'Chemical plant explosion briefly threatens airport', *Taipei Times.* See: www.taipeitimes.com/News/front/archives/2013/08/08/2003569183

[296] 'Explosion rocks metal painting plant in Byram', *WAPT.* See: www.wapt.com/news/central-mississippi/explosion-rocks-metal-painting-plant-in-byram/-/9156946/21363372/-/gl7uk2/-/index.html

[297] 'Lightning Strike Sparks Fire at Venezuela's Puerto La Cruz Refinery', *Wall Street Journal.* See: www.online.wsj.com/article/BT-CO-20130811-700902.html

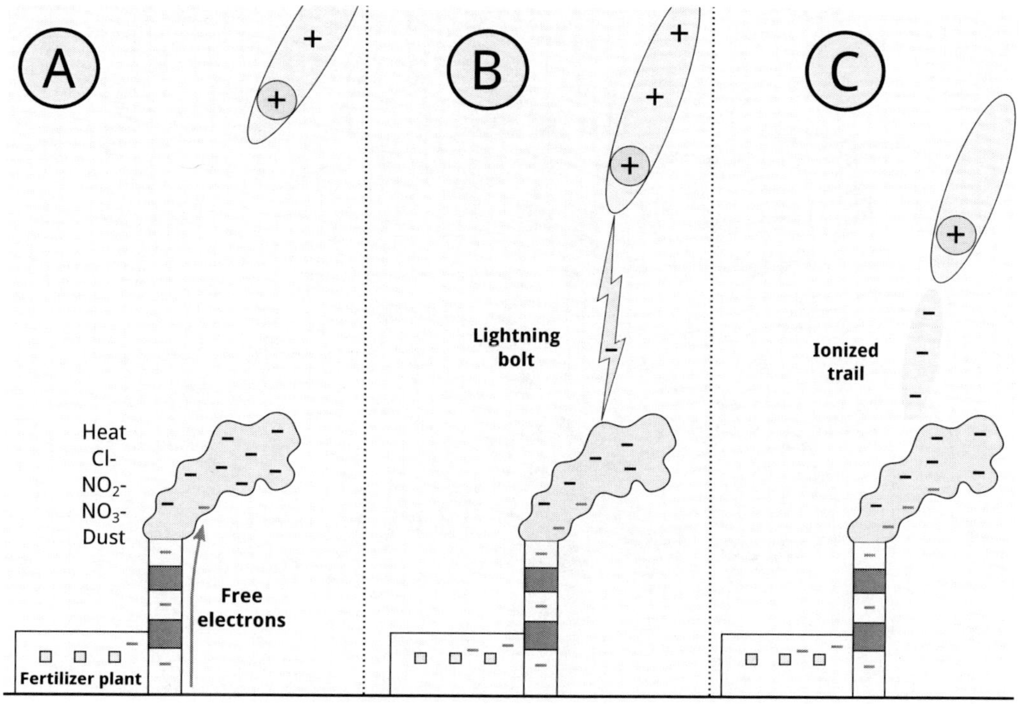

Figure 85: Electric interactions between a meteor and a power plant. (© Sott.net)

causes each time: chemical reactions, accidents, defective equipment, human error, etc. However, I would like to propose that cometary discharge or impact may be the cause for at least some of these explosions and fires. Figure 85 shows how an industrial plume can act as an attractor for cometary discharge.

A (left illustration): A meteor enters the atmosphere. Its (relative) positive charge pulls negative charges (free electrons) up from the plant and the metallic chimney. Those electrons are captured by the low-mobility smoke particles of the plume, making even more negative the charge of fumes that are already loaded with negative ions.[298] The positive meteor and the negative plume attract each other.

B (center illustration): When the electric charge difference is high enough and/or the distance small enough, a massive discharge occurs. An ascending lightning bolt (upward electrons transfer) directly connects the chimney and its plume to the meteor.

C (right illustration): The lightning bolt leaves an ionized and negatively charged atmospheric trail that attracts what remains of the positively charged meteor towards the vicinity of the plant. Note that the meteor may have fragmented before reaching the ground.

Fertilizer plants release chlorine (Cl),[299] nitrites (NO2)[300] and nitrates (NO3).[301] These three

[298] Fertilizer plants eject various kinds of negative ions. In particular Cl-, NO2- and NO3-. See: URS, 'Évaluation de l'impact environnemental et social de l'usine d'engrais de Port-Gentil', *EIES*, p. 94

[299] EPA, *Guidance Manual Alternative Disinfectants and Oxidants*, chapter IV

[300] WHO Regional Office for Europe, *Air Quality Guidelines, Second Edition*, chapter VII

[301] Goeschen C., *et al.*, 'Damage of aromatic amino acids by the atmospheric free radical oxidant NO3 in the presence of

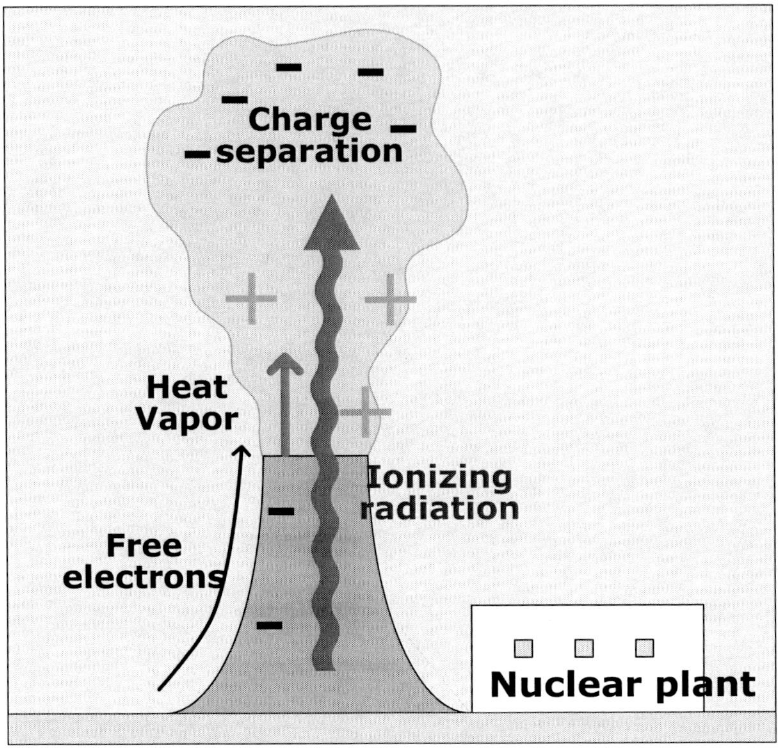

Figure 86: Charge separation in the plume of a nuclear plant. (© Sott.net)

molecules are powerful oxidants; they have a strong tendency to acquire electrons. If they are not already released in negative ionic form, they can be ionized later in the plume. In more general terms, industrial plants tend to eject 'electron-capturing' particles and free radicals (oxidants) into the atmosphere.

That's probably the reason why industrial plumes have been found to hold a negative electric charge.[302] Plumes could therefore act as electric attractors for meteor discharges and possibly meteor/meteorite impacts. They literally behave like lightning rods because they are high up in the sky and they hold an electric charge opposite to that of the space rocks. This explanation may account for some of the explosions and fires listed in the table above, particularly where lightning and/or sonic booms were reported by witnesses.

Nuclear plants exhibit similar properties. They emit ionizing radiation which 'scrape away' electrons from molecules, creating positive and negative ions.[303] In the conductive plume, the negative ions are attracted to the top of the plume by the positive ionosphere, while the positive ions are attracted to the bottom of the plume by the Earth's negative surface.

NO2, N2O4, O3 and O2', *Org Biomol Chem.* (7 May 2011) 9(9): 3380–5

[302] Negative charge has been measured in plumes generated by chemical factories and gasworks. See: Jones, C. D., 'Large electric fields due to industrial chimney stack plumes', *Atmospheric Environment* (1967) 11(12): 1197

[303] The lightning-attracting properties of radioactive material are well known. In 1913, the first radioactive lightning rods (radium, americium) were introduced. Their ionizing properties induce charge separation, which increases the local charge during stormy conditions, enabling them to channel lightning strikes.

Figure 87: Lightning strike on the cooling tower of Salem nuclear plant (19 August 2011). (© Adapted from 'Eagle 9359'-Flickr). The contours of the cooling tower have been enhanced for better clarity.

The fact that nuclear plant plumes are mostly composed of water vapor, a good electric conductor, eases the ionic movement described above and the upward flow of free electrons from the ground to the top of the plume.

On April 18th, 2013, the LaSalle nuclear plant in Illinois experienced an unusual incident: two of its reactors shut down and a radioactive venting procedure was carried out when it was struck by lightning.[304] However, from 1992 to 2003, U.S. nuclear plants were struck by lightning 66 times, yet none of those strikes caused equipment damage or radioactive leakage.[305]

So, what really happened in LaSalle? Was the accident simply due to an insulator defect, as claimed by the U.S. Nuclear Regulatory Commission (NRC),[306] or was the source of this 'lightning' something out of the ordinary? Nuclear plants are very well protected with lightning-prevention equipment,[307] but cometary discharges exhibit two major differences when compared to 'normal' lightning bolts: their polarity is reversed[308] and their intensity can be much higher. The steady increase in cometary activity may cause more such events in the near future.

---

304 'NRC: LaSalle Nuclear Plant Under 'Special Inspection', *Intellihub.* See: intellihub.com/2013/04/28/nrc-lasalle-nuclear-plant-under-special-inspection/

305 Oak Ridge National Laboratory, *Technical Basis for Regulatory Guidance on Lightning Protection in Nuclear Power Plants*, U.S. Nuclear Regulatory Commission, p. 6

306 'Exelon-LaSalle's Two Plant Trip Is A Mess Over A Lightening Strike (really, lightning)', *The Popperville Town Hall blog.* See: steamshovel2002.blogspot.fr/2013/04/exelon-lasalles-two-plant-trip-is-mess.html

307 Lightning rods, inverter fuses, impedance relays, breakers, oversized ground wires and grids, surge arrestors, etc. See: Oak Ridge National Laboratory, 'Technical Basis for Regulatory Guidance on Lightning Protection in Nuclear Power Plants', *USNRC*, pp. 25–31

308 The impedance relays protecting nuclear plants are one-directional, so they don't work when reverse discharges occur. See: *Ibid.*, p. 11

Figure 88: Electron microscope image showing a fossilized diatom embedded in a meteoric rock. (© Wales News Service)

Space-bound bio-hazard

Actual impacts, overhead explosions and their associated shockwaves and electric discharges are not the only threats posed by earthbound asteroids. Ironically, some of the most deleterious effects of comets might well be those least reported and least studied. Fred Hoyle and Chandra Wickramasinghe have shown how viruses can be spread in the Earth's atmosphere by dust in the debris stream of comets.[309] When Earth passes through this stream, the dust and viruses enter the upper parts of our atmosphere,[310] where they can remain suspended for years until gravity finally pulls them down.[311]

Microbes can also be brought by meteorites directly to Earth's surface. That's the conclusion of the same Professor Chandra Wickramasinghe, who published a paper showing the presence of fused fossilized microorganisms in a meteorite that recently hit the ground.[312] Wickramasinghe conducted extensive tests that ruled out any possibility of terrestrial contamination.

A cosmic origin for some microbes may explain why so many new viruses emerge in Asia. The Earth's atmosphere is thinnest at the Himalayas and its surrounding region;[313] therefore 'drop-downs' should take less time there. It might also explain why new strains of viruses usually *affect*

[309] Hoyle, F. & Wickramasinghe, C., 'SARS – a clue to its origins?', *The Lancet* (24 May 2003) 361(9371): 1832

[310] Segura, G., 'New Light on the Black Death: The Viral and Cosmic Connection', *Dot Connector Magazine* (2011) 13

[311] Ulanowski, Z., 'Alignment of atmospheric mineral dust due to electric field', *Atmos. Chem. Phys.* (2007) 7: 6161–6173. Because of their small size, dust particles are very sensitive to wind and not very sensitive to gravity. In addition some electrical phenomenon might explain why dust settles even more slowly than predicted by simulations. This is examined later in Chapter 27: 'Hurricanes, lightning, and tornadoes'.

[312] Wickramasinghe, C., *et al.*, 'Fossil Diatoms In A New Carbonaceous Meteorite', *Journal of Cosmology* (2013) 22

[313] Hoyle & Wickramasinghe, *op. cit.*

birds first, as we see with the numerous avian flus.[314] During their drop-down, microbes are first present in the sky, where they can contaminate birds before eventually reaching the ground. Virus-carrying meteoric material might also explain why Justinian's plague[315] began just four years after a theorized massive overhead cometary explosion in 536 AD, which produced an 18-month-long dust veil,[316] and why the Black Death in 1347 was preceded by intense cometary activity.[317]

[314] Apart from the famous H5N1 strain, there are more than 30 other subtypes of avian influenza. Most of them emerged over the last few years. See: www.cdc.gov/flu/avianflu/influenza-a-virus-subtypes.htm

[315] See Chapter 34: 'Historic evidence of a human-cosmic connection' for more details about this period of history.

[316] Michael the Syrian, *Chroniques Universelles*, Book IX, Chapter XXVI

[317] Baillie, M., *New Light on the Black Death*, pp. 145–155

Chapter 22.

The slowdown of the Earth

As explained in Part I, the spin of stars and planets is electrically driven. The Earth is no exception. It functions as a rotor (with its negative charge) driven by a stator (the ionosphere – which is positively charged, relative to the planet). The ionosphere,[318] or more exactly the magnetosphere, acts as a stator (as in 'stationary') because of its strong asymmetry, as depicted in figure 89.

Indeed, whereas the dayside of the magnetosphere extends only 65,000 km away from Earth,[319] on the nightside, the magnetic field in the magnetotail (the tail of the magnetosphere) extends over 6,300,000 km.[320] The highly asymmetric shape of the magnetosphere keeps its grand axis 'locked' in line with the direction of the solar wind; hence Earth's magnetotail being fixed on its nightside as the planet spins in its orbit around the Sun.

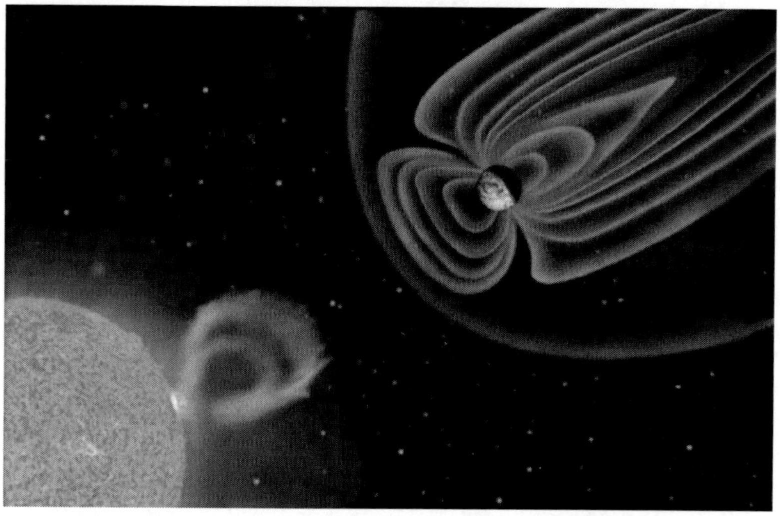

Figure 89: Artist's rendition of the Earth's magnetosphere.

Figure 90 shows the electromotive force (F – black arrow) which, as you'll recall from Chapter 12, is the 'Lorentz force', and is proportional to the vertical atmospheric current (I – light grey arrow). Therefore, *the Earth's rate of rotation is proportional to the intensity of the electric current* between the ionosphere and the planet itself. Consequently, a decrease in this current intensity leads to a decrease in the Lorentz force and therefore a decrease in the Earth's rate of rotation. Thus, the ongoing decrease in solar activity should cause a *slowdown, however small, in the planet's rotation.*

In 1979, Robert G. Curie confirmed that Earth's spin was linked to solar activity and concluded that the "solar sunspot cycle in Earth has been detected."[321] Figure 91 shows the correlation between the Earth's rotation cycle and the 11-year solar activity cycle.

---

[318] In this book, for simplicity's sake, we consider the ionosphere to be equivalent to the magnetosphere.

[319] 'Magnetosphere', *Wikipedia*. See: en.wikipedia.org/wiki/Magnetosphere

[320] *Ibid.* For comparison, that is more than 15 times the Earth–Moon distance.

[321] Curie, R. G., 'Detection of the 11-yr sunspot signal in Earth rotation' *Science* (23 January 1981) 211(4480): 386–389

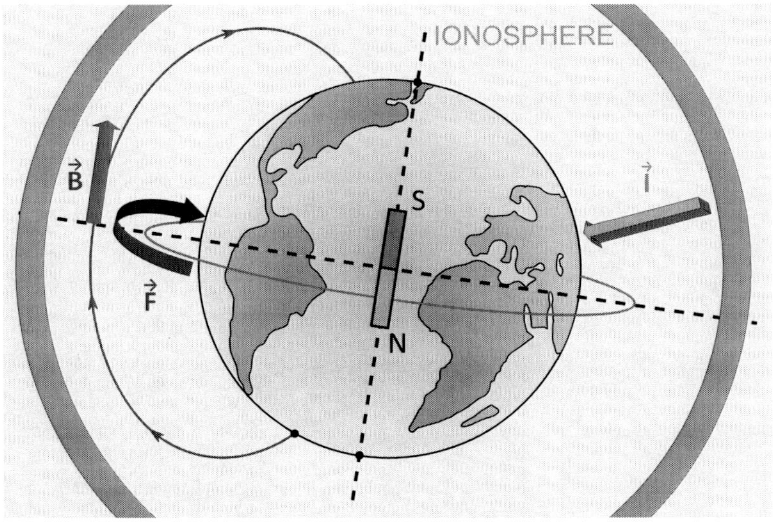

Figure 90: The electrically driven spin of the Earth. (© Sott.net)

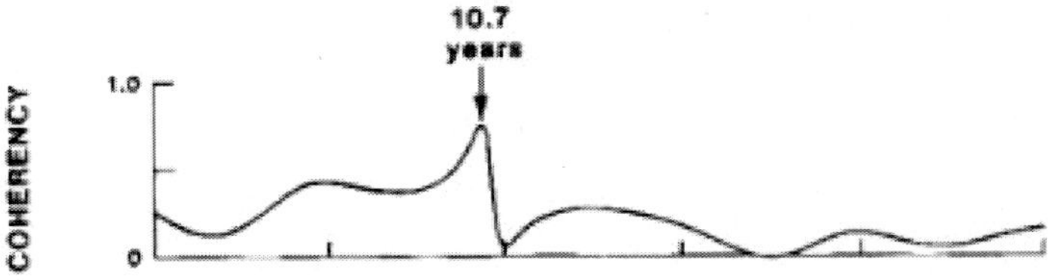

Figure 91: Coherency level between the Earth's rotation cycle and the solar cycle. The variation in Earth's rotation follows a 10.7-year cycle similar to the Sun's cycle. (© Curie)

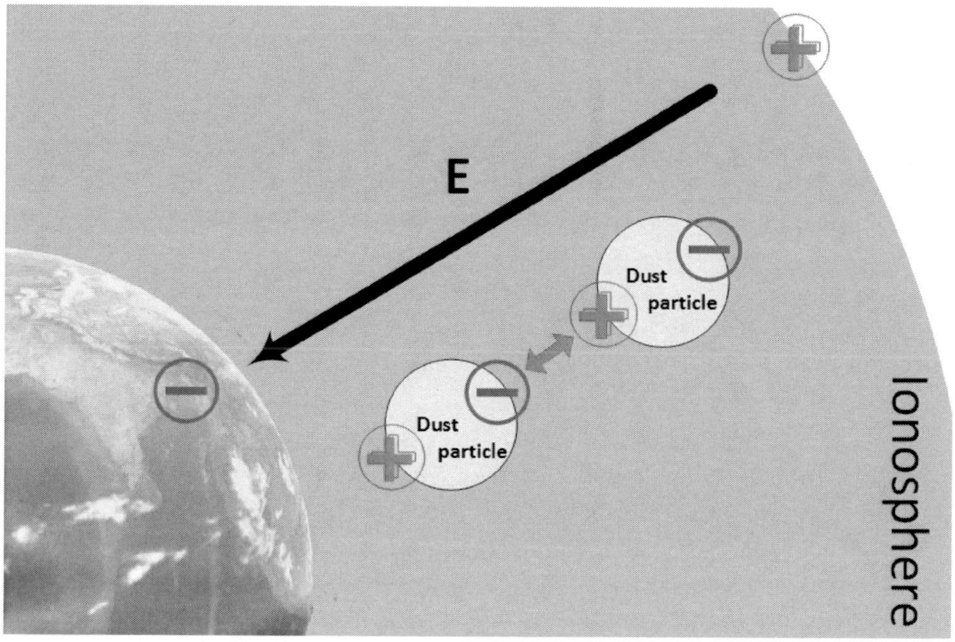

Figure 92: Polarization of dust particles in the Earth's atmosphere. (© Sott.net)

A higher concentration of cometary dust in the atmosphere[322] may also contribute to the Earth's slowdown by increasing the drag exerted by the atmosphere over the planet, i.e. the drag between the stator and the rotor.

Here we are not so much talking about mechanical drag *per se* (friction, for example) as about electrical drag. When dust is present, the small atmospheric ions attach to these relatively large dust particles and lose mobility.[323] In addition, because of the atmospheric vertical electric field, these now charged particles tend to be polarized and align themselves with the vertical atmospheric E-field.[324] Their positive side points down towards the Earth's negative surface and their negative side points upwards towards the positive ionosphere, as depicted in figure 92.

If the Earth was not spinning, opposite charges would attract and the polarized particles would stay perfectly aligned along the atmospheric E-field. But the planet's spinning disrupts this alignment and tends to disorganize the particles. The energy necessary to induce this particle agitation is equal to the amount of kinetic energy lost by our planet, hence the contribution to the abovementioned infinitesimal slowdown. The higher the electric drag, the stronger the slowdown.

It's conventionally thought that the Earth's slowdown is due exclusively to tidal effects[325] exerted by the Moon on the planet's oceans. But if that is the case, how are we to reconcile it with the

[322] See previous chapter: 'Increase in fireballs and atmospheric cometary dust'
[323] Schneider, H.S., *et al.*, *Encyclopedia of Climate and Weather*, p. 67
[324] Ulanowsky, Z., *et al.*, 'Alignment of atmospheric mineral dust due to electric field', *Atmos. Chem. Phys.* (2007) 7: 6161–6173
[325] 'Tidal acceleration', *Wikipedia*. See: en.wikipedia.org/wiki/Tidal_acceleration

concomitant slowing down of the rotation of other planets including Venus[326] and Saturn,[327] which has been widely acknowledged?

> Interestingly, the **Earth's rotation is also slowing down**, but scientists attribute this to tidal acceleration, frictional 'drag' caused by the Moon's gravitational pull. **This explanation cannot apply to Venus' slowing spin, however, because Venus has no moon of its own.**[328]

The fact that Venus is also slowing down suggests that the Moon's tidal force, while being one possible contributing factor to Earth's slowdown, is not the *sole* one. In addition, the previously mentioned[329] increased eccentricity of the Moon's orbit should lessen the tidal force and, therefore, reduce the lunar influence on the Earth's slowdown.

The fact is, the concomitant slowdown of Saturn, Venus and Earth is consistent with the Electric Universe hypothesis, whereby the Sun's electric activity drives the planets' spin, and therefore, reduced solar activity induces a slowdown of the planets in the solar system, including Venus, Saturn and the Earth.

That the speed of the planet's rate of spin varies was first established long ago,[330] and several forces seem to be at work in directing it. It's interesting to note, however, that data for the period 1960–2012 show that the slowdown in the rate of spin is not constant. Figure 93 shows the deviation between the 'theoretical' 24-hour day length (represented by the horizontal black straight line), and the actual measured day length (the dark grey curve). As you can see, the measured day length (jagged light grey curve) throughout this period is longer than the theoretical day length by up to four milliseconds/day (see year 1971). The ascending medium grey curve shows the cumulative time difference between the theoretical day duration and the observed day duration (+25 seconds between 1970 and 2010).

Notice also that, since 2003, the average measured day length has been getting longer (dark grey ascending line on the right of the chart), from an average of +0.3 ms per day in 2003 to an average of +1.0 ms per day in 2011. Since 2003, the rate at which the Earth's spin is slowing down seems to be increasing.

A lengthening of the day by one millisecond per day leads to an actual year that is 0.365 seconds longer than the theoretical one. Some adjustments are therefore required to avoid an ever-increasing drift and people in the northern hemisphere eventually having to celebrate Christmas in the middle of summer. Since 1972, an extra second, also known as a leap second, has been added about once a year. This extra second was initially added because the duration of one mean solar day is slightly

326 'Could Venus be shifting gears?', *European Space Agency* (10 February 2012). See: www.esa.int/Our_Activities/Space_Science/Venus_Express/Could_Venus_be_shifting_gear

327 Barry, P., 'Saturn's rotation puts astronomers in a spin', *New Scientist* (3 May 2006). See: www.newscientist.com/article/dn9100-saturns-rotation-puts-astronomers-in-a-spin.html#.UryPbbTDviE

328 Nelson, B., 'Scientists baffled to discover that Venus' spin is slowing down', *MNN* (17 February 2012). See: www.mnn.com/earth-matters/space/stories/scientists-baffled-to-discover-that-venus-spin-is-slowing-down-0

329 Chapter 16: 'Radio carbon dating 'anomalies''

330 As far back as 140 AD, Ptolemy acknowledged the difference between true day duration and theoretical day duration. See: Toomer, G. J., *Ptolemy's Alemagest*, pp. 6–7

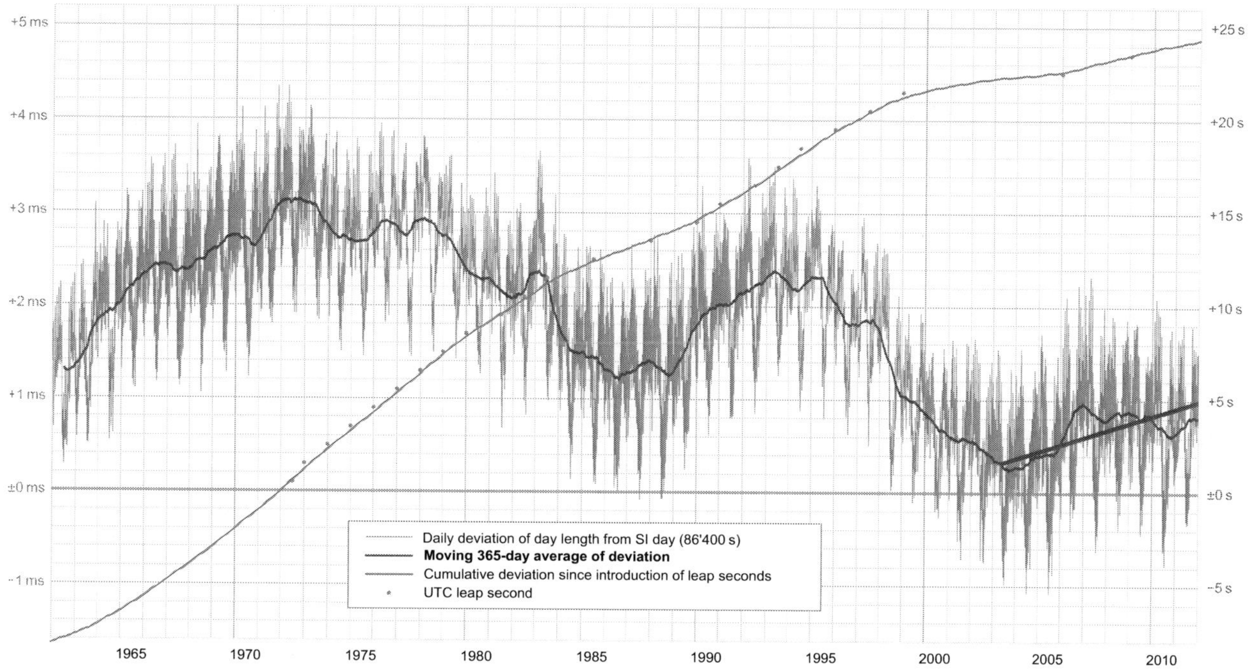

Figure 93: Theoretical day length vs. observed day length. (© Sott.net, adapted from U.S. Naval Observatory data)

longer than the UTC[331] day, which counts 24 hours, i.e. 86,400 SI seconds.[332] However, the Earth's slowdown now requires the addition of an ever greater number of leap seconds,[333] since leap seconds have to compensate not only for the UTC day – which is already too short – but also for the fact that actual days are getting longer and longer.

Another piece of evidence for the electrically driven spin of the Earth is what scientists call 'glitches'. In August 1972, an exceptional solar outburst hit the Earth, provoking one such 'glitch': a sudden variation in the length of the day.[334] The electric disturbance from the solar flare affected the spin of the Earth such that, for a few days, it spun more slowly than usual. Because of its unusual power, the solar outburst flooded the Earth's double layer, diminishing its effectiveness, leading to a reduced Lorentz force and the subsequently lowered spin rate.[335]

This last point might seem like a paradox since we've already established that higher solar activity in an electric model would tend to make Earth spin faster. This is true until the capacitor (Earth's surface ionosphere) reaches the 'breakdown' voltage. Beyond this point *atmospheric discharges occur*, the atmospheric electric field drops and so does the subsequent Lorentz force, hence the

[331] UTC stands for Coordinated Universal Time. It is based on atomic clocks and doesn't take into account any slowdown of Earth.

[332] International System Unit.

[333] McCarthy, D., *et al.*, 'The Physical Basis of the Leap Second', *Astronomical Journal* (2008) 136: 1906–1908

[334] Gribbin, J., *et al.*, 'Discontinuous Change in Earth's Spin Rate following Great Solar Storm of August 1972', *Nature* (4 May 1973) 243

[335] De Grazia, A. & Milton, E., *Solaria Binaria*, p. 99

lowered spin rate.

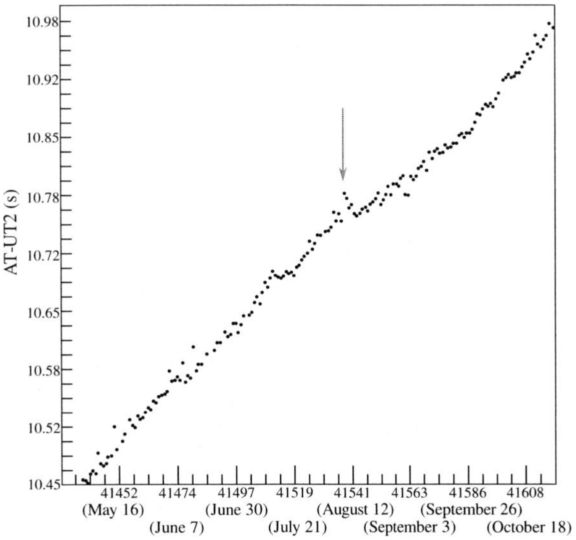

Figure 94: Change of slope in the day-length curve after the massive 1972 solar outburst. (© Dachille)

Note that a slowdown of the Earth, even if infinitesimal,[336]can produce devastating effects, particularly earthquakes and volcanic eruptions, as explained by Caltech seismologist Don L. Anderson:

> There is an immense reservoir of energy stored in the Earth's rotation, so any change in the rotation rate could dump an enormous amount of energy in the crust and conceivably trigger earthquake[s] in regions where the crust has already been stressed to its breaking point ... Volcanoes sense pressure changes below the lithosphere; when the pressure increases, so does their venting. In places like Japan and Chile we know that earthquakes and volcanoes are intimately related.[337]

The mainstream media and mainstream science don't mention any possible adverse effect caused by this slowdown; they don't even mention the acceleration of the slowdown. But in the following chapters, we will see how an infinitesimal slowdown of the Earth may be responsible for inducing effects that are not exactly negligible, to understate the matter.

[336] A lengthening of 1 ms/day means a 0.000 001% slowdown in Earth's spin frequency.

[337] Anderson, D., 'Giant quakes increase: Seismic storm Lull Ends', *Los Angeles Times* (29 August 1976)

Chapter 23.

Earth 'opening up': Earthquakes, eruptions and sinkholes

Figure 95 shows the Earth and some of the effects we might expect from reduced spin. On the left, the planet is represented with a normal rate of rotation. Note that it's slightly broader at the level of the equator (ellipsoidal shape) because of the centrifugal force. On the right, Earth is depicted with a reduced rate of rotation. This slowdown induces a decrease in centrifugal force, which results in mechanical stress exerted on the lithosphere (Earth's crust). Compression forces (white arrows) occur at low latitudes, while extension forces (grey arrows) occur at higher latitudes. Thus, the planet is deformed: its shape becomes less ellipsoidal and more spheroidal. Of course, this deformation has been exaggerated here (oval shape on the left vs. circular shape on the right) in order to make the effect more visually apparent.

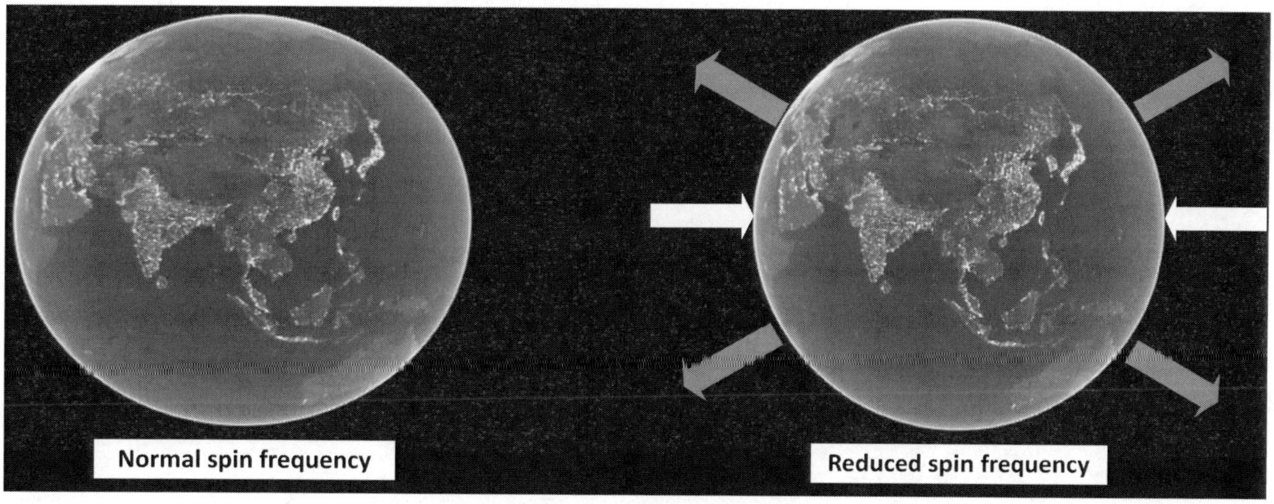

Figure 95: Effects of the Earth's slowdown (tension, compression and shape). (© Sott.net)

We've already seen how reduced solar activity decreases the Earth's surface–core E-field.[338] This electric field acts as a powerful 'binder' of the planet, with the (relatively) negative surface being attracted to the (relatively) positive core and therefore literally **pulling the surface and core together**. Reduced binding means a 'looser' planet with its 'parts' not as tightly held together.

To illustrate this, let's use a simple example. Imagine an electromagnet (magnet powered by

[338] See Chapter 20: 'Impact of the Sun's reduced activity on planet Earth'

electricity) on which you drop various parts of steel: washers, nuts, filings... All this steel material of various size and shape will stick strongly to the magnet and form a relatively coherent strong coat. Now, imagine you switch off your magnet. All of the steel will suddenly become loose, the individual parts free to move in any direction, not held together any more, losing their coherency and bonds. That's what happens to some extent to the Earth's crust when solar activity drops: the heterogeneous parts constituting the crust (soil, rocks, mountains, continental plates...) are not tightly bound together anymore and become free to move away from each other.

These two factors mentioned above – the Earth's slowdown and the reduced surface–core E-field – work together to trigger a literal 'opening up' of the planet, which in turn might be the cause for several earthly effects that have recently begun to happen or have increased in frequency, i.e. formation of new islands, new geysers, sinkholes, vents and rifts, landslides, etc.

Below, we will focus on a few effects induced by the 'opening-up': first, macro effects that are earthquakes and volcanic eruptions and then a more localized effect, namely sinkholes.

Figure 96: The magnet attracts, holds and binds together the otherwise loose and incoherent iron filings.

Earthquakes and volcanic eruptions

Figure 97 shows the annual frequency of earthquakes[339] (white curve) registering a magnitude of 6.0 or higher, and the yearly number of volcanic eruptions[340] (black curve) worldwide.

From 1973 to 1996, earthquake and eruption frequencies were almost stable, increasing only slightly year after year, but from 1996 onwards, an acceleration is noticeable. Volcanic eruptions show an increase from about 59 eruptions per year at the end of the 1990s to roughly 75 eruptions per year in the period 2007–2010 (+30%).

Today, the increase in volcanic activity has reached such a level that, by late November 2013, 35 volcanoes were actively erupting,[341] including volcanoes that had been dormant for decades.[342]

It could be argued that the increase in both the frequency and intensity of earthquakes and volcanic eruptions is, at least partly, a result of the slowdown and 'opening up' processes:

1. The Earth's minute slowdown exerts mechanical stress on the crust (compression at low latitudes and extension at high latitude). This stress deforms the crust. This deformation is more pronounced and can even lead to partial ruptures around the weakest spots of the crust, i.e. the fault

[339] The earthquake frequency curve is based on data provided by the USGS earthquake database.

[340] The eruption frequency curve is based on the Smithsonian volcanic eruptions database.

[341] Mahapatra, L., 'Mount Sinabung is just one of 35 currently erupting volcanoes', *International Business Times* (26 November 2013). See: ibtimes.com/volcano-indonesia-one-35-currently-erupting-volcanoes-heres-where-they-are-map-1486866

[342] The usual level of activity for the Ring of Fire is more like 0 to 3 simultaneous eruptions.

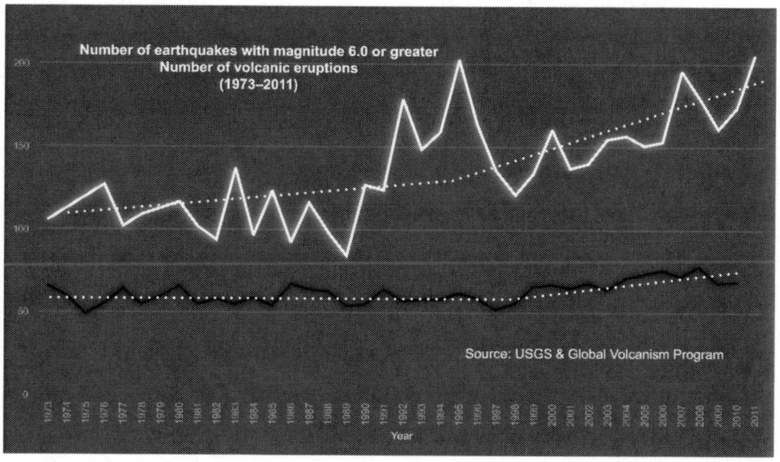

Figure 97: Number of earthquakes (white curve) and number of volcanic eruptions (black curve) over the period 1973–2011. (© Sott.net)

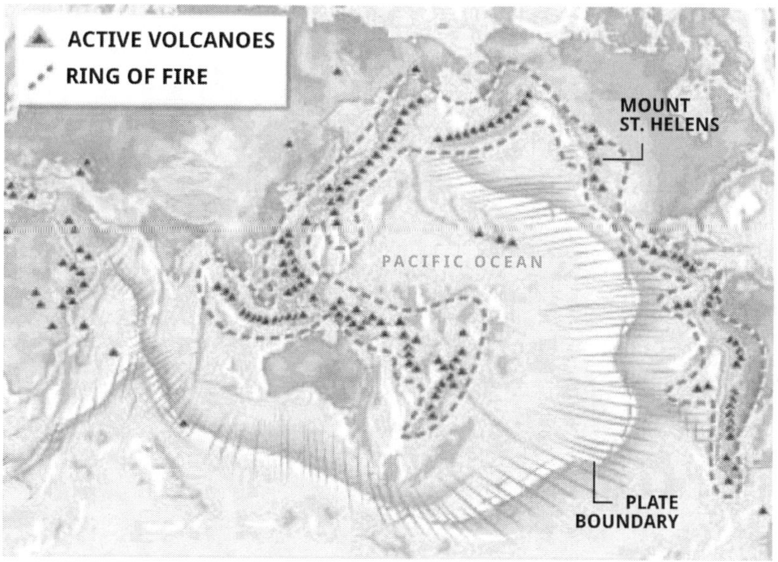

Figure 98: The Pacific Ring of Fire. (© MSNBC)

lines (boundaries between tectonic plates) which are the typical location of seismic and volcanic activity.

2. The mantle has a higher density than the crust[343] and, therefore, it also has a higher momentum and won't slow down as fast as the crust. This difference in rotation between the crust and the mantle is called crustal slippage. The fluidity of the mantle enables slippage induced by the different momentum carried by the crust,[344] the upper mantle and the core. This speed difference can cause friction at the interface between the crust and the mantle. This friction can locally deform the crust and cause earthquakes and eruptions.[345]

> [Change] in Earth's speed of rotation would induce changes in the magma tide as it adjusted to the new equator or altered rotational speed. Such changes, however, might not be uniform throughout, owing to a 'drag' factor deep in the magma itself, although, overall, they would certainly **impose terrible strains on the lithosphere generally**.[346]

3. The decrease in the surface–core E-field reduces the binding force and loosens the tectonic plates relative to each other. The plates are then free to move relative to each other. It is this very relative movement (divergence, convergence or sliding) which is one of the main causes of earthquakes and volcanic eruptions.

4. A final factor involved in earthquakes and volcanic eruptions is electromagnetism:

> Some scientists have become aware of a correlation between sunspots and earthquakes and want to use sunspot data to help predict earthquakes. The theory is that an intensification of the magnetic field can cause changes in the geosphere [i.e. crust]. NASA and the European Geosciences Union have already put their stamp of approval on the sunspot hypothesis, which suggests that certain changes in the Sun-Earth environment affect the magnetic field of the Earth, which can then trigger earthquakes in areas prone to them. It is not clear how such a trigger might work.[347]

As a matter of fact, several strange earthquake precursors have been noted for decades: low-frequency electromagnetic emission,[348] magnetic field anomalies, earthquake lights from ridges and mountaintops, temperature anomalies over wide areas as seen in satellite images, and changes in the plasma density of the ionosphere.[349]

The problem with the electric nature of earthquakes is that rocks are very poor electric conductors. So how could such poor conductors manifest any electromagnetic phenomena?

---

[343] The density of the crust (mostly rocky material) is comprised between 2.7 and 3.3. The density of the mantle (mostly iron and silicates) is comprised between 3.3 and 5.7. See: Geoman, 'What is the best estimate of the densities of the various layers of the Earth?', *University of Oregon*. Jersey.uoregon.edu/~mstrick/AskGeoMan/geoQuerry57.html

[344] The crust is composed mainly of rocks and is 30 km thick, on average. The mantle is about 3,000 km thick, composed mainly of olivine-rich rock. In the upper mantle the rock is cool and brittle, while in the lower mantle, the rock is warm and soft. See: 'Three parts of the Earth's interior', *Geology*. geology.com/nsta/earth-internal-structure.shtml

[345] Rankin, I., *This Erratic Planet: What Happens When the Earth Changes Its Axis of Rotation*, p. 33

[346] Allan, D. & Delair, J., *Cataclysm!*, p. 180

[347] 'Sunspots and Earthquakes', *Thunderbolts.info* (21 December 2005).
See: www.thunderbolts.info/tpod/2005/arch05/051 221earthquake.htm

[348] Hattori, K., 'ULF Geomagnetic Changes Associated with Large Earthquakes', *TAO* (September 2004) 15(3): 329–360

[349] Freund, T. F., 'Rocks That Crackle and Sparkle and Glow: Strange Pre-Earthquake Phenomena', *Journal of Scientific Exploration* (2003) 17(1): 37

Type of Margin	Divergent	Convergent	Transform
Motion	Spreading	Subduction	Lateral sliding
Effect	Constructive (oceanic lithosphere created)	Destructive (oceanic lithosphere destroyed)	Conservative (lithosphere neither created or destroyed)
Topography	Ridge/Rift	Trench	No major effect
Volcanic activity?	Yes	Yes	No

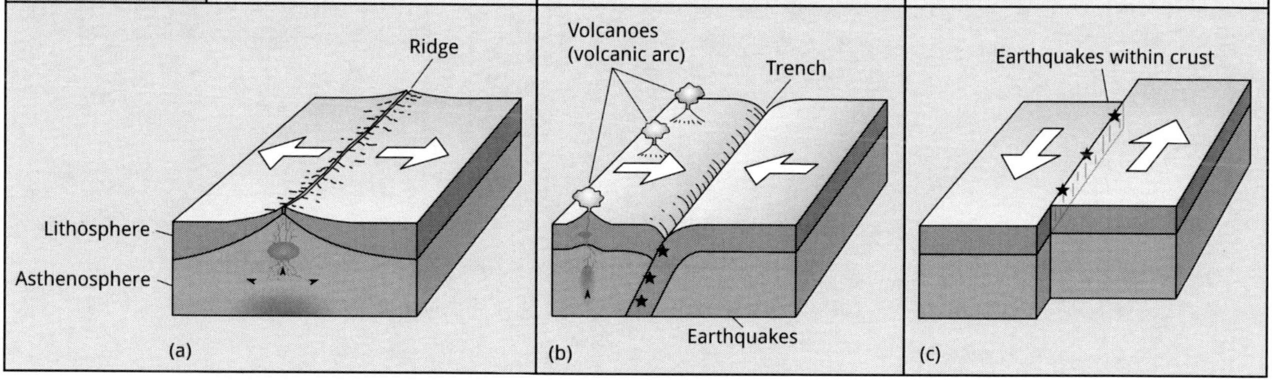

Figure 99: Types of tectonic plate movement and subsequent seismic and volcanic activity. (© age-of-the-sage.org)

NASA senior researcher Friedemann T. Freund has shown that rocks are, in fact, not always poor conductors. When subjected to mechanical shocks or mechanical stress, the conductivity of rocks increases dramatically and they suddenly become very good conductors:

> ... igneous and metamorphic rocks, which make up a major portion of the Earth’s crust, **contain electric charge carriers, which have been overlooked in the past. These charge carriers are defect electrons in the valence band, i.e., positive holes. Under normal conditions they are dormant, but when they ‘wake up’, the rocks begin to sparkle and glow.**[350]

Figure 100 shows the behavior of a sample of rock (granite – large grey rectangle on the right) subjected to a mechanical shock (‘impact’ – light grey arrow). Electric current as shown by the curve on the left titled “Top Electrode” is measured through a top electrode and three coils (top, middle and bottom).

Before the impact, the electrode and the three coils return a zero signal. No electric current occurs within the rocks. After the impact (about 50 microseconds later), the granite sample starts to conduct electricity. All three coils and the top electrode detect a sudden surge in electric current traversing the rock.

So, most of the *Earth’s crust can become highly conductive* if subjected to mechanical stress/shock, for example the type of stress due to the slowing down and opening up of the Earth’s lithosphere described above. When high conductivity is reached, *electric current can occur between ground*

[350] *Ibid.*, p .53

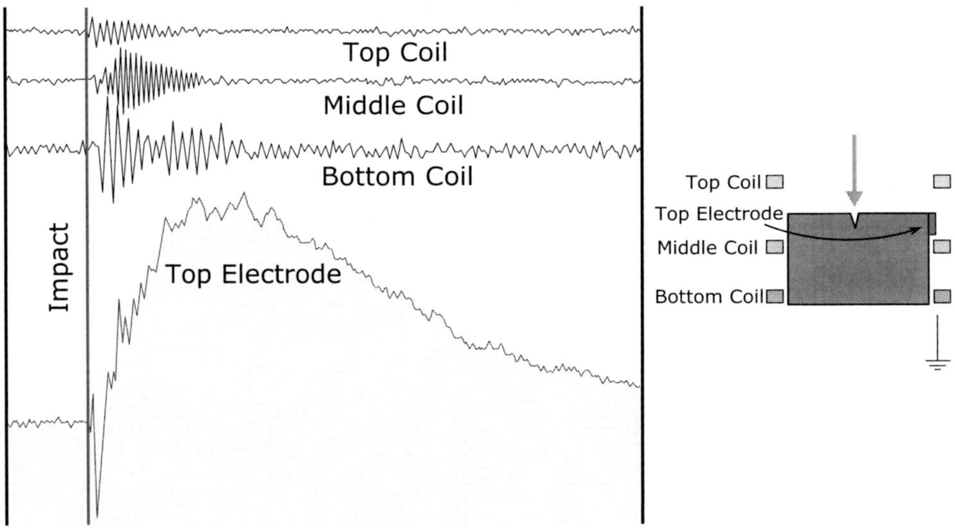

Figure 100: Electric effects of a mechanical impact on granite recorded by means of three magnetic pick-up coils and one electrode. (© Sott.net, adapted from F. T. Freund)

regions of different electric potential.[351] This current is fed, among other factors, by the change in surface–core E-field mentioned previously.

At this point a second phenomenon called piezoelectricity might intervene. Some crystals, in particular quartz, which is very frequent in granite rocks,[352] will deform if subjected to electric current (that's almost the reverse of the above-described phenomenon, where mechanical deformation triggers electric currents).

One could thus hypothesize that earthquakes are not unlike underground lightning. Earthquakes being to underground electric phenomenon what lightning is to atmospheric electric phenomenon: a simple charge rebalancing process generating some mechanical side effects: air waves (thunder) for lightning and crustal waves (seismic tremors) for earthquakes.

Figure 101 shows how those various phenomena, all being consequences of reduced solar activity, might interact and contribute to earthquakes and volcanic eruptions.

Sinkholes

Along with earthquakes and volcanic eruptions, sinkholes could be another consequence of the Earth's 'opening up' phenomenon. Sinkholes are conventionally defined as follows:

> A sinkhole is basically any collapsed or bowl-shaped feature that's formed when a void under the ground creates a depression into which everything around it drains.[353]

[351] Such effects, combined with emissions of gasses from inside the Earth, another consequence of 'opening up', could be another factor in the occurrence of anomalous fires and explosions.

[352] Interestingly most mountain ranges are made of granite. Concomitantly, mountain ranges are also some of the most quake prone geological formations.

[353] Berlin, J., 'Sinkhole Science: A primer', *National Geographic* (5 March 2013). See: news.nationalgeographic.com/news/2013/03/130 305-florida-sinkhole-science-causes-world/

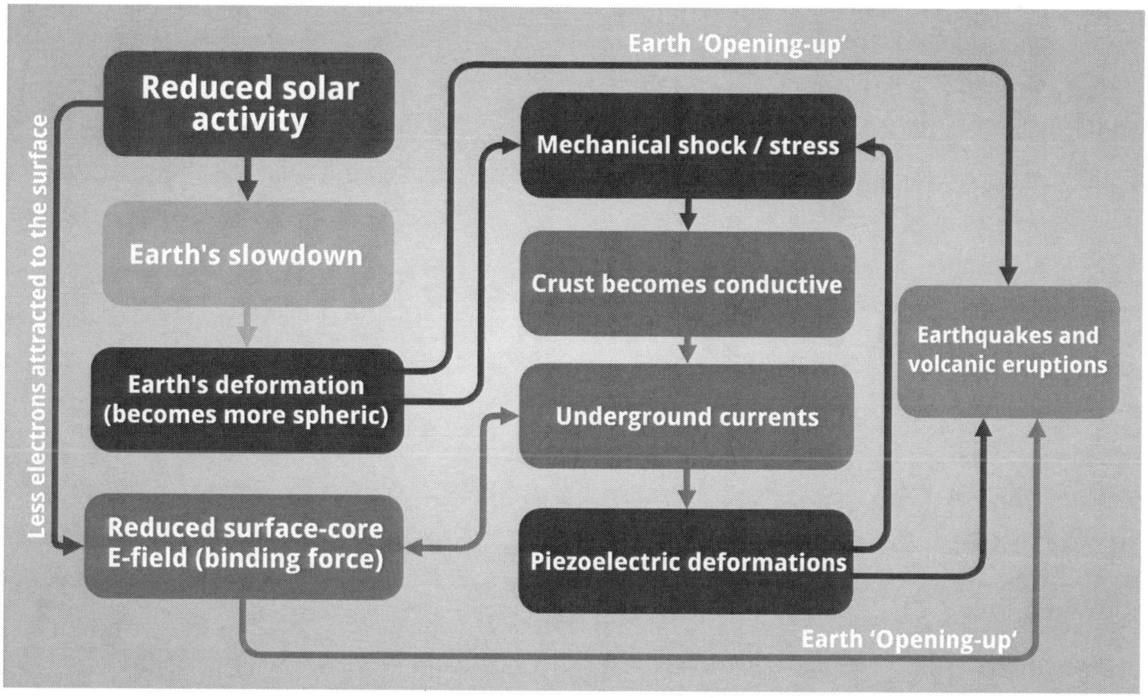

Figure 101: Hypothesized mechanisms through which reduced solar activity can increase earthquakes and volcanic eruptions. (© Sott.net)

Sinkholes come in all kinds of shapes and sizes. Their volume can vary from a few liters to about 100 million cubic meters.

The largest sinkhole in the world is the Xiaozhai Tiankeng,[354] located in China. It is up to 662 meters deep and its volume exceeds 119 million cubic meters. It has been well known to local inhabitants of the area since ancient times.[355] The Devil's Millhopper sinkhole (Florida), which formed nearly 20,000 years ago,[356] is another large one, and the Sima de las Cotorras sinkhole[357] (Chiapas, Mexico) even displays ancient cave-type rock paintings, a living testimony of its venerable age.

Indeed, sinkholes are not a new phenomenon; they have conventionally been studied as old geological formations, often having their own names, and are usually filled with water because of years of exposure to rainfall and/or underground water.

Since most 'old' sinkholes are located in rural areas and/or rocky terrain, human industrial causes can be excluded. The progressive dissolution of the underground rock as the most likely cause for the majority of old sinkholes can stand uncontested for the most part.[358]

[354] 'Xiaozhai Tiankeng', *WonderMondo.* See: www.wondermondo.com/Countries/As/China/Chongqing/Xiaozhai.htm
[355] *Ibid.*
[356] Ulmer, G. L., *The Florida Landscape: Revisited*, The Polk Museum, 1992
[357] 'Sima de las Cotorras', *SustainableTrip.* See: www.sustainabletrip.org/profile/sima-de-las-cotorras
[358] Lard, L., Paull, C. & Hobson, B., 'Genesis of a submarine sinkhole without subaerial exposure', *Geology* (1995) 23(10): 949–951

Figure 102: Xiaozhai Tiankeng, the largest sinkhole in the world. (© Public domain)

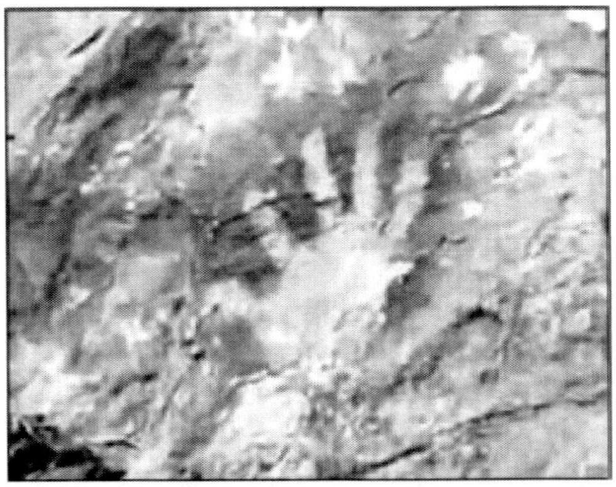

Figure 103: Cave-type painting on a cliff of the Sima de las Cotorras sinkhole. (© Mayantravelguide.com)

Figure 104: Picture of the giant sinkhole that opened up in Guatemala City, swallowing an intersection and a 3-story building on May 30th, 2010. (© Reuters/Daniel LeClair)

However, this 'landscape' began to change recently, around 2004, with what appears to be one of the first modern reports concerning the sudden appearance of a new generation of sinkholes:

> To folks around Wildwood, it is nothing but freaky: an entire 23-acre lake vanished in a matter of days, as if someone pulled the plug on a bathtub. Lake Chesterfield went down a sinkhole this week, leaving homeowners in this affluent St. Louis suburb wondering if their property values disappeared along with their lakeside views.[359]

Looking back through reports over the last ten years, it seems that the appearance of new sinkholes accelerated in 2007 when a giant sinkhole opened up in Guatemala City.[360] A second monster hole appeared in 2010 (figure 104).

The frequency of sinkholes over the last three or four years has increased to the point that people are being unexpectedly 'swallowed' and even killed in urban areas.[361] Homes and vehicles have also been gobbled up in ever-larger numbers.[362] An increasingly wide range of explanations for this startling phenomenon has been put forward by experts, including man-made factors such as damage to underground water pipes or sewer collapse leading to the dissolution of underground

[359] 'Woe! Lake begone!', *Associated Press* (11 June 2004). See: http://carour.com/trucking-safety-3/safety-3-2378.html

[360] 'Guatemala homes swallowed up by hole', *BBC*. See: news.bbc.co.uk/2/hi/americas/6391 117.stm

[361] 'Massive sinkhole kills five in south China town of Shenzhen', *News* (21 May 2013). See: www.news.com.au/world/massive-sinkhole-kills-five-in-south-china-town-of-shenzhen/story-fndir2ev-1226647965 016

[362] 'Woman escapes after giant sinkhole swallows her car on US street', *The Telegraph* (4 July 2013). See: www.telegraph.co.uk/news/worldnews/northamerica/usa/10159 096/Woman-escapes-after-giant-sinkhole-swallows-her-car-on-US-street.html

Figure 105: Harrisburg's sinkhole map. (© New York Daily News)

rock,[363] the depletion of aquifers due to fracking,[364] loose soil as a result of flooding (that is in turn invariably attributed to 'man-made global warming'), hastily or improperly constructed buildings or infrastructure, abandoned coal mines, heavy rain following severe drought[365] and so on.

While some sinkholes may indeed be due to the above-mentioned causes, how is it possible to explain the overall, sudden global increase in newly formed sinkholes? Is rock suddenly dissolving faster?[366] Have sewers all of a sudden decided to collapse at the same time in widely divergent locations? Even 'fracking', which has been pushed strongly in the U.S. in recent years, can't account for *all* the new sinkholes, especially those that have appeared where fracking isn't practiced.

Notice that most causes mentioned above are based on rock dissolution, but only three common rocks are water soluble, namely rock-salt, gypsum and limestone. Therefore, dissolution can't explain sinkholes that suddenly appear in non-soluble rocks like granite, for example.

The formation of new sinkholes, which were almost unheard of before 2004,[367] has become so

[363] Cochran, M., 'What causes sinkholes to form', *Accuweather* (22 April 2013). See: www.accuweather.com/en/weather-news/sinkholes-proven-to-be-dangero/10802 869

[364] Hydraulic fracturing involves forcing chemicals deep into the ground under high pressure in order to access natural gas deposits.

[365] Orndorff, R., 'The Science of Sinkholes', *WAMU 88.5*. See: thedianerehmshow.org/shows/2013-03-20/science-sinkholes/transcript

[366] Natural Environment Research Council, 'Soluble rocks', *British Geological Survey*. See: www.bgs.ac.uk/products/geosure/soluble.html

[367] Except on a small scale, usually in Florida, a state sitting on top of limestone 'Swiss cheese'.

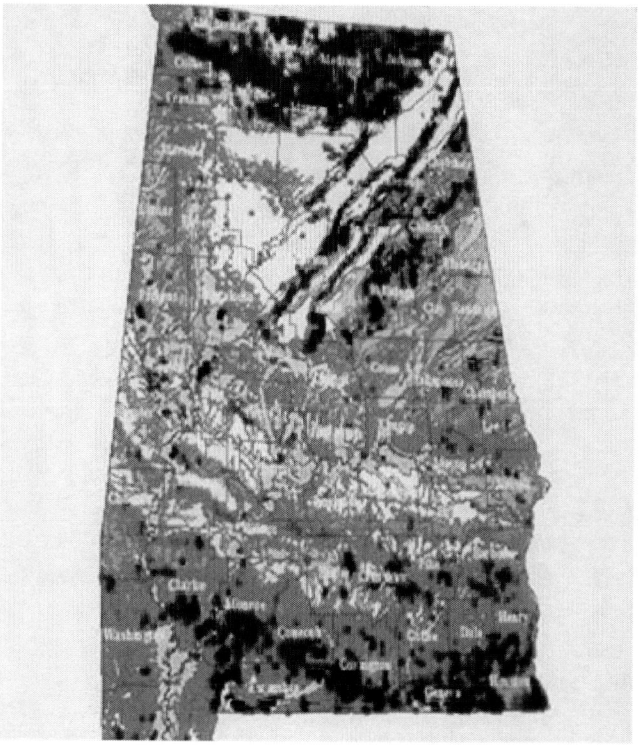

Figure 106: The black dots on this map show thousands of known sinkholes in Alabama. (© WSFA)

common that they are being reported on an almost daily basis.[368] An extraordinary outbreak of 41 sinkholes in the town of Harrisburg, Pennsylvania, in early 2013 (see figure 105)[369] was matched by a similar outbreak in the Russian city of Samara just two months later, when *dozens* of sinkholes suddenly opened in a matter of a few weeks, swallowing vehicles as they crossed busy intersections.[370]

In the state of Alabama alone more than 6,000 known sinkholes have been identified (see figure 106).[371]

In Missouri, the USGS listed hundreds of sinkholes, most of them having appeared recently (see figure 107).

Unfortunately, no single source extensively lists the thousands of sinkholes that have appeared around the world over the last few years. As with the other phenomena we're looking at here, as a general rule, the mainstream media downplay and 'normalize' the phenomenon, presenting it as

[368] Emric, A., 'Bosnians panic as sinkhole swallows village pond', *Associated Press* (26 November 2013). See: news.yahoo.com/bosnians-panic-sinkhole-swallows-village-pond-072935 587.html

[369] Harrisburg counts about 50,000 inhabitants. Caulfield, P., 'Bankrupt Harrisburg, Pa., can't fix the 41 sinkholes plaguing its streets', *New York Daily News* (1 February 2013). See: www.nydailynews.com/news/national/photos-harrisburg-pa-plauged-sinkholes-article-1.1253 254

[370] Williams, A., 'The Russian city being 'eaten alive': Cars, buses, and trucks disappear beneath the earth as they are swallowed by giant sinkholes', *Daily Mail* (9 April 2013). See: www.dailymail.co.uk/news/article-2306 085/Samara-The-Russian-city-eaten-alive-giant-sinkholes.html#ixzz2ooaMcTm8

[371] Henri, B., 'Sinkhole Shock', *WSFA* (30 April 2013). See: wsfa.com/story/22111 363/bryans-sinkhole-story

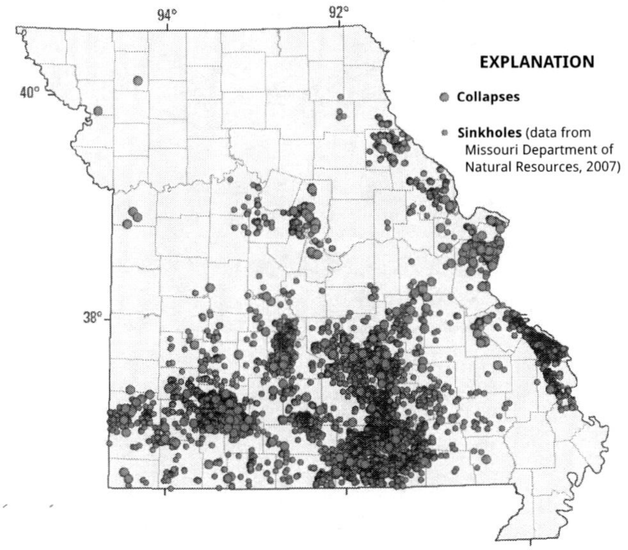

Figure 107: Missouri sinkhole map – 2007. (© USGS)

harmless, mostly man-made and common.

New sinkholes have increased not only in number, but in severity too. If ten years ago you were told that a sinkhole had literally swallowed human beings alive,[372] you would probably have dismissed it as the plot from a bad horror movie. Well, that is today's reality. In the last few years, over 20 individuals have experienced 'death by sinkhole'.[373]

Since none of the invoked causes can explain the sudden appearance of so *many* new sinkholes in so *many* different locations, we're left to consider that some *new* factor must underpin the sharp increase. If makes us wonder if the 'opening up' of the Earth is not this new factor. Interestingly, the previously mentioned Guatemala sinkhole appeared at the same time as the massive eruption of the Pacaya volcano[374] located just 30 miles south of Guatemala City, suggesting a correlation between volcanic activity and sinkhole formation, as posited by our Earth 'opening up' hypothesis.

---

[372] Pearson, M. & Zarrella, J., 'A loud crash, then nothing: Sinkhole swallows Florida man', *CNN* (5 March 2013). See: www.cnn.com/2013/03/01/us/florida-sinkhole

[373] Campbell, A., 'Killer Sinkholes: Unexpected Holes Swallow Entire Families, Cars And Homes', *Huffington Post* (1 March 2013). See: www.huffingtonpost.com/2013/03/01/killer-sinkholes-deaths_n_2791 634.html

[374] 'Pacaya volcano erupts in Guatemala', *tvnz.co.nz* (31 May 2013). See: tvnz.co.nz/world-news/pacaya-volcano-erupts-in-guatemala-5452 571

Chapter 24.

Global cooling

Evidence

According to Official Science, Earth is currently suffering from 'anthropogenic global warming'. Human activity, however, contributes just 5% of atmospheric emissions of CO2,[375] and CO2 is only a minor contributor (3%) to greenhouse gas emissions.[376] 5% of 3% means that man-made CO2 accounts for just **0.15%** of the 'greenhouse effect'. By comparison, water vapor – itself also of natural origin – accounts for a whopping 95% of the greenhouse effect.[377]

Climate change propaganda aside, it's clear that by the end of the 20th century, the 'warming' data that **global warming climatologists had been using to advance their thesis had given way to lower average temperatures**, and that this cooling (like the warming that preceded it) is not of human but cosmic origin. If, as claimed by proponents of anthropogenic global warming, human carbon dioxide emissions are the main cause of recent warming on Earth, how could that explain why other planets in our solar system have been warming up too? Over the last 50 years or so, Earth's average temperature has, by some accounts, risen about one degree. During this same time period, global warming has been observed on Mars,[378] Neptune[379] and Pluto.[380] Is this just coincidence?

Figure 108 shows IPCC[381] temperature predictions (four grey curves) compared to the observed temperatures (black smoothed curve and light grey jagged square curve). Notice that the surface warming (Y axis) represents the deviation from the average global temperature over the period 1960–1990. For example, the measured temperature for 1998 was 0.55 degrees higher than the 1960–1990 average. In 2011 it was only 0.35 degrees warmer. This means that **between 1998 and 2011, the average temperature dropped by 0.2 degrees** according to the very data provided

[375] Segalstad, T., 'Correct Timing is Everything - Also for CO2 in the Air', *CO2 Science* (5 August 2009) 12(31). See: www.co2science.org/articles/V12/N31/EDIT.php

[376] Tomkin, J., 'What gas lies behind greenhouse effect?', *Phys. Today* (December 1992): 13–15

[377] Freidenreich, S., *et al.*, 'Solar radiation absorption by CO2, overlap with H2O, and a parameterization for general circulation models', *Journal of Geophysical Research: Atmospheres* 98: 7255–7264

[378] Taylor, J., 'Mars Is Warming, NASA Scientists Report', *The Heartland Institute* (1 November 2005). See: news.heartland.org/newspaper-article/2005/11/01/mars-warming-nasa-scientists-report

[379] Hammel, H. & Lockwood, G., 'Suggestive correlations between the brightness of Neptune, solar variability, and Earth's temperature', *Geophysical Research Letters* (8 April 2007) 34. See: www.agu.org/pubs/crossref/2007/2006GL028 764.shtml

[380] 'Pluto is undergoing global warming, researchers find', *MIT News* (9 October 2002). See: web.mit.edu/newsoffice/2002/pluto.html

[381] Intergovernmental Panel on Climate Change.

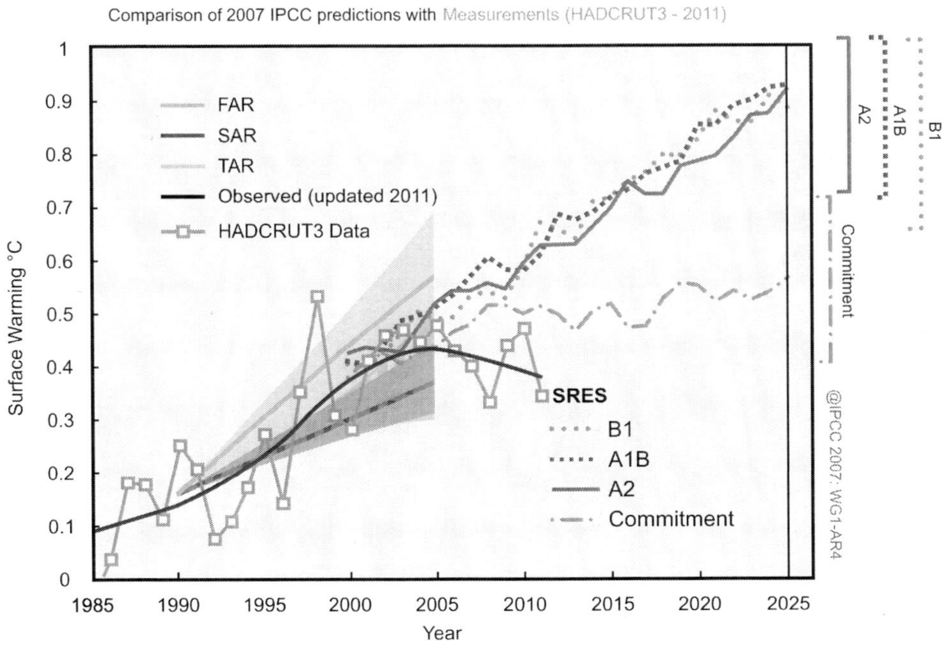

Figure 108: Warming as predicted by the IPCC vs. observed cooling. (© IPCC)

by the IPCC. How's that for "hiding the decline"?[382]

While all IPCC models predicted a strong and steady increase in temperature (about one degree in 40 years), measurements actually show that, since the 1998 maximum, *observed temperatures have been consistently lower* than predicted. The black curve, which displays the 3-year average observed temperature, reveals a steady cooling since at least 2003, over 10 years ago. This overall downward trend was reluctantly and partly acknowledged by the IPCC in its September 2013 Fifth Assessment Report.[383]

As mentioned previously,[384] the current solar cycle, SC24, took unusually long to start and didn't reach a clear maximum, exhibiting a double peak, which is an unusual feature. In addition, SC24 passed through its maximum at the beginning of 2014.[385] The Sun's activity has now begun to drop even more. The observed average maximum for SC24 is *below* 70,[386] making **SC24 the weakest cycle since SC14**, which displayed a maximum of 64 in February 1906.[387] SC14 lasted from 1902 to 1913, some of the coldest years of the 20th century.

[382] Black, Richard, ''Hide the decline' revisited', *BBC News* (2 November 2011). See: www.bbc.co.uk/news/science-environment-15538845

[383] Idso, C. D., Carter, R. M., Singer, S. F., & Soon, W., 'Scientific Critique of IPCC's 2013 'Summary for Policymakers'', *Watts Up With That?* (19 October 2013). See: wattsupwiththat.com/2013/10/19/scientific-critique-of-ipccs-2013-summary-for-policymakers/

[384] See Chapter 14: 'Variations in solar activity'

[385] 'Monthly sunspot numbers', *Australian Government – Bureau of Meteorology*. See: ips.gov.au/Solar/1/6

[386] Smoothed monthly values barely reached 70 sunspots/month at the beginning of 2012 and at the end of 2013. See: sidc.oma.be/sunspot-index-graphics/sidc_graphics.php and solarscience.msfc.nasa.gov/predict.shtml

[387] Adams, M., 'MSFC Solar Group', *NASA*. See: solarscience.msfc.nasa.gov/predict.shtml

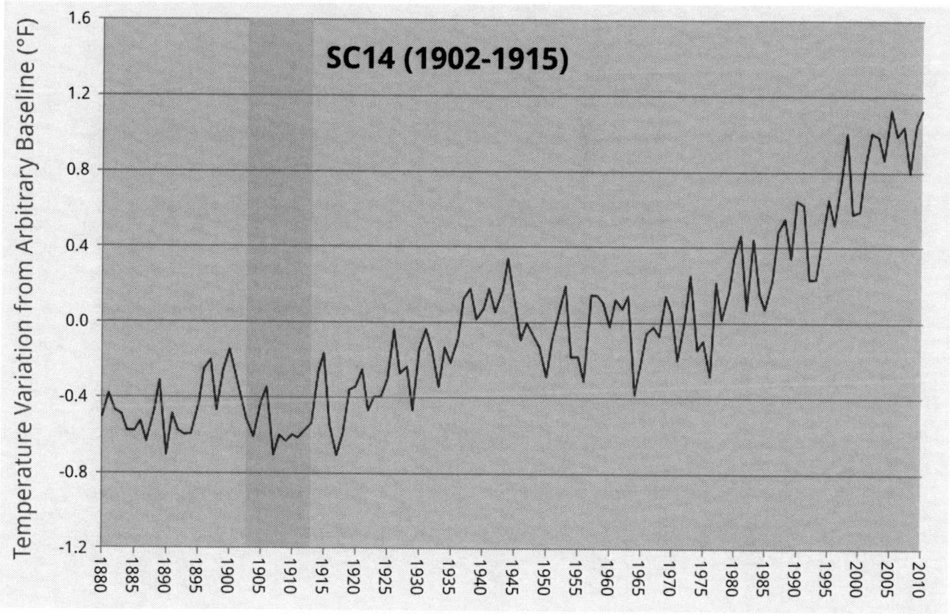

Figure 109: Average annual global surface temperature and SC14. (© Sott.net, adapted from Goddard Institute for Space Studies)

Before SC14 there was another recorded drastic drop in solar activity and temperatures, known as the 'Maunder Minimum',[388] a 70-year period that began around 1645, during which sunspots virtually vanished from the Sun's surface.

Although official observation and recording of sunspots only began in 1760 (see the grey jagged curve in figure 110), the low sunspot numbers characterizing the Maunder Minimum were not due to a lack of observations since astronomers Cassini, de La Hire, Hevelius and Picard[389] were already conducting systematic solar observations at the time (see grey dots in figure 110). The low sunspot number was simply due to the low number of sunspots actually occurring.

One positive consequence of the 'Little Ice Age', which in its broadest extent lasted from about 1350 to about 1850, was the exceptional quality of Stradivarius violins, which were made of wood that grew during this cold period. Because of the biting cold, this wood exhibited very small tree rings, making it very dense and providing it with unique acoustic properties.[390] This positive note aside, the Little Ice Age was a gloomy time that brought with it low temperatures, bitter winters, crop failure, famine, plague and social unrest.[391] According to historians Elizabeth Ewan and Janay

---

[388] Soon, W. & Yaskell, S., *Maunder Minimum And the Variable Sun-Earth Connection*, pp. 8–15. Actually it was not Walter Maunder but German astronomer Gustav Sporer who retrospectively identified, in the middle of the 19th century, the period known as 'Maunder Minimum'. This latter term was only coined in 1976 by Jack Eddy of the High Altitude Observatory in Colorado in reference to Walter Maunder, superintendent of the London Observatory who wrote about the topic in the 1890's, several years after Sporer's work. See: Svensmark, H. & Calder, N., *The Chilling Star*, p. 16

[389] *Ibid.*, p. 21

[390] Ben-Menahem, A., *Historical Encyclopedia of Natural and Mathematical Sciences* 1: 1207

[391] Soon & Yaskell, *op. cit.*, p. 28

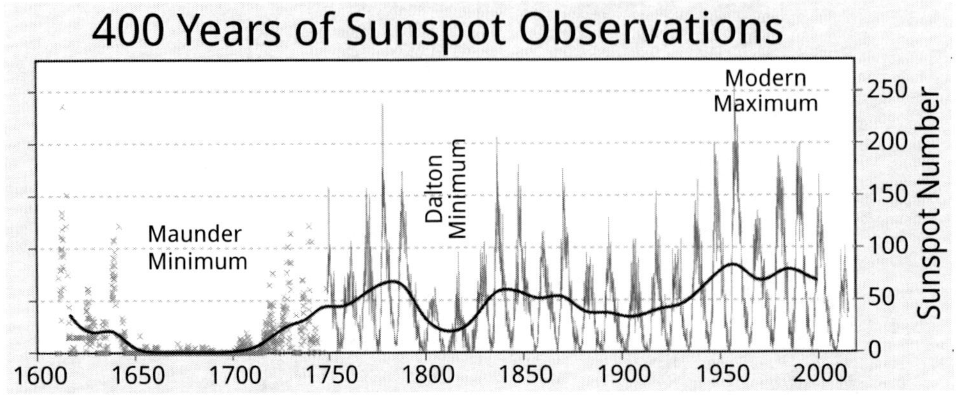

Figure 110: Solar activity during the Maunder Minimum. (© Wikimedia Commons/Global Warming Art)

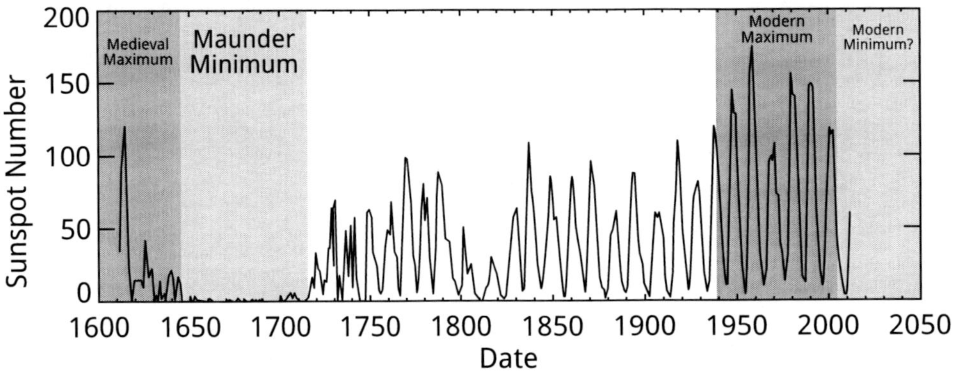

Figure 111: Similarities between the late Middle Age and the modern era. (© Sott.net, adapted from NASA / MSFC)

Nugent,[392] the famines that struck France in 1693, Norway in 1695 and Sweden in 1696 killed about 10%[393] of the population of each country.

Not only is the decreased solar activity observed over the last few years similar to the lack of activity seen around 1650, the Maunder Minimum was also preceded by a period of increased solar activity (the 'Medieval Maximum').[394] As depicted in figure 111 (medium grey area), most of the 20th century was a period of exceptionally intense solar activity,[395] known as the 'modern maximum' or 'grand solar maximum'.[396]

For years now, the mainstream media have fiercely supported the man-made warming dogma while clear evidence of global cooling has been piling up. Despite 10 years of denial, independent scientists have persuaded many to acknowledge the global cooling that began at the turn of the century. The Space and Science Research Center, an independent U.S. climate research company,

392 Elizabeth, E. & Janay, N., *Where is the family in medieval and early modern Scotland?*

393 Notice that this figure may actually be quite conservative.

394 Soon & Yaskell, *op. cit.*, p. 17

395 Solanki, S., Usoskin, I., Kromer, B., Schussler, M. & Beer, J., 'Unusual activity of the Sun during recent decades compared to the previous 11,000 years', *Nature* 431: 1084–7

396 Abreu, J., *et al.*, 'For how long will the current grand maximum of solar activity persist?', *Geophys. Res. Lett.* (2008)

forecasts a future climate scenario quite different from that predicted by the IPCC:

> The Space and Science Research Center (SSRC) announces today that the most recent global temperature data through January 31, 2011 using NASA and NOAA weather satellites supports the previous forecast from the SSRC that a **historic drop in global temperatures is under way and that the previously predicted climate change to one of a long and deep global cooling era has begun.**
>
> SSRC Director John L. Casey explains, 'Based on the data from the AMSR-E instrument on board the NASA Aqua satellite, sea surface temperatures just posted this week showed their steepest decline since the satellite was made operational in 2002. This major drop from the warm temperature levels seen in 2010 is also echoed by a dramatic decline in atmospheric temperatures in the lower troposphere, where we live, with the data coming from NOAA satellites. At present rates of descent, both ocean and atmospheric temperatures are likely to soon surpass the temperature lows set in the 2007–2008 period. Even with a small correction that is usually seen after such a rapid drop, **there is no doubt that the Earth is entering a prolonged global cooling period and will soon set another record drop in temperatures** by the November–December 2012 time frame as was forecast in the SSRC press release from May 10, 2010.'[397]

Causes

As demonstrated above, the Sun has been unusually quiet for several years. The Sun rises every morning and, sure enough, every evening it sets. However, solar activity fluctuates along with the temperature of our planet (figure 112).

The above only shows a *correlation* between solar activity (number of sunspots) and Earth's surface temperature. Further down, we will explain how reduced solar activity (and also an increase in atmospheric cometary dust) causes Earth's global cooling.

According to mainstream science, the huge variation in solar activity as reflected by sunspot counts has little if any impact on the energetic output of the Sun. Indeed, solar radiation remains almost stable with a mere 0.1% variation over the solar constant.[398] However, such measurements only take into account **irradiance**, i.e. measurable solar radiations. Notice also that irradiance is measured during fair weather. Therefore, it doesn't take cloudiness into consideration.

While the Sun does indeed emit radiations – ultraviolet, visible spectrum, infrared, gamma rays, X-rays – it also ejects massive quantities of particles through its solar wind.[399] This outflow of ionized particles (mostly protons and electrons, with an overall positive charge) plays an important role on the electromagnetism of the solar system[400] including the Earth's and the Sun's magnetic shields.

[397] 'Global Cooling Begins and Global Warming Ends with Record Drop in Temperatures', *Space and Science Research Center* (4 February 2011). See: www.spaceandscience.net/sitebuildercontent/sitebuilderfiles/ssrcpressrelease2-2011globalcooling.pdf

[398] 1.3 watts/m^2 variations compared to 1366 watts/m^2 for the average solar irradiance. IPCC, *Third Assessment Report - Climate Change 2001*, Working Group I: The scientific basis, chapter 6.11

[399] McCanney, J. M., *Atlantis to Tesla: The Kolbrin Connection*, p. 7

[400] Gerd, P., *Physics of the Earth's Space Environment: An Introduction*, pp. 281–282

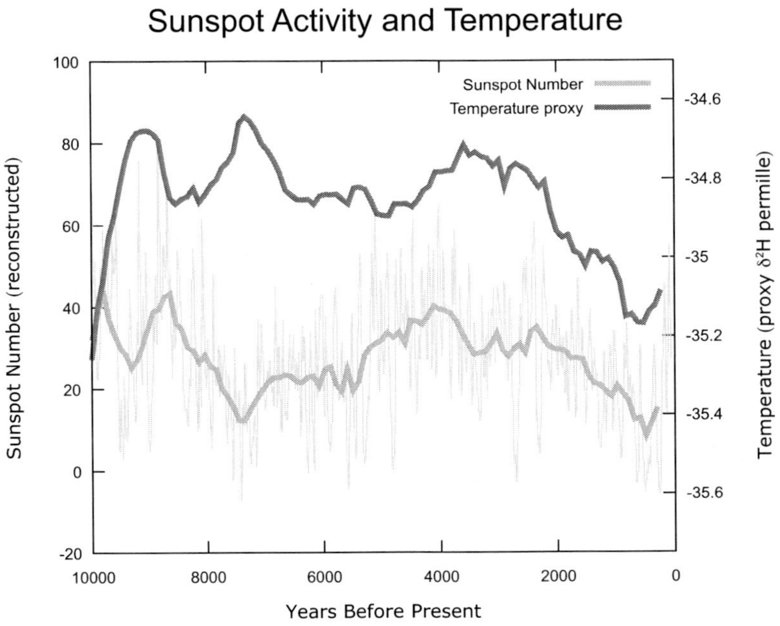

Figure 112: Sunspot number vs. average temperature over a 10,000-year span. (© NOAA)

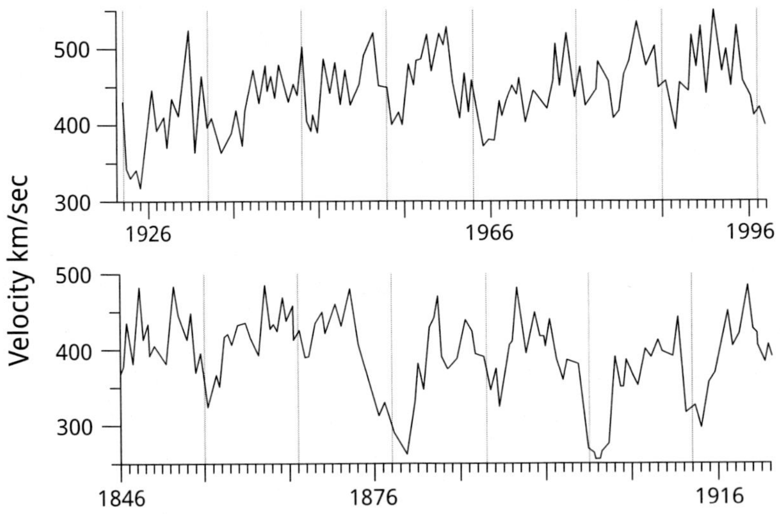

Figure 113: Monthly value of solar wind speed. The vertical lines correspond to solar minimums (1845–2000). (© Rangarajan et al.)

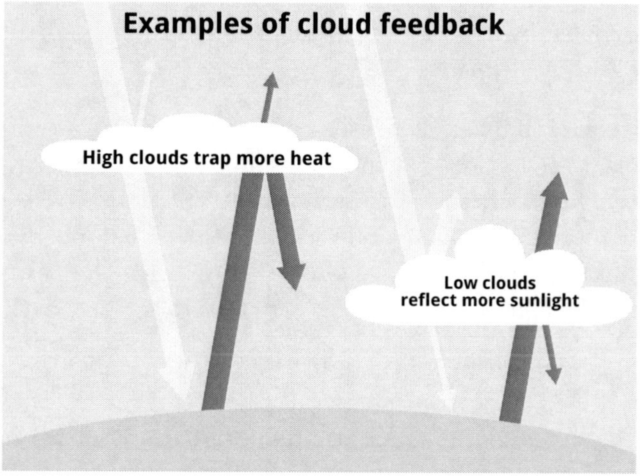

Figure 114: Clouds reflect solar radiation more than Earth radiation. Overall, clouds induce a net cooling. (© Sott.net, adapted from skeptikalscience.com)

Now, if the *minute decrease in solar irradiance* due to reduced solar activity is not the cause for global cooling, how does reduced solar activity lead to global cooling? One key factor seems to be cloud formation.

Clouds have both a cooling and a warming effect. Clouds have a cooling effect because they scatter back into space about half of the incoming sunshine that would otherwise warm the Earth (that's why cloudy days are colder than sunny ones).[401] Clouds can also have a warming effect (greenhouse effect) by trapping the heat escaping from the Earth's surface (that's why cloudy nights are warmer than starry nights).

The radiation 'budget' of clouds (are clouds net warmers or net coolers?) was a matter of scientific conjecture until three dedicated satellites were sent into space during the 1980s to measure incoming sunlight and outgoing infrared radiation.[402]

The results were clear. Overall,[403] clouds have a strong net cooling effect.[404] For instance, if nothing else changed, removing the cloud canopy would increase the Earth's surface temperature by about 10° Celsius.[405]

Clouds are made of water droplets in suspension in the air. For these droplets to form, three main factors are involved:

1. The **temperature** must drop low enough (below dew point) for condensation to occur. During

[401] Committee on Scientific Accomplishments of Earth Observations from Space, *Earth Observations from Space: The First 50 Years of Scientific Achievements*, p. 26

[402] The Earth Radiation Budget Experiment (ERBE), *NASA*. See: science.larc.nasa.gov/erbe/

[403] Apart from high-altitude cirrus which radiate into space less heat than they block going out from the Earth. See: 'Clouds and radiations', *NASA Earth Observatory*. earthobservatory.nasa.gov/Features/Clouds/

[404] Ramanathan, V., *et al.*, 'Cloud-Radiative Forcing and Climate: Results from the Earth Radiation Budget Experiment', *Science* (6 January 1989) 243(4887):v57–63

[405] This can be seen intuitively when observing that usually cloud tops are white (very reflective) and cloud bottoms are grey (absorbent). See: Svensmark & Calder *op. cit.*, p. 67

condensation, atmospheric water vapor (water in gaseous form) is transformed into droplets (tiny drops of water in liquid form).

2. Also, atmospheric **particles** must be present to help the droplets form.[406] These particles are called 'cloud condensation nuclei'. Without them, even if temperatures were very low there would be no condensation, and therefore no clouds.[407] As discussed previously,[408] Earth has experienced a steep increase in atmospheric cometary dust over recent years. These dust particles act as potential cloud condensation nuclei. Notice that even when atmospheric dust doesn't generate clouds, it has a net cooling effect as described earlier (global dimming).[409]

3. **Cosmic rays** accelerate cloud formation. In the following we'll describe this process.

As previously mentioned,[410] the Sun's magnetic field and the Earth's magnetic field are powered by the Sun's activity. Those fields literally act as magnetic shields against incoming cosmic rays; the Sun's magnetic field (heliosphere) deflects about 50% of the cosmic rays.[411] Comparatively speaking, the Earth's magnetic field isn't as effective at shielding us from cosmic rays. Even if it disappeared entirely, the number of incoming cosmic rays would only increase by 3%.[412] That is, the Sun does most of the 'protective' work.

Cosmic rays are charged particles (mostly protons) that can almost reach the speed of light if highly energetic. They are generated by stars and supernovae.[413] Our Sun produces cosmic rays too (solar winds) but they are energetically weak (only about 700 km/s).[414]

Thus, **when solar activity decreases, solar winds (low energy cosmic rays) decrease** (see figure 113), **but also the Sun's magnetic shield weakens, allowing a greater amount of high energy cosmic rays to reach the solar system and ultimately our planet**. The negative correlation between solar activity and cosmic ray flux has been evidenced by several researchers.[415][416]

When reaching the Earth's atmosphere, cosmic rays combine and interact with the particles of the Earth's atmosphere, leading to swiftly moving particles called 'secondary cosmic rays'.[417] This particle mayhem occurs mainly between 15 and 25 km above our heads[418] and leads to only one

[406] Cloud seeding has been used successfully since the 1950s. The basic principle is to release particles (from aircraft or rockets) into the atmosphere in order to stimulate the formation of water droplets.

[407] Apart from a very specific kind of nucleation called 'homogeneous nucleation', where no particle is necessary and water vapor (gaseous phase) directly transforms into ice crystals (solid phase) without going through the liquid phase. See: Chandrasekar, *Basics of Atmospheric Science*, p. 127

[408] See Chapter 21: 'Increase in cometary activity'

[409] See section 'Global dimming' in Chapter 21: 'Increase in cometary activity'.

[410] See Chapter 7: 'The insulating bubble'

[411] Svensmark & Calder, *op. cit.*, p. 48

[412] *Ibid.*, p. 61

[413] *Ibid.*, p. 41

[414] *Ibid.*, p. 42

[415] Kirkby, J., *et al.*, 'Cloud formation may be linked to cosmic rays', *Nature* (2011) 476: 429–433

[416] Svensmark, H., *et al.*, 'Variation of cosmic ray flux and global cloud coverage: A missing link in solar climate relationships', *J. Atmos. Solar-Terr. Phys.* 59: 1225–1232

[417] Svensmark & Calder, *op. cit.*, p. 52

[418] *Ibid.*, p. 53

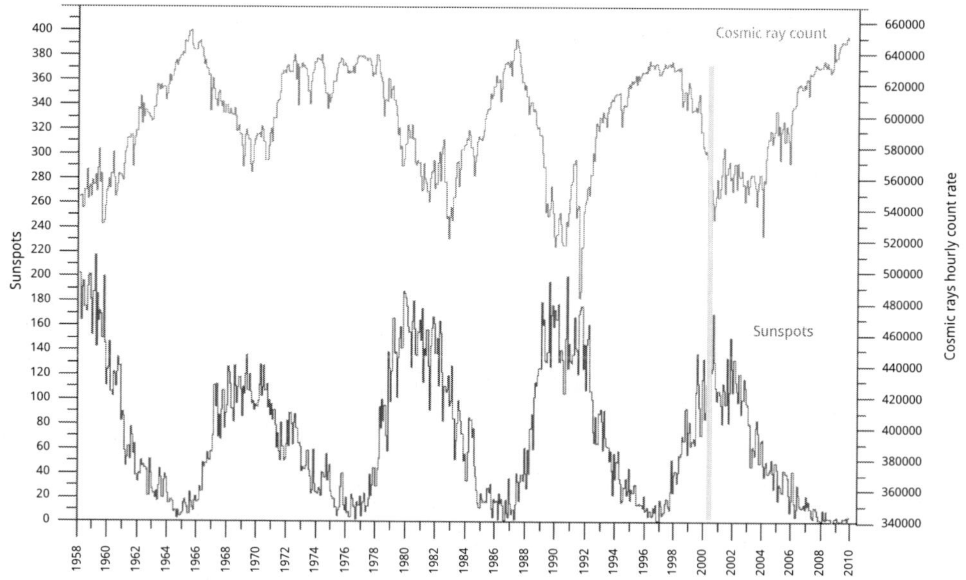

Figure 115: Number of sunspots (dark grey curve) vs. cosmic ray count (light grey curve) over the 1958–2010 period. Notice that after 2000 (vertical light grey line), sunspots count dropped sharply while cosmic rays experienced a drastic increase. (© Climate4you.com)

kind of particle that is able to reach the Earth's surface in large numbers and without substantial loss of energy: muons. (See figure 116.)

Muons are like electrons except for their mass: they are 200 times heavier than electrons. Also, muons have an extremely short life[419] and they quickly transform into electrons.[420] But, because of their very high speed, muons have enough time to penetrate deep into our atmosphere and spread electrons like 'collateral damage' along their entire atmospheric journey.

In figure 117 we can see from top to bottom how cosmic rays (black arrow) generate (via muons) electrons (black circles). Electrons accelerate the formation of clusters (medium grey circles) made of positively charged molecules in suspension in the atmosphere like clay dust, carbon, sulfur dioxide (white circles), eventually leading to stable and electrically neutral clusters (light grey circles) which act as condensation nuclei (dark grey circle) around which a water droplet can form.

The catalyzing effect of electrons has been repeatedly demonstrated experimentally in cloud chambers.[421]

Knowing that the main causes of cloud formation are cosmic rays and atmospheric dust, and that

[419] 2 millionths of a second. *Ibid*, p. 55

[420] *Ibid.*, pp. 54–55

[421] Cosmic ray electrons combine with O2 molecules: e- + O2 -> O2-
O2- attracts water molecules (Wn) and through ligand switching leads to O3 water cluster:
O2-.Wn + O3 -> O3-.Wn-1 + O2 + W1
O3- oxidizes SO2: O3- + SO2 -> SO3- + O2
Then SO3- reacts with water molecules leading to sulphuric acid: SO3- + Wn -> H2SO4 + Wm.
At this point the electron has catalyzed the formation of a stable molecule that can act as a nucleation agent.
See: Svensmark, H., *et al.*, 'Experimental evidence for the role of ions in particle nucleation under atmospheric conditions', *Proc. R. Soc. A* (February 2007)

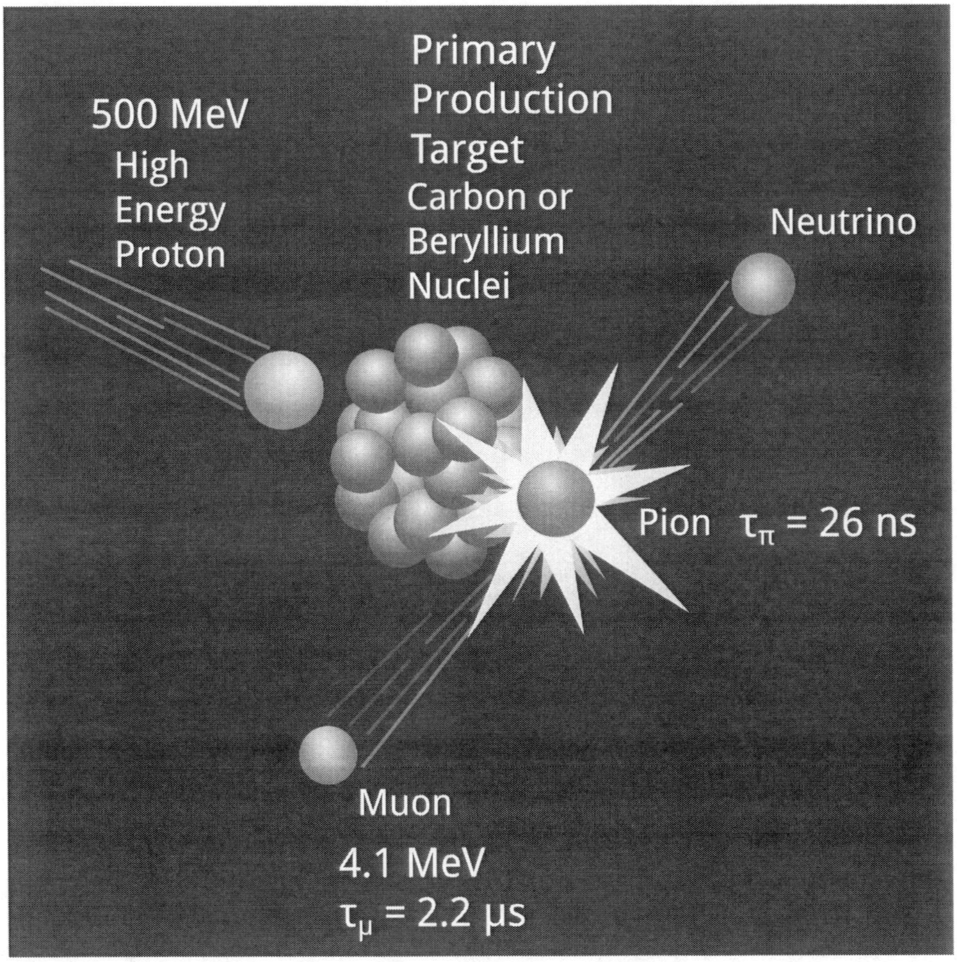

Figure 116: A collision between a proton (top left – primary cosmic ray) and an atmospheric particle (central cluster – carbon or beryllium) forms a muon (bottom). (© Triumf.ca)

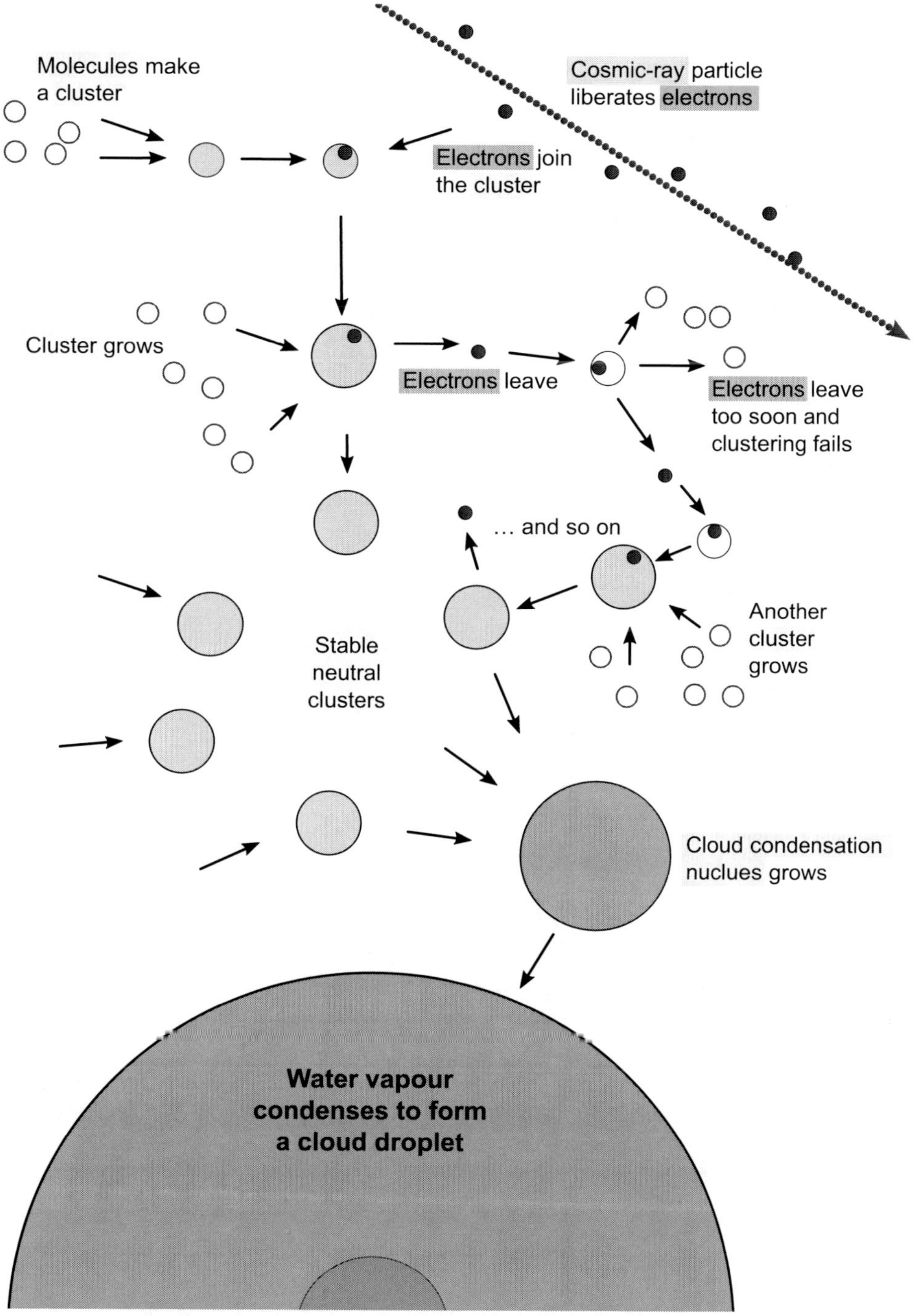

Figure 117: Electrons are the main catalysts of molecular clustering, i.e. cloud droplet formation. (© Sott.net, adapted from Svensmark)

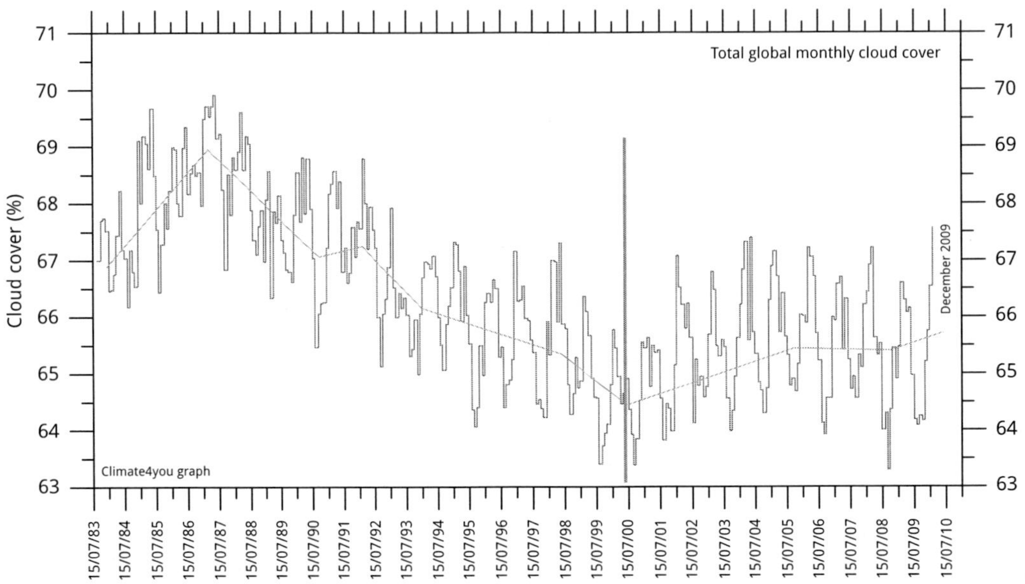

Figure 118: Overall cloud cover (%) between 1983 and 2010 (jagged line). The middle line show linear regressions. The vertical line shows the year 2000, when the cloud cover started to increase. (© Sott.net, adapted from Climate4you)

these two factors are on the rise because of the hypothesized approach of Nemesis (grounding the Sun and reducing its activity) and its cometary swarm, we can expect an overall increase in cloud cover in the future and the resulting cooling effects.

Actually, this trend started several years ago. Around 2000, when solar activity started to weaken and atmospheric cometary dust began to increase, the overall cloudiness started increasing (see figure 118).[422] [423]

Notice that all the factors and mechanisms involved in global cooling might not act in a strictly linear way. Weather science was one of the first disciplines to acknowledge the concept of 'the butterfly effect'.[424] **Small causes can indeed lead to great effects, and two seemingly independent causes can act synergistically to bring disproportionate consequences.** Weather phenomena include threshold effects and feedback loops. One well-documented non-linear weather phenomenon is the ice-albedo effect, where the cause and the effect amplify each other:

> Snow and ice are white and very reflective. They have what scientists call a very high **albedo – that's a measure of how much light a surface reflects**. Between 70 and 80 percent of the sun's rays that hit this kind of frozen surface are bounced right back out into space. So the land or water beneath the snowy blanket doesn't get a chance to absorb much of that solar radiation ... whereas the darker sea water and land readily absorb the sunlight.

[422] High-level cloud amount started to increase in 2000, middle-level clouds in 1996 and low-level clouds in 2009. See *NASA/ISCCP*, 'Cloud analysis – Part 7': isccp.giss.nasa.gov/climanal7.html

[423] Unfortunately, the ISCCP stopped collecting data about cloud cover in 2010

[424] This expression was first expressed in the title of a talk given by American mathematician and meteorologist, and pioneer of chaos theory, Edward Lorentz: 'Does the flap of a butterfly's wings in Brazil set off a tornado in Texas?' See: Ghys, E., 'The butterfly effect', *12th International Congress on Mathematical Education*, 8-15 July 2012, COEX, Seoul, Korea

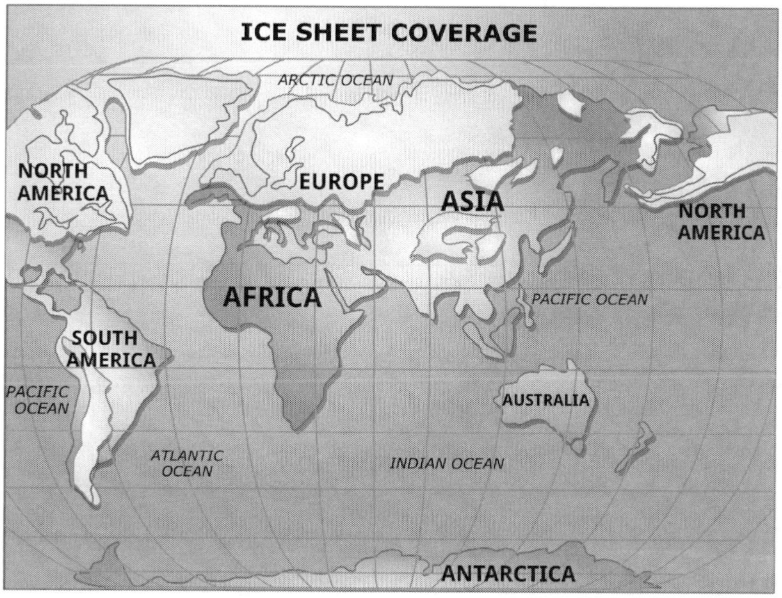

Figure 119: Permanent ice sheet cover during the last ice age – the Younger Dryas – which occurred 12,800 years ago. (© How Stuff Works)

> Now imagine that a little bit of heat is added to the system. That's exactly what is happening in the real world; scientists say that the average temperature in Alaska has risen 4 degrees Fahrenheit since the 1950s. With warmer temperatures, some of the snow and ice melts, exposing the darker land or water underneath. These surfaces have much lower albedos – open water reflects less than 10 percent of the solar energy that hits it, for example. So more heat is absorbed by the landscape.
>
> Then a **feedback loop kicks in**. More heat is absorbed by the darker surface, so more snow and ice melt. More of the darker surface is exposed, leading to the absorption of even more heat, more snow and ice melt, and so on. Just a small temperature rise can set this feedback cycle into motion. **The opposite effect is possible too; a small temperature decrease would lead to more snow and ice, would lead to more solar radiation being bounced back to space, would lead to colder temperatures, would lead to more snow and ice, and so on. Scientists describe the onset of past ice ages in this way.**[425]

The ice-albedo effect may also be exacerbated by a 'warming/cooling rebound' effect because of the (non-anthropogenic) global warming our planet underwent during most of the 20th century.[426] [427] This warming has led to the evaporation of massive quantities of water from oceans, seas, lakes, snow, ice and rivers. Global cooling started around the turn of the century,[428] but average

[425] Villiger, M., 'The Arctic – Our Global Thermostat', *Scientific American* (15 June 2004). See: www.pbs.org/saf/1404/features/thermostat.htm

[426] Solanki, S. K., Usoskin, I. G., Kromer, B., Schussler, M. & Beer, J., 'Unusual activity of the Sun during recent decades compared to the previous 11,000 years', *Nature* 431: 1084–7

[427] Abreu, J. A., *et al.*, 'For how long will the current grand maximum of solar activity persist?', *Geophys. Res. Lett.* (2008)

[428] Lockwood, M., Rouillard, A. P. & Finch, I. D., 'The rise and fall of open solar flux during the current grand solar

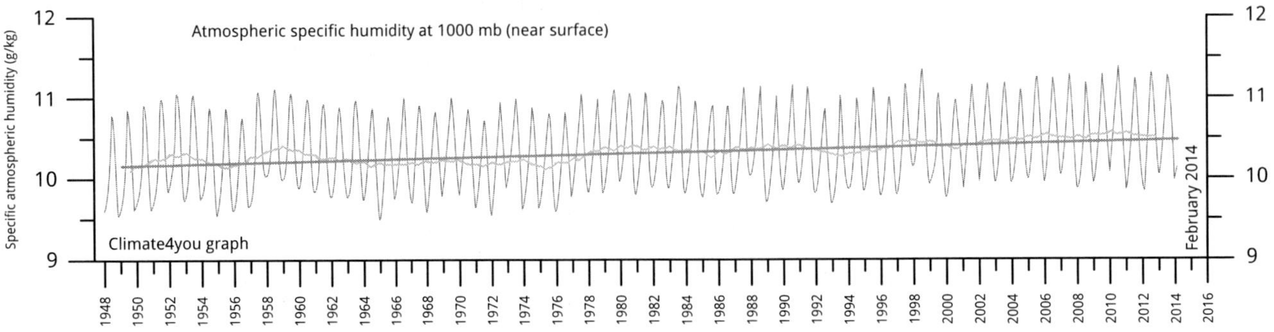

Figure 120: Atmospheric humidity at sea level since 1948. Notice that, concomitant with global warming, humidity has steadily increased between 1948 and 2004 (grey linear regression). Since then this increase, like global warming, has stopped. (© Kalnay et al.)

temperatures are still high and there are still massive quantities of water vapor in the atmosphere (figure 120).

With such large amounts of humidity in the atmosphere, a severe winter could generate massive precipitation, leading to more extensive ice and snow cover that might increase the ice-albedo effect to a point where the planet could not warm up again when 'spring' comes and consequently experience a winter lasting several years.

In addition, water vapor being, by far, the main 'greenhouse gas',[429] cooling and the subsequent condensation means there would be less of a greenhouse effect. Less greenhouse effect in turn leads to more cooling. In figure 121 we have two feedback loops (greenhouse loop and albedo loop) that feed each other and are potential contributors to global cooling.

Feedback loops and non-linear systems may partly explain why ice ages initiate much faster than previously expected. Until recently, scientists believed that the last ice age – the one that wiped out the mammoths and other species about 12,800 years ago, and which presents very strong evidence of cometary impact[430] [431] – took about 10 years to settle over Europe.[432]

Additional research by William Patterson of the University of Saskatchewan involving the analysis of mud deposits from Lough Monreagh in County Clare, Ireland, revealed that the northern hemisphere could be forced[433] into an ice age in *as little as three months.*[434] Patterson described

maximum, *Astrophys. J.* (2009)

[429] Depending on sources, the contribution of water vapor to the greenhouse effect is comprised between 66% (Fleagle, 1994) and 95% (Freidenreich, 1993).

[430] Mahaney, W., *et al.*, 'New Evidence from a Black Mat Site in the Northern Andes Supporting a Cosmic Impact 12,800 Years Ago', *Journal of Geology* (July 2013) 121(4): 309–325

[431] Wittke, J., *et al.*, 'Evidence for deposition of 10 million tons of impact spherules across four continents 12,800 y ago', *PNAS* (2013)

[432] Taylor, K., 'Rapid climate change', *American Scientist* (July–August 1999) 87(4)

[433] For Patterson it was the stalling of the Gulf Stream that caused this sudden ice age. We'll deal with this topic in Chapter 27: 'The Gulf Stream'.

[434] Frozen woolly mammoths are not necessarily evidence of 'quick freeze'. Most specimens were found in mud deposits in a standing or sitting position. In addition, they had several broken bones. This suggests that the mammoths were swept up by a major flood, possibly caused by the close approach of a comet that may also have caused the Ice Age itself. In this sense, mammoths are not victims of the Ice Age so much as victims of what caused the Ice Age. See:

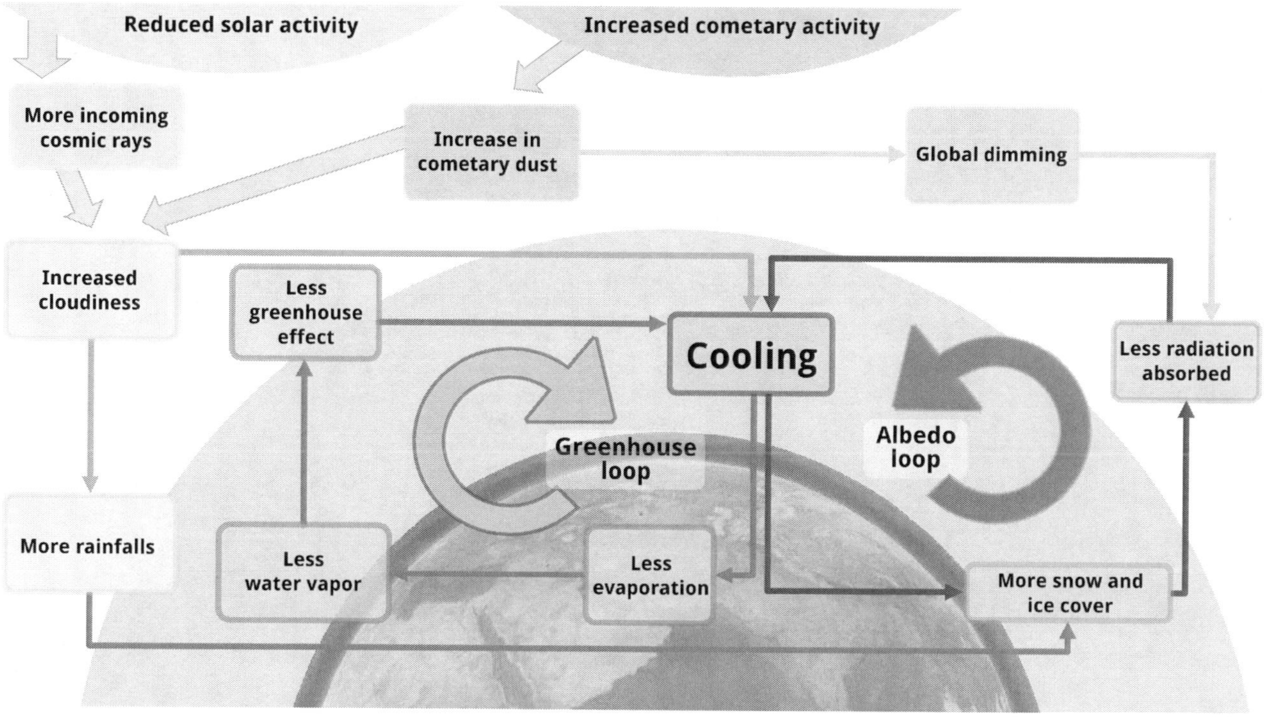

Figure 121: The interacting greenhouse feedback loop and albedo feedback loop. (© Sott.net)

this unexpectedly sudden shift in the following terms:

> It would be like taking Ireland today and moving it up to Svalbard, **creating icy conditions in a very short period of time.**[435]

The nonlinear nature of Earth's weather system is such that the effects caused by all the changes listed above are difficult to predict and probably interact synergistically with each other. It is therefore highly possible that the sum effect of those changes far exceeds the effect of each individual change. For this reason, global cooling may progress much more strongly and much quicker than expected.

Allan, D. & Delair, J., *Cataclysm!*, pp. 126–128

435 'Big Freeze Plunged Europe Into Ice Age in Months', *Science Daily* (30 November 2009). See: www.sciencedaily.com/releases/2009/11/091130112421.htm

Chapter 25.

Strange clouds

Now that we know a bit more about clouds, let's turn our attention to the variety of 'strange' clouds that have recently appeared in our skies. Clouds come in many strange forms – such as roll clouds, mammatus clouds and wave clouds – but in this chapter we're just going to focus on two kinds, namely **noctilucent** clouds (also known as 'night-shining' clouds) and **contrails**.

Let's keep in mind what is required for clouds to develop: temperature must drop low enough for condensation to occur and atmospheric particles – condensation nuclei – must be present to help droplets form.

Noctilucent clouds

Noctilucent clouds were first noticed[436] in 1885, quite recently relative to other kinds of clouds, and they appear at *very high altitudes*, around 80 kilometers (50 miles). They are actually the highest known clouds. According to mainstream science, the cause for noctilucent clouds remains unclear:

Figure 122: Noctilucent cloud over Estonia – 59° north. (© Martin Koitmae)

> Noctilucent clouds are **not fully understood** and are a recently-discovered meteorological phenomenon. Noctilucent clouds can form only under very restrictive conditions; their occurrence can be used as *a sensitive guide to changes in the upper atmosphere*. They are a relatively recent classification. The **occurrence of noctilucent clouds appears to be increasing in frequency, brightness and extent**. It is theorized that this increase is connected to climate change.[437]

[436] First reported by Leslie in 1885. See: Leslie, R., 'First Sighting of a Noctilucent Cloud', *Nature* (16 July 1885) 32(245)

[437] 'Noctilucent cloud', *Wikipedia*. See: en.wikipedia.org/wiki/Noctilucent_cloud

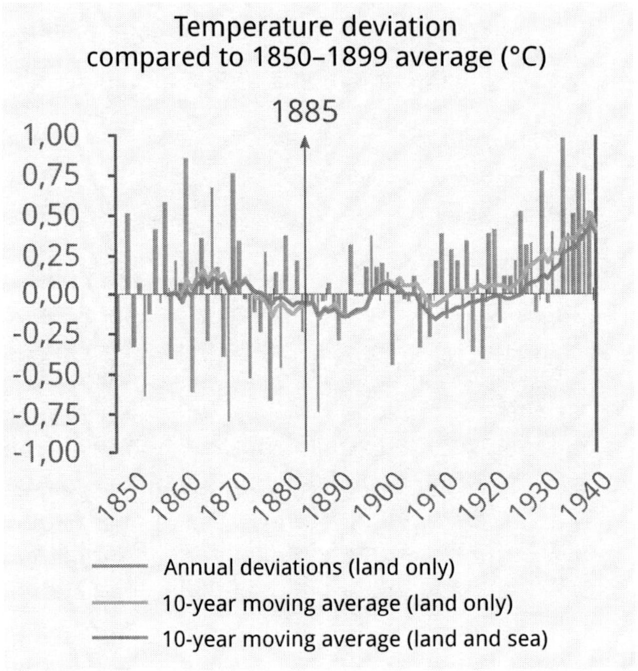

Figure 123: Average temperatures and deviation (1850–1940). (© Sott.net)

One popular claim is that noctilucent clouds are caused by global warming.[438] This sounds like a dubious explanation for several reasons, the first being that, as shown previously,[439] global temperatures have not been increasing but *decreasing* since 2000, while *the appearance of noctilucent clouds is on the rise.*

Also, the first observations of noctilucent clouds were made in 1885, a year that was colder than the previous 20 years (see figure 123). If warming was the cause, why didn't the first observations of noctilucent clouds occur during the warmer years prior to 1885?

A more likely cause of noctilucent clouds is atmospheric dust. Indeed, the first observation of noctilucent clouds in 1885 was **just two years after the eruption of the Krakatoa** volcano[440] that released an estimated 11 cubic miles (45 cubic km) of dust into the atmosphere. This release was so severe that the dust cloud darkened the sky 250 miles (400 km) away.[441] The ash deposit on nearby islands was so thick that for several years it was difficult for plants to grow. Until 1886, ashes suspended in the upper atmosphere caused exceptional red sunsets all around the world.[442]

The role of dust in noctilucent clouds was shown in 2009 when they were artificially created with

[438] Madrigal, A., 'Mysterious, Glowing Clouds Appear Across America's Night skies', *Wired* (16 July 2009). See: www.wired.com/wiredscience/2009/07/nightclouds/

[439] See Chapter 20: 'Impact of the Sun's reduced activity'

[440] Bagley, M., 'Krakatoa Volcano: Facts About 1883 Eruption', *Livescience* (26 March 2013). See: www.livescience.com/28 186-krakatoa.html

[441] 'The Volcano That Shook the World: Krakatoa 1883', *Scholastic*, 26 August 1883. See: teacher.scholastic.com/activities/wwatch/volcanoes/witnesses.htm

[442] Holt, L., *Daniel's Key*, p. 40

Figure 124: Krakatoa volcano eruption on September 2009 displaying lightning along the plume. (© John Seach)

exhaust particles released by a space rocket:

> The Charged Aerosol Release Experiment (CARE) was conducted by the Naval Research Laboratory and the Department of Defense Space Test Program using a NASA four-stage Black Brant XII suborbital sounding rocket. Using ground based instruments and the STP/NRL STPSat-1 spacecraft, scientists will study an **artificial noctilucent cloud formed by the exhaust particles of the rocket's fourth stage** at about 173 miles altitude.[443]

With dust identified as the cause of noctilucent clouds, some sources attributed their appearance to man-made pollution.[444] But remember, noctilucent clouds are very high-altitude clouds. For industrial dust to reach high enough concentrations at these altitudes, it should literally saturate the lower atmosphere (as with the Krakatoa eruption). This dust saturation of the lower atmosphere should make the Sun barely visible, which is obviously not the case. So if the dust found in the higher atmosphere doesn't come from below, it must come from above.

After beating around the bush for years, in August 2012 NASA finally acknowledged the presence of meteor dust in noctilucent clouds. Moreover, it recognized that meteor dust is *the* nucleating agent around which noctilucent clouds form:[445]

> "We've detected bits of **'meteor smoke' imbedded in noctilucent clouds**," reports James Russell of Hampton University, principal investigator of NASA's AIM mission to study the

[443] 'Night Time Artificial Cloud Study Using NASA Sounding Rocket', *NASA* (19 September 2009). See: www.nasa.gov/centers/wallops/CARE.html

[444] 'Noctilucent clouds', *Weather FAQ UK*. See: http://weatherfaqs.org.uk/book/export/html/175

[445] Interestingly, this observation doesn't seem to have reached the mainstream media, which, after the July 2012 NASA paper, continued to blame global warming and man-made pollution as the cause of noctilucent clouds.

Figure 125: Rocket used to conduct the CARE experiment. (© NASA)

> phenomenon. **'This discovery supports the theory that meteor dust is the nucleating agent around which NLCs form.'**[446]

So noctilucent clouds are not a product of industrial pollution. They are not 'anthropogenic'. NASA quantified the sharp increase in noctilucent cloud frequency in the following excerpt:

> **Since 1980, satellite observations show the number of noctilucent clouds increasing about 28 percent per decade**. The ice crystals in the clouds also appear to be getting bigger, with the brightness of the clouds rising about 7 percent per decade.[447]

At this point, NASA has acknowledged that meteor dust is the cause of noctilucent clouds and that the number of noctilucent clouds has increased. **The obvious conclusion would be that meteor activity is on the rise**. Why doesn't NASA acknowledge this important piece of data?

The increase in noctilucent clouds is one of the effects – among others – of increased dust concentration in the atmosphere in general, and in the upper atmosphere in particular. We suspect that most of this atmospheric dust is of cometary origin, while some of it might be due to the recent increase in volcanic activity.[448]

In addition to being more frequent overall, noctilucent clouds can also be observed at lower latitudes.[449]

[446] Philips, T., 'Meteor Smoke Makes Strange Clouds', *NASA* (7 August 2012). See: science.nasa.gov/science-news/science-at-nasa/2012/07aug_meteorsmoke/

[447] Chang, K., 'First Mission to Explore Those Wisps in the Night Sky', *NY Times* (24 April 2007)

[448] Cometary activity and volcanic activity are positively correlated. See Chapter 23: 'Earth opening up: earthquakes,

Figure 126: July 14th, 2013 – Bright and dense display of noctilucent clouds over Paris (48° north). (© Myriam Kieffer)

According to several researchers[450] the cause of this low latitude occurrence is *severe cooling of the upper atmosphere*, which is much more dramatic than the limited cooling observed at sea level:

> 'The sun is in a very unusual period,' said Marty Mlynczak, a TIMED team member at NASA Langley in Virginia. 'The Earth's thermosphere[451] is responding remarkably up to **an order of magnitude decrease in infrared emission/radiative cooling** by some molecules.' Less radiation in both directions means that this layer of the atmosphere also cools substantially. In fact, the **thermosphere has cooled by a factor of 10 since the last solar maximum in early 2002**.[452]

Observations also show that noctilucent clouds are more frequent when solar activity is low.[453] During solar minimum, **cosmic rays generated by the Sun (solar winds) decrease**, unlike **high energy cosmic rays** generated by stars and supernovas.[454] Because of their low energy, solar winds **can't** penetrate deep into our atmosphere but **they do reach the upper atmosphere**[455] in massive quantities.

Although the total solar irradiance varies by only ~0.1% during a solar cycle, nearly 6% changes are observed in the ultraviolet spectrum at high altitude, which has been shown to be **enough to change the temperature in the upper atmosphere**.[456]

Thus, when solar activity is low, less solar wind reaches the higher atmosphere, which then

eruptions and sinkholes'

449 For decades noctilucent clouds were only observed close to the poles, but recently sightings have occurred at latitudes as low as 40° north. See: Russell III, J. M., 'Analysis of low latitude Noctilucent Cloud occurrences using satellite data and modeling', *Aeronomy of Ice in the Mesosphere* (March 2013)

450 *Ibid*, p. 18

451 Noctilucent clouds are found at 80 km (50 miles) altitude. The thermosphere is right above, beginning at about 85 km (53 miles) altitude.

452 Thompson, A., 'Earth's Upper Atmosphere Cooling Dramatically', *Space* (17 December 2009). See: www.space.com/7685-earth-upper-atmosphere-cooling-dramatically.html

453 'Mysterious Glow-in-the-Dark Arctic Clouds Invade USA', *Daily Kos* (29 July 2009). See: www.dailykos.com/story/2009/07/29/759 373/-Mysterious-Glow-in-the-Dark-Arctic-Clouds-Invade-USA

454 See Chapter 24: 'Global cooling'

455 Svensmark, H. & Calder, N., *The Chilling Star*, pp. 76–77

456 Haigh, J. D., 'Climate variability and the influence of the sun', *Science* (2001) 294: 2109–2111

Figure 127: Condensation trails left behind the 4 engines of a Boeing 747. (© Wiki Commons)

receives substantially less ultraviolet radiation, which causes the dramatic cooling of the higher atmosphere.

Contrails

So-called 'chemtrails' are often described as evidence of aerosol spraying of anthropogenic toxic substances on a mass scale from *high altitudes* as part of a government plot to poison the population.[457] This is claimed in spite of the fact that far simpler and more efficient means of delivering toxic substances to masses of people have long existed (and been applied), such as the fluoridation of 'drinking' water,[458] the genetic manipulation of the food supply[459] and the nuclear explosions conducted around our planet,[460] just to name a few.

It should also be pointed out that **high-altitude spraying is an extremely inaccurate delivery method**. Permanent winds at this altitude (about seven miles up) can reach hundreds of miles per hour and aerosols can take up to two years[461] to reach Earth's surface.[462] So, when you spray from airliner cruising altitude, you can't know where and when the released aerosols will

[457] Steiger, B. & Steiger, S., *Conspiracies and Secret Societies: The Complete Dossier*, pp. 91–93

[458] 'Fluorine Compounds Make you Stupid - Why is the Government not merely allowing, but promoting them?', *Sott.net* (4 February 2008). See: www.sott.net/article/148 371-Fluorine-Compounds-Make-you-Stupid-Why-is-the-Government-not-merely-allowing-but-promoting-them

[459] Renter, E., 'Another Study Finds GMO Compounds in 100% of Pregnant Women and Fetuses', *Activist Post* (27 October 2012). See: www.activistpost.com/2012/10/another-study-finds-gmo-compounds-in.html

[460] Since 1945, more than 2,000 nuclear explosions have been conducted. Their total yield is estimated to be approximately 510 megatons, equivalent to about 40,000 Hiroshima-size atomic bombs. See: 'General overview of the effects of nuclear testing', *CTBTO*. www.ctbto.org/nuclear-testing/the-effects-of-nuclear-testing/

[461] 'Atmospheric aerosols: what are they and why are they so important', *NASA* (August 1996). See: www.nasa.gov/centers/langley/news/factsheets/Aerosols.html

[462] Fall speed can also be drastically reduced because of electromagnetic forces exerted on the particles. See Chapter 26: 'Hurricanes, lightning and tornadoes'

land, and you might actually end up spraying yourself.

That being said, there is evidence that the ruling elites in various countries have, indeed, conducted aerial spraying of toxic substances, but these were **low-altitude spraying** that allowed greater control over where and when the toxic agent would reach the ground. For example, the CIA sprayed infectious agents over Cuba:

> Fort Detrick SO [Special Operations] Division microbiologists assisted in several covert attacks against rural and agricultural areas in Cuba. These attacks involved aerial spraying with swine flu virus, dengue, and other lethal infectious agents. As a result, hundreds of farm animals and several humans died.[463]

High-altitude spraying has also been tested by releasing sulfur dioxide and other aerosols for the purpose of cloud-seeding,[464] but this doesn't require control over the time and location of the landing, and has been shown to be generally ineffective.

Another factor that makes the 'chemtrails' theory iffy at best is the 'evidence' collected by its proponents. They point to high levels of aluminum, barium and strontium found in soil and water samples.

Aluminum is the **most abundant metal in the Earth's crust**. Every year about 100 million tons of aluminum are produced.[465] It is extensively used in the car, plane, boat and building industries. From this perspective, it's not surprising to find high levels of aluminum in most if not all soil and water samples. Such levels do not imply that this aluminum comes from jet spraying.

Barium[466] is found in nature as a free element.[467] It is used for numerous industrial applications: glassmaking,[468] fluorescent lamps, and screens.[469] Barium is also used as an 'atmospheric aerosol spray for enhancing/refracting the signaling of radio/radar waves along military jet flight paths, missile test ranges.'[470] So, toxicology results indicating high levels of barium in some water sources near barium production and usage sites are hardly surprising.

Finally, strontium is a natural element, the fifteenth most common element on Earth. It is also massively produced (about 300,000 tons a year)[471] and used in screens for TVs, computers and phones. In addition, one radioactive isotope of strontium, called **strontium 90, is present in radioactive fallout**. Since 1945, 2,051 nuclear explosions have occurred on planet Earth.[472] Many of these explosions happened in the atmosphere and were conducted in places as far apart as the Pacific Islands, China, the U.S., Algeria, Australia, Russia and Kazakhstan. Thus, it is not surprising to find high levels of strontium (radioactive or not) all over the planet.

---

463 Albarelli, H. P., *A Secret Order*, p. 13

464 Steiger & Steiger, *op. cit.*, pp. 105–107

465 'List of countries by aluminium production', *Wikipedia.* See: en.wikipedia.org/wiki/List_of_countries_by_aluminium_production

466 'Barium', *Wikipedia.* See: en.wikipedia.org/wiki/Barium

467 It makes up 0.0425% of Earth's crust.

468 Lide, D., *CRC Handbook of Chemistry and Physics*, p. 4

469 Ullman, F., *Encyclopedia of Industrial Chemistry*

470 Purdey, M., 'Chronic barium intoxication disrupts sulphated proteoglycan synthesis: a hypothesis for the origins of multiple sclerosis', *Medical Hypotheses* (2004) 62(5): 746–54

471 'Strontium', *Wikipedia.* See: en.wikipedia.org/wiki/Strontium

472 'International Day Against Nuclear Tests: Translating Words Into Action', *Arms Control Association* (2 September 2011). See: www.armscontrol.org/events/International-Day-Against-Nuclear-Tests-Translating-Words-Into-Action

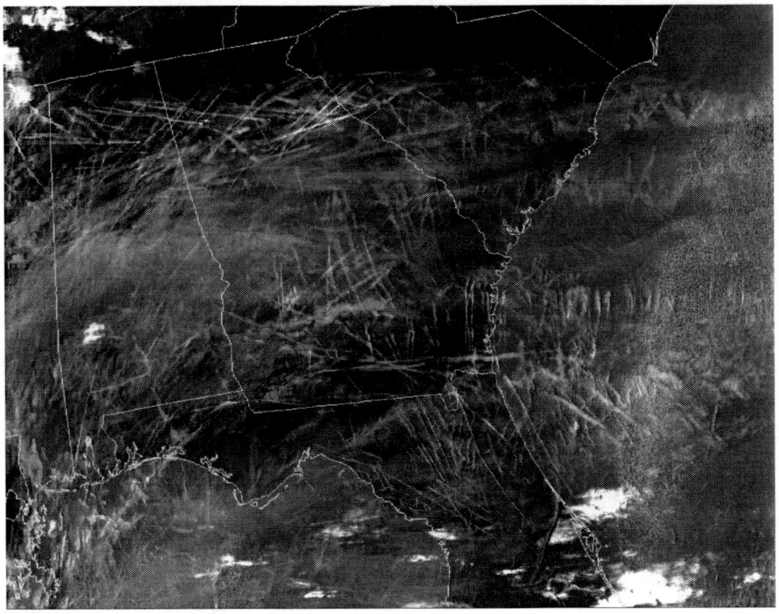

Figure 128: Contrails over the southeastern U.S. sky.

In short, the elements claimed to be evidence of 'chemtrail spraying of populations' have numerous industrial sources.

Condensation trails have been observed since WWI, when bombers managed to reach high enough altitudes for cold air to allow contrails to form.[473] The phenomenon was so obvious that bomber pilots complained that contrails were giving away their position.[474] Chemtrailers claim that 'chemtrails' are thicker and longer lasting compared to contrails. So, basically the only valid piece of evidence brought by 'chemtrailers' is the fact that jet planes did not previously leave any persistent and thick condensation trails behind them and *now they do.* This observation is in fact valid. Indeed, around the end of the 1990s, something changed and thicker contrails started being observed more frequently. Incidentally, the term 'chemtrails' was first introduced by journalist William Thomas in 1997.[475]

A condensation trail, or contrail, is the result of the condensation of jet fumes. Here, unlike noctilucent clouds, atmospheric dust doesn't play a major role, since jet engines already release massive quantities of particles,[476] particularly sulfur molecules[477] that are some of the most efficient condensation nuclei.[478]

---

473 About 33,000 feet, the altitude at which airliners fly today.

474 Steiger & Steiger, *op. cit.*, p. 98

475 Eden, D., 'Chemtrails, what's going on?', *Viewzone* (July 2013). See: www.viewzone.com/chemtrails.html

476 15,000 particles / cm^3 vs. background concentration equal to 6–18 particles / cm^3. See: Fahey *et al.*, 'Emission Measurements of the Concorde Supersonic Aircraft in the Lower Stratosphere', *Science* (6 October 1995) 270(5233): 70–74

477 Panel on Atmospheric Effects of Aviation, *Atmospheric Effects of Aviation: A Review of NASA's Subsonic Assessment Project*, p. 14

478 Svensmark & Calder, *op. cit.*, pp. 126–131

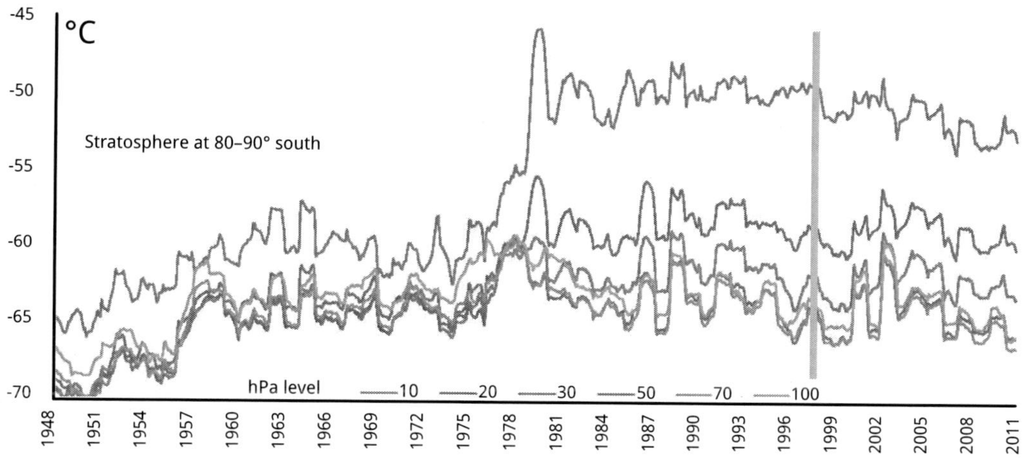

Figure 129: Temperature of the stratosphere over the period 1948–2011. At the beginning of the century (light grey vertical line), the temperature of the stratosphere started to decrease. (© Climate Change Blog, adapted by Sott.net)

However, *what does play a major role is temperature.* The cooler the ambient air, the stronger and longer lasting the condensation. The cruising altitude for an airliner is about 10 km (7 mi); this is just below the stratosphere,[479] which has cooled down *'by several degrees since 2002'*.[480]

Far from the conspiracy theory proposed by the 'chemtrailers', data strongly suggests that persistent contrails are a direct consequence of the cooling of the stratosphere, a far more troubling state of affairs than 'chemtrails' released in upper atmospheres where the winds aloft make it unlikely that anything released in a given location will filter to the ground anywhere near, or anytime soon.

[479] The stratosphere covers the 20–50 km altitude range.

[480] Laštovička, J., *et al.*, 'Global Change in the Upper Atmosphere', *Science* (24 November 2006) 314(5803): 1253–1254

Chapter 26.

Hurricanes, lightning and tornadoes

Introduction

The accumulation of cometary dust in the Earth's atmosphere plays an important role in the increase of tornadoes, cyclones, hurricanes and their associated rainfalls, snowfalls and lightning. To understand this mechanism we must first take into account the electric nature of hurricanes, tornadoes and cyclones, which are actually manifestations of the same electric phenomenon at different scales or levels of power. Because of this similarity, we will refer to these three phenomena collectively as 'air spirals' in the following discussion.

McCanney describes the electric nature of hurricanes in these terms:

Figure 130: A waterspout associated with lightning. (© Flickr)

> A simple model showed that these [tropical] storms formed when **electrical currents connected between the ionosphere and the top of the clouds.** ... the reason **hurricanes lost power when they approached land was that the powering electrical current from the ionosphere to the cloud tops and to the Earth's surface had no connection (anode) while over the ocean** so it drew up vast surface areas of ionized air from the ocean surface and sucked them up a central column (the spinning vortex was caused by the moist air rising 'up the drain') whereas the land provided a 'ground' for the current and therefore it shunted out the storm's power source. ... I also calculated that **the warm water theory for hurricane development lacked sufficient energy to account for the energy in these massive storms. We later witnessed hurricanes on Mars where there is no water** at all. Clearly, the warm water concept did not work ...[481]

From this perspective, air spirals are simply the manifestation of electric discharges between the ionosphere and the Earth's surface. Figure 130 shows a waterspout and a lightning bolt occurring in the same place at the same time, suggesting that, indeed, the electric potential difference between

[481] McCanney, J., *Planet X, comets and earth changes*, pp. 70, 71

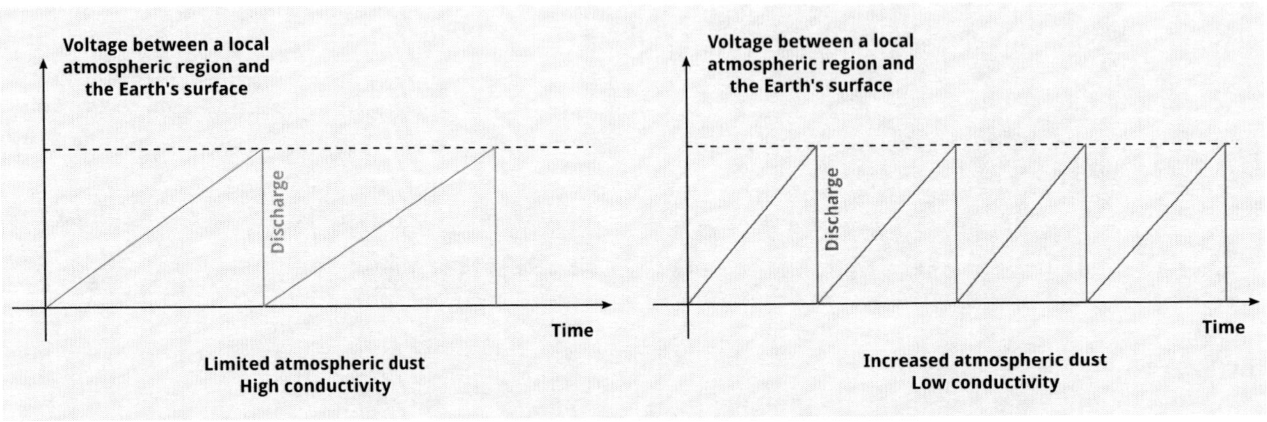

Figure 131: Influence of atmospheric dust on frequency of 'air spirals' and lightning. (© Sott.net)

the clouds at the top of the picture and the ground at the bottom is what powers both the lightning and the tornado.

Discharge frequency

If air spirals are electrically driven, how, then, can we explain an increase in their frequency when the Sun's activity has dropped and the atmospheric E-field has therefore weakened?[482] While the *overall* atmospheric E-field has indeed weakened, another factor must be taken into account: The **increase in atmospheric dust concentration**[483] **reduces the electric conductivity of the atmosphere.**[484] Conductivity in the atmosphere is due to the mobility of small ions. When dust is present, these ions, instead of moving freely, attach to the relatively large dust particles and lose mobility, hence the decrease in atmospheric conductivity.[485]

The two graphs in figure 131 compare the effects of low levels of atmospheric dust (left) and high levels of atmospheric dust (right). On the left, the increase in voltage between a local atmospheric region and the Earth's surface grows slowly because of the **fair weather circulation: electrons move up freely to balance the charge differences**. On the right, the fair weather circulation is restricted because of the dust-induced reduced conductivity. Electrons can't move freely; they are captured by dust particles. Thus local electric charges build up faster, hence the increase in discharge frequency.

This additional feature of dust particles – their ability to carry an electric charge – means that dust accumulation enables any given area of the atmosphere to carry potentially massive electric charges, which can differ from the charge of adjacent regions, from the charge of the ionosphere and from the charge of the Earth's surface. This suggests that lightning and air spirals of all kinds

[482] Because of reduced solar activity, the ionosphere is less (positively) charged, therefore the atmospheric E-field decreases.

[483] See Chapter 21: 'Increase in cometary activity'

[484] Schneider, H., *et al.*, *Encyclopedia of Climate and Weather*, p. 67

[485] Ulanowsky, Z., *et al.*, 'Alignment of atmospheric mineral dust due to electric field', *Atmos. Chem. Phys.* (2007) 7: 6161–6173

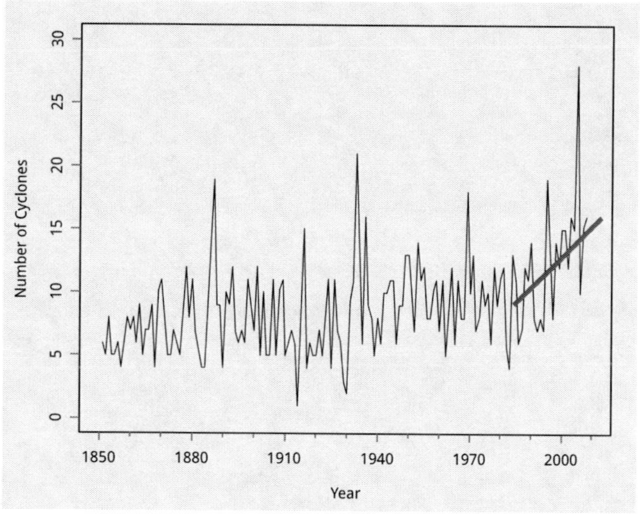

Figure 132: Tropical storm frequency (1851–2008). (© Clemson)

should increase in frequency and geographic scope.

Figure 132 is adapted from a scientific paper published by Robert Lund, professor of mathematical sciences at Clemson, who studied changes in the tropical storm-cycle record in the North Atlantic between 1851 and 2008.[486] His study shows an increase in tropical-storm frequency that started in the mid-'80s (dark grey ascending line) and which continues today.

Discharge magnitude

While air spirals (tornadoes and hurricanes) have been more frequent since the end of the 1980s and should be more frequent in coming years because of the ongoing increase in atmospheric cometary dust, very powerful discharges, like Category 5 hurricanes, should not necessarily increase.

Category 5 hurricanes last for days and enable massive transfers of energy (charge rebalancing) between the ionosphere and the Earth's surface. So the main energetic source for major hurricanes is the electric potential difference between the Earth and its ionosphere. Reduced solar activity means a less (positively) charged ionosphere, which means reduced electric potential difference *between the ionosphere and the Earth*, which in turn leads to less frequent Category 5 hurricanes. After studying the tropical-storm records made over the last 150 years, Robert Lund reached a similar conclusion concerning the strength of the storms:

> 'This is a hot button in the argument for global warming,' said Lund. 'Climatologists reporting to the U.S. Senate as recently as this summer testified to the exact opposite of what we find. **Many researchers have maintained that warming waters of the Atlantic are increasing the strengths of these storms. We do not see evidence for this at all, however we do find that the number of storms has recently increased.** ' The **study represents one of**

[486] Lund, R., 'Study in the changes in the tropical cycle record in the North Atlantic between 1851 and 2008', *Journal of the American Statistical Association* (2009)

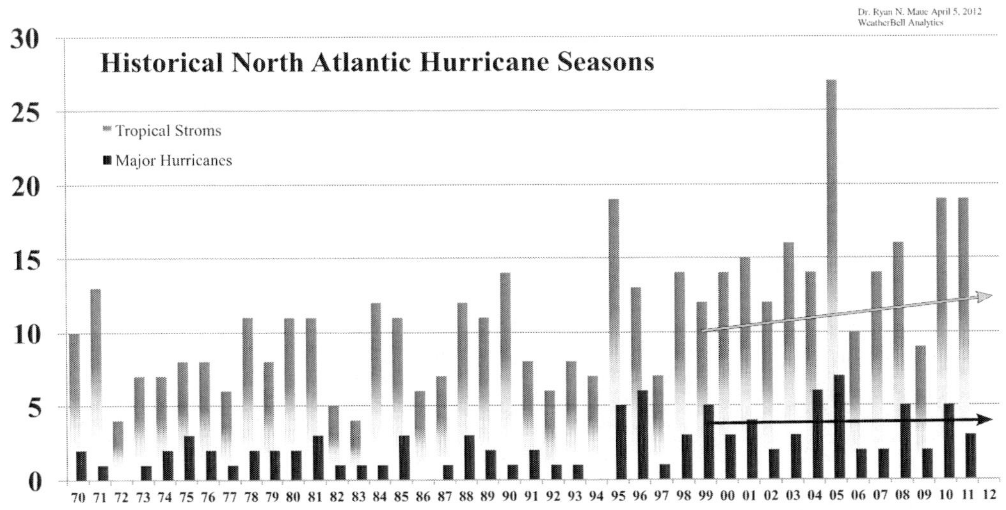

Figure 133: Annual number of tropical storms (light grey) vs. annual number of major hurricanes (black) (1970–2012). (© Ryan N. Maue)

the first rigorous statistical assessments of the issue with uncertainty margins calculated in.[487]

Lund's statement seems to be confirmed by recent statistical data compiled by Dr. Ryan N. Maue:

Figure 133 confirms Lund's observation of an increase in tropical storm frequency (light grey ascending arrow – linear regression) since the '80s, while the frequency of major hurricanes seems relatively stable (black horizontal arrow – linear regression).

The study of tornado statistics in the U.S. leads to a similar result. Figure 134 was compiled from data provided by NOAA.[488] While the **overall number of tornadoes shows a steady increase since 2002** (light grey ascending bars and curve), the number of powerful tornadoes (dark grey bar and line) has been relatively constant. As a result, the **percentage of violent tornadoes is decreasing.**[489]

As seen previously,[490] two opposing factors may modulate solar activity. **On one side, an approaching Nemesis would tend to reduce the Sun's activity because of 'grounding'; on the other side, an accompanying cometary swarm would tend to increase solar activity** because of comet-triggered solar discharges.

[487] 'Hurricane Frequency Is Up But Not Their Strength, Say Researchers', *Science Daily* (23 September 2009). See: www.sciencedaily.com/releases/2009/09/090922112 207.htm

[488] National Oceanic and Atmospheric Administration. See: www1.ncdc.noaa.gov/pub/data/cmb/images/tornado/2011/dec/tornado-counts-0112-2011.png and: www1.ncdc.noaa.gov/pub/data/cmb/images/tornado/clim/EF3-EF5.png

[489] For example, in 1974, 45 F3+ tornadoes were officially counted out of a total of 950. In 2010, there were 30 F3+ tornadoes and a total of 1300. So, over the period 1974–2010, the percentage of violent tornadoes was approximately halved.

[490] See Part II: 'The Sun's companion and its accompanying cometary swarm'

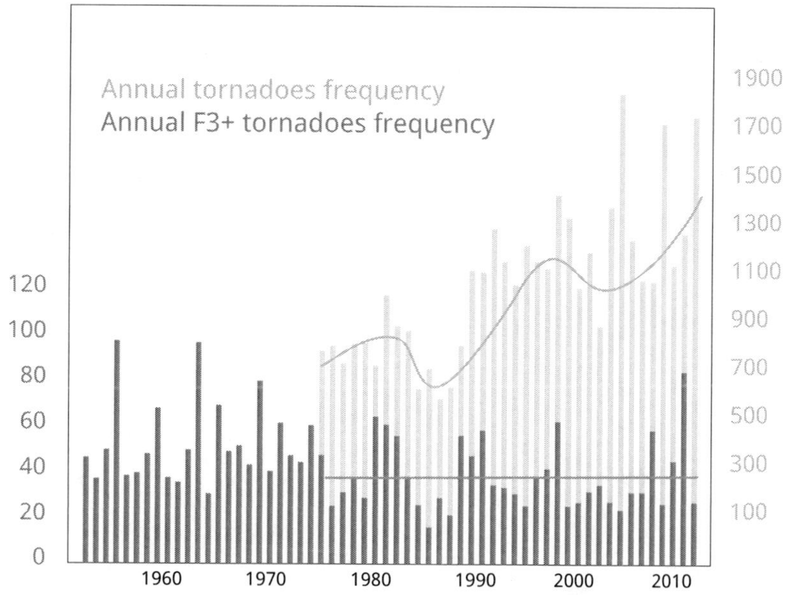

Figure 134: Tornado annual frequency (1975–2011) and F3+ tornado annual frequency (1950–2011). (© Sott.net)

In recent years, the Sun has been unusually quiet, which suggests that, overall, the Nemesis grounding factor overpowers the comet-discharge factor. Hence a reduced frequency of major hurricanes (since, as seen above, solar activity is the main driver for major hurricanes). However, this is not always the case: the Sun can still experience major activity spikes.

In November 2013 the Sun exhibited a period of intense activity, with eight X-class[491] solar flares[492] between October 25th and November 19th, and yet, over the preceding two years the Sun had experienced an average of just 0.5 X-class solar flares per month.[493]

The development of Typhoon Haiyan, which devastated parts of the Philippines in November 2013, seems to synch with the high and unusual solar activity that was happening at the time.

Knowing that solar winds take two to four days to reach Earth, Haiyan originated as a tropical storm on November 2nd (possibly triggered by the X2.3 flare that occurred on October 29th). By November 5th, it was a Category 5 typhoon. On November 7th it was still intensifying (possibly strengthened by the X3.3 flare of November 5th) with one-minute sustained winds reaching 315 km/h (196 mph). This very same day, Haiyan made its first landfall without any change in intensity, becoming the most powerful storm in recorded history to make landfall.[494]

[491] Solar flares are classified in 5 categories – A,B,C, M or X – according to their power or 'peak flux' (measured in w/m^2). An X-class flare is ten times more powerful than an M-class flare, which in turn is ten times more powerful than a C-class flare. The classification letter is usually followed by a number. For example an X2 flare is twice as powerful as an X1 flare.

[492] X1.7 flare on October 25th, X2.1 flare on October 25th, X1.0 flare on October 28th, X2.3 flare on October 29th, X3.3 flare on November 5th, X1.1 flare on November 8th, X1.1 flare on November 10th, X1 flare on November 19th.

[493] Between March 7th, 2011, and October 25th, 2013, the Sun has experienced 17 X-class solar flares. See: 'Solar Flare list', *My solar alerts.* mysolaralerts.blogspot.fr/p/solar-flare-list.html

[494] Fischetti, M., 'Was Typhoon Haiyan a Record Storm?', *Scientific American* (12 November 2013) See: blogs.scientificamerican.com/observations/2013/11/12/was-typhoon-haiyan-a-record-storm/

Figure 135: Satellite picture of Typhoon Haiyan. (© NOAA/AFP/Getty Images)

What could explain this spike in solar activity overpowering an otherwise quiet Sun due to the theorized Nemesis-induced grounding? As a matter of fact, there were a **high number of active comets observed in the Sun's vicinity** at that time. On November 21st, **17 comets**[495] were visible with a small telescope, of which five[496] exhibited a magnitude between 4 and 7.5 and could be seen with binoculars or the naked eye.[497]

The situation was quite unusual because, amongst the seventeen comets mentioned above, the two brightest ones, ISON (2012 S1) and Lovejoy (2013 R1), did not follow the usual Jupiterian orbit. ISON and Lovejoy are two very long-period comets. Lovejoy's period is 10,371 years and ISON's period is still unknown but its eccentricity suggests that it might be a 'one-journey' comet, never to be seen again in the solar system.[498]

Comets with elongated trajectories are the most electrically active. In contrast to comets of Jupiterian orbits, they are highly negatively charged because they come from the negatively charged outer regions of the solar system and they rush in almost a straight line towards the positively charged Sun and its surrounding space, hence the comets' brightness (due to electric stress) and the spikes in solar activity (due to electric discharging of the Sun triggered by these very active comets).

Figure 136 is a screen-capture from a video recorded by NASA's Stereo probe. If you watch this video,[499] which spans a three-day period from November 19th to November 22nd, you will see how

[495] 'Comet list', *British Astronomical Association – Comet section.* Check observable comets for 21 November 2013 at www.ast.cam.ac.uk/~jds,/

[496] ISON (2012 S1), Lovejoy (2013 R1), 2P/Encke, Linear (2012 X1), P/Nevski (2013 V3).

[497] 'Comet chasing – visual observation of telescopic comets', *Cometchasing.* See: cometchasing.skyhound.com

[498] See Chapter 18: 'Comets or asteroids?'

[499] Battams, K., 'ISON and Encke in STEREO: Nov19-23, 2013', *NASA Comet ISON Observing Campaign.* See:

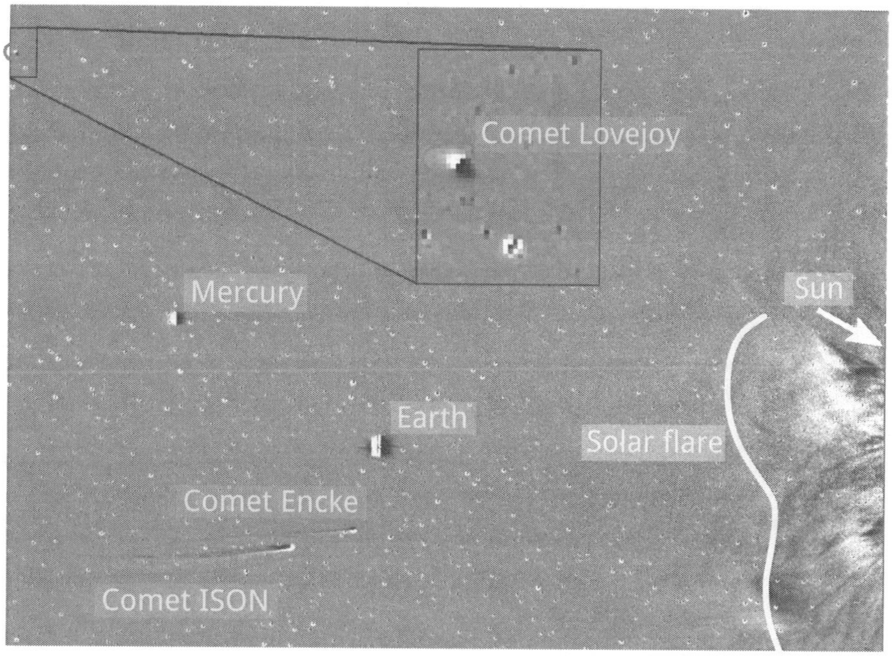

Figure 136: Image from STEREO showing 3 active comets in the vicinity of the Sun: 2P/Encke, ISON and Lovejoy triggering massive solar flares. (© NASA-CIOC)

incoming comets can trigger a substantial solar flare.

Now that we have a better understanding of the possible cause for this sudden spike in solar activity and its effects on weather phenomena, let's go back to hurricanes and tornadoes.

Despite the ferocious Typhoon Haiyan, even after taking its energy into account, the Accumulated Cyclone Energy (ACE) index for 2013 – at 74% – was still below normal, the lowest it has been since the 1970s.[500] As you can see in figure 137, both global (light grey line) and northern hemisphere (dark grey line) data show an overall downward trend in ACE since around 1998 and a continuous drop since 2006.

According to mainstream science, hurricanes are due to moist warm air that rises from the Earth's surface, which then begins to spiral because of the Coriolis effect,[501] finally condensing to create clouds. But, as is often the case with mainstream explanations, there are some 'anomalies' that don't fit the proposed model. **According to the dominant weather science, no hurricane can form on the equator**[502] because: 1) the Coriolis effect is very weak at such low latitudes and 2) the Coriolis effect in the southern hemisphere is opposite to the Coriolis effect in the northern hemisphere. This opposition would tend to split apart any cyclone spreading over the line of the equator.

---

www.ISONcampaign.org/ISON-encke-nov19-23

[500] Lomborg, B., 'Bjørn Lomborg on Haiyan/Yolanda', *wattsupwiththat* (10 November 2013). See: wattsupwiththat.com/2013/11/10/bjorn-lomborg-on-haiyanyolanda

[501] See Chapter 27: 'The Gulf Stream'

[502] Barnes-Svarney, P. & Svarney, T. E., *Skies of Fury: Weather Weirdness Around the World*, p. 173

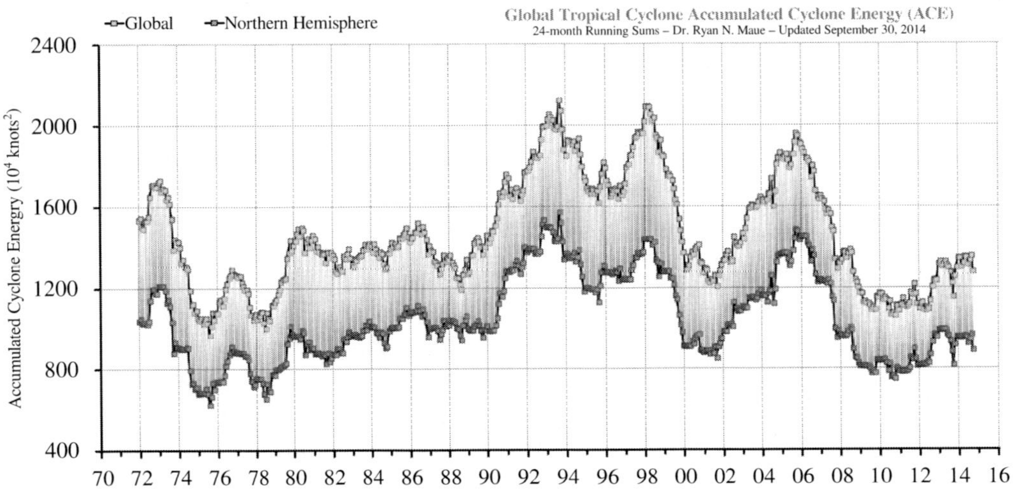

Figure 137: Accumulated Cyclone Energy (1972–2014). (© Weatherbell.com)

However, on December 27th, 2001, Typhoon Vamei[503] formed on the equator and exhibited counterclockwise circulation **on both sides of the equator**. If the Coriolis effect was the only factor causing hurricanes to spin, Typhoon Vamei and its equatorial position would be simply impossible.

Let's now deal with moist air, which is considered the energy source of hurricanes. The second hurricane of the 1971 season (unnamed, hereafter referred to as 'Hurricane #2') formed east of Bermuda on July 29th, 1971.[504] It moved northeast and started gaining power, even though it was moving over ever cooler waters. As the storm passed a drilling rig 200 nautical miles from Newfoundland,[505] its winds reached 127 mph (204 km/h). The radius of maximum winds was about 35 miles (50 km) wide. Hurricane #2 had all the characteristics of a hurricane when it was at latitude 46° north (on the east coast of Canada),[506] where the water temperature in July averages 12° Celsius.[507] At that time, the water was actually *five degrees colder than the air.*[508] According to the textbooks, 'Hurricane #2' should never have formed.

An even more striking example comes from Mars, a planet devoid of any sea which, nonetheless, regularly experiences hurricanes. They were first observed in 1979 by the Voyager satellite.[509] In 1999, Hubble shot pictures of a massive Martian hurricane that lasted for months. Its size was nearly four times larger than the state of Texas (see figure 139).

In conclusion, Typhoon Vamei shows that the Coriolis effect is probably not the main force

[503] Tropical cyclones have several names depending on their location and intensity. In Asia they are called 'typhoons', while in the U.S. they are called 'hurricanes', 'tropical storms' or 'tropical depressions', depending on their strength.

[504] Unlike other hurricanes (Wilma, Katarina, etc.), this hurricane wasn't named. See: Simpson, R. & Hope, J., *'Atlantic hurricane season 1971'*, NOAA, 1971

[505] Newfoundland is a Canadian island located along the eastern Canadian coast.

[506] 'New Brunswick map', *Geology.* See: geology.com/canada/new-brunswick.shtml

[507] 'Sea temperature map', *NOAA.* See: www.ospo.noaa.gov/data/sst/contour/global.fc.gif

[508] Simpson & Hope, *op. cit.*

[509] Sincell, M., 'Hubble Snaps Martian Hurricane', *Science* (19 May 1999). See: news.sciencemag.org/1999/05/hubble-snaps-martian-hurricane

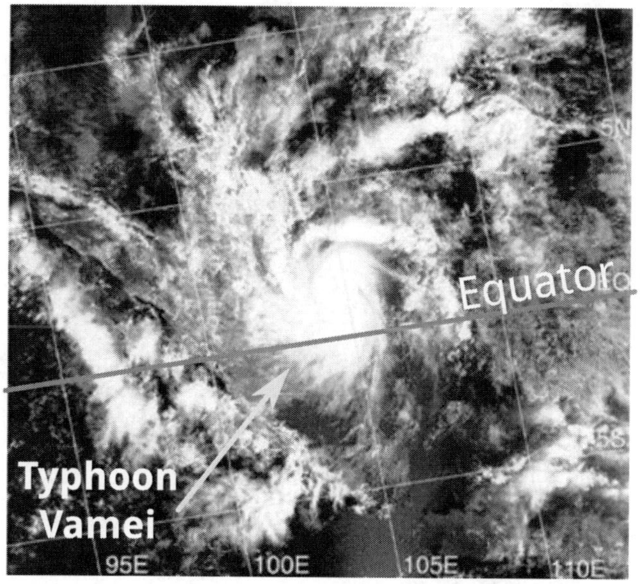

Figure 138: Satellite image of Typhoon Vamei on December 27th, 2001. (© CRISP/National University of Singapore)

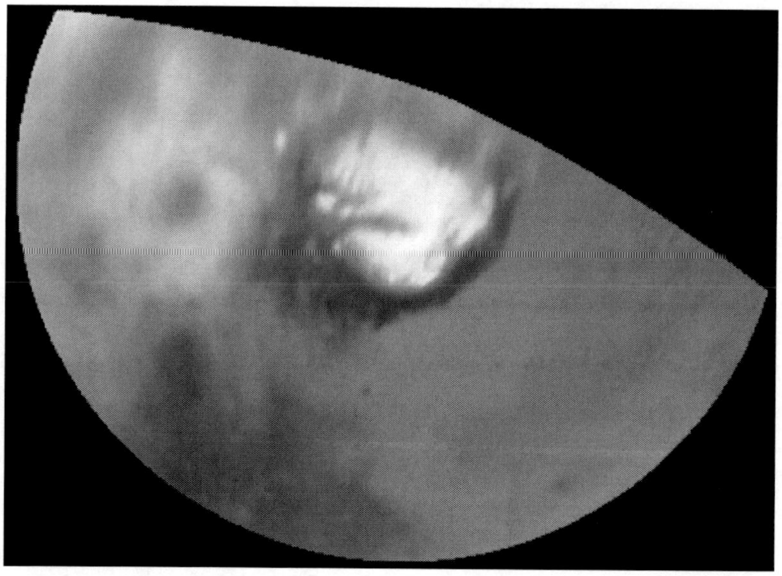

Figure 139: Hurricane snaking around Mars's North Pole on April 27th, 1999. (© Hubble/NASA)

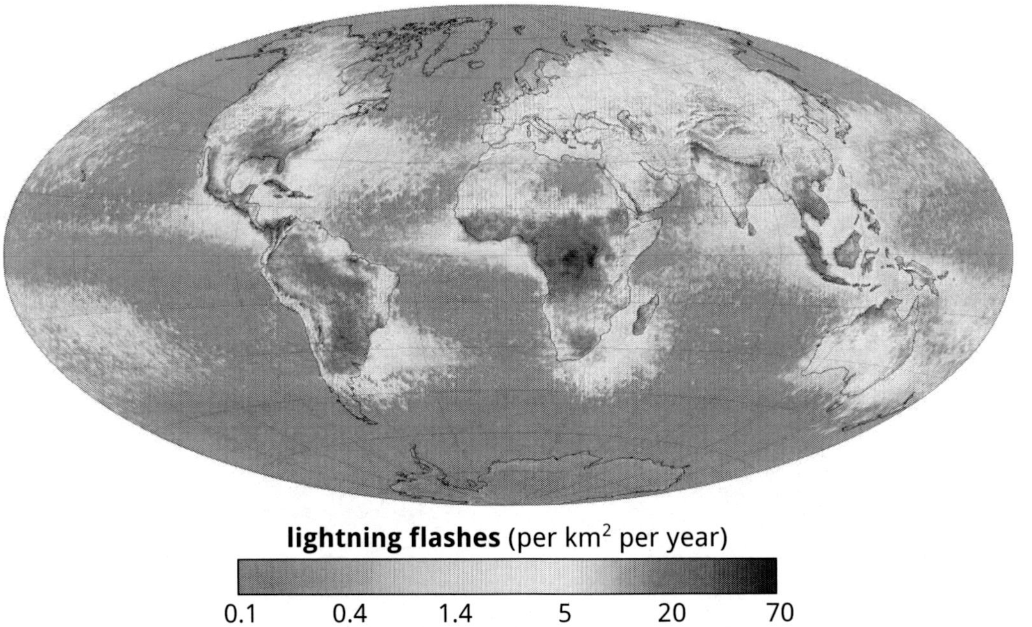

Figure 140: Global lightning frequency. (© NASA/GHRC/NSSTC)

causing hurricanes to spin and 'Hurricane #2', plus the Martian hurricanes, show that warm water is not the primary factor fueling hurricanes.

The electric dimension of hurricanes and lightning

Electricity seems to play a major role in air-spiral phenomena, including hurricanes. To better understand these electric phenomena, let's first consider lightning, which, as we will soon see, is closely related to hurricanes, depressions, tornadoes and their accompanying cloud masses.

The highest lightning frequency occurs around the equator, towards the end of the day.[510] This time and place correspond to maximum ionosphere electric charge. Indeed, the **intertropical zone directly faces the solar wind** that hits the ionosphere during the day, until it reaches a **maximum charge in the evening after one full day of direct solar exposure**. In addition, the intertropical zone is where atmospheric conductivity is the lowest.[511] So, in this region, the atmosphere can hold a very high charge difference between the ionosphere and the Earth's surface, which leads to violent discharging (lightning bolts) instead of fair-weather charge rebalancing.

The two drawings in figure 141 depict cloud formation followed by a cloud-to-ground lightning discharge. The 6 steps of the process listed below correspond to the numbers in light grey.

Top drawing:

1) The proton-rich solar wind projects positively charged particles into the Earth's ionosphere.

2) As the positive charge of the ionosphere increases, it begins to attract electrons located on

[510] Christian *et al.*, 'Global frequency and distribution of lightning', *Journal Of Geophysical Research* 108

[511] Plotkin, V., 'Earth Charge Produced By Thunderstorms', *Russian Geology and Geophysics* (2003) 44(3): 260–267

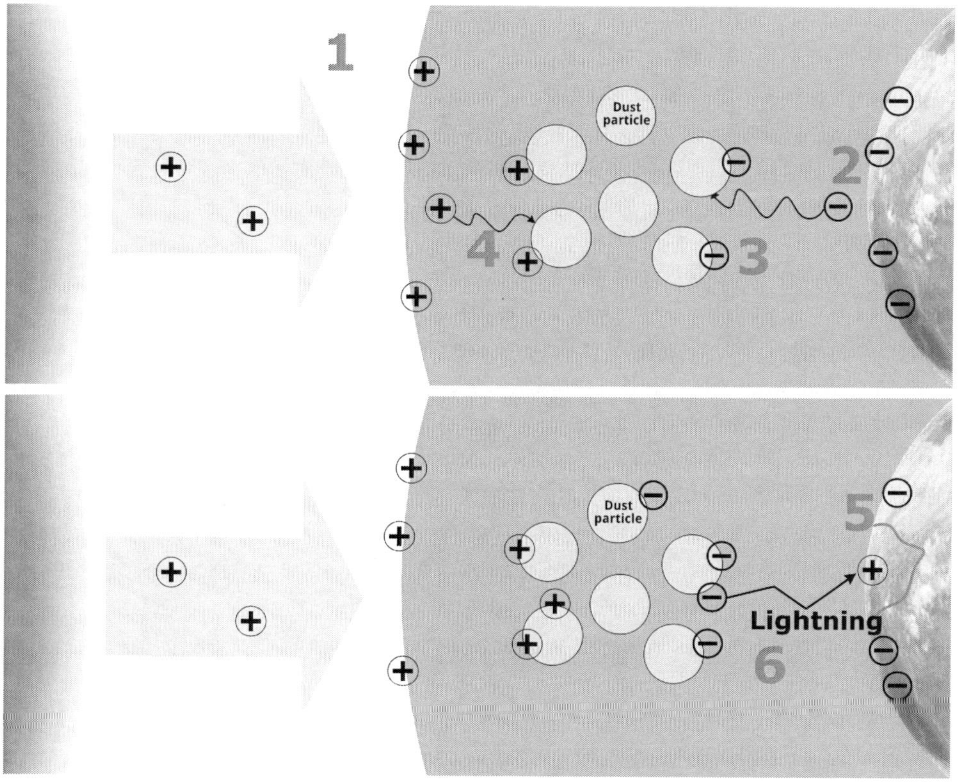

Figure 141: Lightning formation process. (© Sott.net)

Earth's surface.[512]

3) The electrons from Earth's surface rise in the air and are captured by low-mobility atmospheric particles (dust, droplets).

4) Similarly, protons from the ionosphere are attracted by the Earth's negative charge and start sinking into the atmosphere, eventually being captured by atmospheric particles.[513]

Bottom drawing:

5) The upward electron flow creates an electron-deficient region on Earth's surface (positive region within the medium grey line) and an electron-rich area at the bottom of the cloud.

6) When the electric potential difference is high enough, lightning strikes, carrying a massive electron flow back to the ground that re-balances electric charges between the bottom of the cloud mass and the locally positive region on Earth's surface.

Lightning and hurricanes seem to be similar charge rebalancing processes. Lightning mostly occurs above continents and is far less frequent above oceans.[514] This may be due to the difference between ground conductivity and sea conductivity. When electrons start flowing upwards from the ocean, the high conductivity of salt water[515] usually prevents the formation of electron-deficient regions, which is one of the causes of lightning. However, when the upward electron flow occurs above a continent, the poor conductivity of the ground[516] enables the formation of electron-deficient pockets that will trigger and receive lightning discharges.

In terms of location, hurricanes are the opposite of lightning bolts: they mostly occur above oceans and usually weaken or die when they reach land. When a massive flow of electrons is pulled up above the ocean, the high conductivity of salt water can provide and conduct free electrons from all adjacent regions, thus offering an almost endless supply of electrons to power the ongoing hurricane. When the hurricane reaches the ground, the electron supply is limited by the poor conductivity of the ground and the hurricane weakens.

The map in figure 142 depicts the trajectories of the 14 listed hurricanes that occurred in 2000. All of those that reached the coast 'died' soon after.

Notice that the rainfall that usually accompanies hurricanes also participates in the charge re-balancing process.

When a raindrop falls to the ground, it can capture electrons from the bottom of the cloud or below it, thus carrying a negative charge to the ground and rebalancing electric potential differences in a manner similar to lightning. From this perspective, lightning and rain are both caused by a strong atmospheric E-field and both lead to a charge rebalancing between the Earth's surface and its atmosphere.

Notice that the atmospheric field has an influence on raindrops' formation and size. In figure

[512] This is comparable to the low, progressive leakage current observed in a capacitor.

[513] See previous note.

[514] 'World lightning map', *Geology.* See: geology.com/articles/lightning-map.shtml

[515] Seawater is an extremely good conductor. Its conductivity is equal to 4.8 siemens/meter. See: Cox *et al.*, *Deep Sea Res.* (1970) 17: 679

[516] Soil conductivity is between 1 and 100 millisiemens/meters. It is several orders of magnitude lower than the conductivity of seawater (4,800 millisiemens/meter).
See: Barbosa, R. & Overstreet, O., 'What Is Soil Electrical Conductivity?', *LSU AgCenter Pub* 318

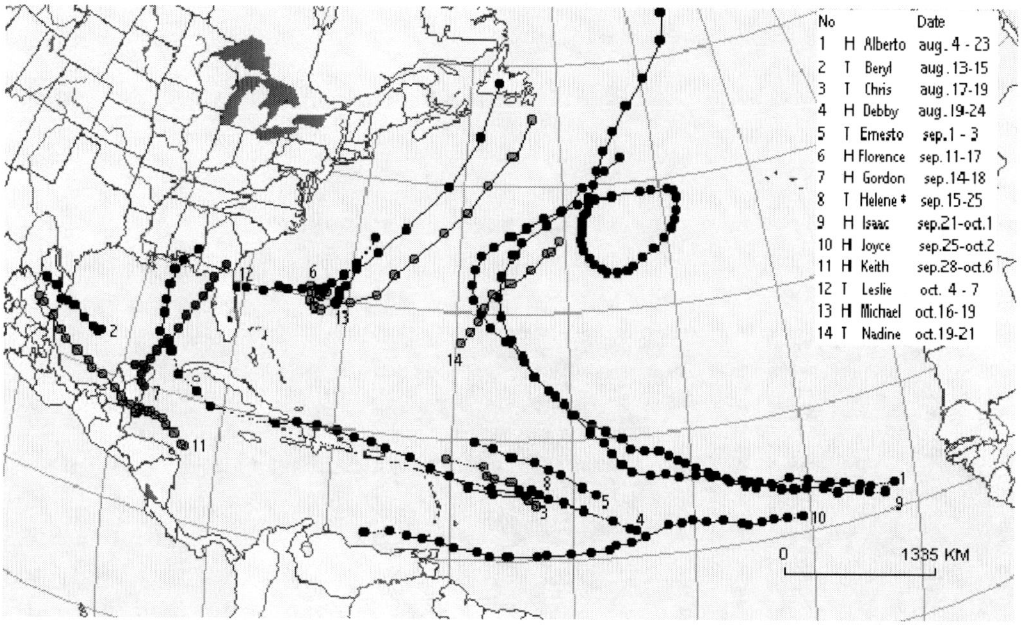

Figure 142: Trajectory of the 14 hurricanes that occurred in 2000. (© met.inf.cu)

143,[517] a thin water jet was created by a hypodermic needle connected to a water faucet. On the left, no electric field was applied. The jet took the form of a mist made of small droplets. On the right, an electric field was applied to the water jet, triggering the binding of droplets to one another and the subsequent formation of large water drops. This experiment is very similar to what occurs in clouds, where *water droplets* tend to align along the atmospheric E-field and attract each other, forming heavier and heavier water drops.

From the above we can see that lightning and hurricanes are very similar electric phenomena. Hurricanes are to sea surface what lightning bolts are to ground surface. They are both caused by upward electron flows and they both rebalance electric charges by returning electrons to the ground: rainfall in the case of hurricanes, lightning in the case of electrical storms.

Before closing this chapter, a few further comments about atmospheric dust are necessary: as we've seen previously, atmospheric dust plays a major role in storm dynamics. On a physical level it acts as a nucleus for the formation of condensed water droplets (clouds). On an electrical level it holds electric charges that can trigger lightning.

Atmospheric dust also seems able to modulate cloud elevation. According to mainstream science, atmospheric dust and water droplets stay in suspension in the atmosphere because of their very small size: low weight and a comparatively large drag.[518] However, some observations don't fit the

[517] Jermacans, K. & Laws, K., 'Coalescence of Raindrops in an Electrostatic Field', *The Physics Teacher* (April 1999) 37

[518] For example, if the radius of a particle is divided by two, its gravity will be divided by eight, while its drag will only be divided by four. Downward gravity depends on the volume of the particle, which is proportional to the cube of the radius of the particle (volume of a sphere is equal to $4/3\ ^{x}\ \pi\ ^{x}\ r^3$), while the drag exerted by upward air convection depends on the surface of the particle, which is proportional only to the square of the radius of the particle (surface of a sphere is equal to $4\ ^{x}\ \pi\ ^{x}\ r^2$).

gravity-drag model and, in some cases, dust clouds settle much slower than predicted:

> Interestingly, it appears that some hitherto **unknown atmospheric process counteracts gravitational settling of larger atmospheric dust particles** (Maring *et al.*, 2003), as models of long-range dust transport often underestimate the larger particle fraction (Colarco *et al.*, 2003, Ginoux *et al.*, 2001), and dust samples collected after fallout events show that large numbers of 'giant' dust particles (larger than 62.5 micrometers) can be carried thousands of kilometers from their source (Middleton *et al.*, 2001).[519] [520]

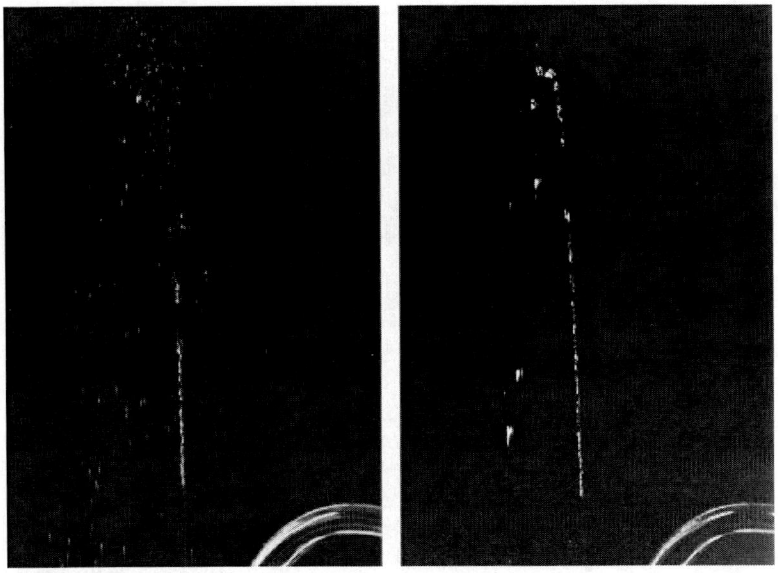

Figure 143: Influence of electric field on the size of water drops. (© Pierce Bounds)

If you remember the Millikan experiment,[521] a droplet charged with only one electron can counteract gravity and literally levitate *when exposed to a vertical electric field.* For this to happen, the vertical electric field has to be 32,100 V/m.[522] Although the atmospheric electric field is normally about 100 V/m at ground level,[523] atmospheric dust or atmospheric droplets can drastically increase this value because they reduce conductivity. Electric fields of 2,000 V/m have been measured under dust storms,[524] up to 20,000 V/m *under* thunderstorms[525] and up to 200,000 V/m *within* thunderstorms.[526] In addition, unlike the Millikan experiment, some particles can carry more than one electron charge.

This means that the atmospheric electric field can play a role in the fall speed, location, movement and elevation of clouds whether they are made of dust or droplets (or both). It can cause the particles to 'levitate' or literally rise up in the air.

---

519 Ulanowski, Z., 'Alignment of atmospheric mineral dust due to electric field', *Atmos. Chem. Phys.* (2007) 7: 6161–6173

520 Note that this point was emphasized in the discussion about 'chemtrails' allegedly laid in the upper atmosphere. It simply makes no sense because there is absolutely no control over where (or even when) any such substances would reach ground.

521 See Chapter 2: 'Electromagnetic forces vs. gravity'

522 For one droplet charged with one electron to levitate, the downward gravitational force must be equal to the upward electric force. If we take a water droplet 10^{-6} meters in diameter, its weight (w) is $4/3 \times \pi \times \rho \times r^3 \approx 5.1 \times 10^{-14}$ N. So the electric field is equal to $w/e = 5.1 \times 10^{-14} / 1.6 \times 10^{-19} = 32{,}100$ V/m.

523 Roble, R., *The Earth's Electrical Environment*, p. 206

524 Kamra, A., 'Measurements of the electrical properties of dust storms', *Journal of Geophysical Research* 77(30): 5856–5869

525 Pawar, S., 'Anomalous electric field changes and high flash rate beneath a thunderstorm in northeast India', *Journal of Earth System Science* (October 2010) 119(5): 617–625

526 Evans, L., 'The growth and fragmentation of ice crystals in an electric field', *J. Atmos. Sci.* 30: 1657–1664

Chapter 27.

The Gulf Stream

In this chapter we will focus on the Gulf Stream, the main current in the North Atlantic Ocean. Notice however that each ocean has a similar current and the principles mentioned below can be applied to any of those ocean currents.

All the main ocean currents in the northern hemisphere, including the Gulf Stream (see figure 144) flow in clockwise loops, while the currents in the southern hemisphere flow counterclockwise. According to mainstream science, this phenomenon is exclusively due to the 'Coriolis effect'.

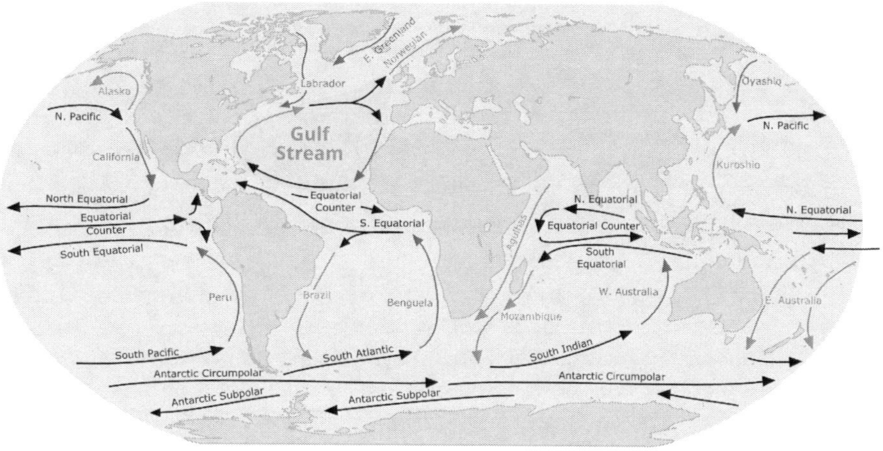

Figure 144: The ocean currents. light grey arrow: warm current. Dark grey arrow: cold current. (© L30nc1t0 – Wikimedia, adapted by Sott.net)

The Coriolis effect states that the movement of a fluid (water or air, for example) will be deflected to the right (facing north) if it occurs in the northern hemisphere. If the fluid is located in the southern hemisphere, then it will be deflected to the left (facing south), hence the spin direction of the oceanic currents. So, in the North Atlantic Ocean, water is deflected to its right, thus the clockwise spin of the Gulf Stream.

If the Earth's rate of rotation decreases, then the Coriolis effect should drop, since the **Coriolis effect is proportional to the rotation rate.**[527] This is the first factor behind a cosmically-induced slowdown of the Earth[528] leading to a weakening of the Gulf Stream. But another factor of an electromagnetic nature may also contribute to this weakening.

The North Atlantic Ocean, like any other water surface, is traversed by the atmospheric vertical E-field and by the Earth's magnetic field. Seawater is partly ionized; about 3.5% of its constituent molecules are in ionic form (mostly sodium and chloride).[529] Because of this ionization,[530] seawater

[527] Lappa, M., *Rotating Thermal Flows in Natural and Industrial Processes*, p. 20

[528] See Chapter 22: 'The slowdown of the Earth'

[529] 'Sea water', *Science Daily*. See: www.sciencedaily.com/articles/s/seawater.htm

[530] For comparison, pure water is a very poor conductor with resistance about one million times higher. See: 'Water

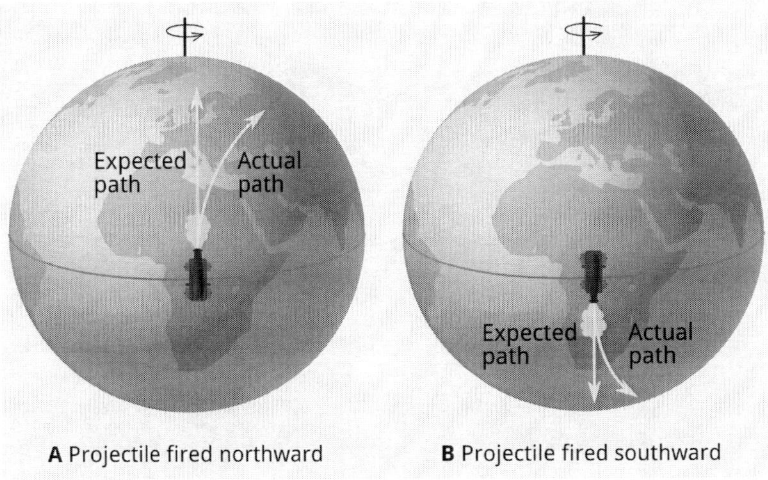

Figure 145: Representation of the Coriolis effect in the northern and the southern hemisphere.

is a very good conductor,[531] meaning that it carries an electric charge similar to the charge of the Earth's surface. This electric charge was illustrated by Peabody,[532] for example, who measured a 30 ampere current crossing the Panama Isthmus, between the Atlantic Ocean and the Pacific Ocean.

Having charged water molecules in movement within an electric field and a magnetic field results in a Lorentz force being exerted on the water. However, the electric field between the ionosphere and the Earth's surface is not perfectly uniform; it changes at different latitudes:

> The electrical conductivity of the atmosphere also exhibits a latitudinal variation owing to the shielding effect of the Earth's magnetic field. The magnetic field deflects incoming cosmic rays away from the equatorial region more effectively than from the polar regions, with the result that the **electrical conductivity is about 50 percent greater at high latitudes than at [low] latitudes.**[533]

Conductivity is the ability of a milieu to conduct electricity; the higher the atmospheric conductivity, the higher the vertical electric current traversing it. Therefore, this variation of atmospheric conductivity relative to the latitude has an influence on the strength of the Lorentz force. **The higher the latitude, the stronger the Lorentz force.**

As depicted in figure 146, at low latitudes, conductivity is low, and therefore the local electric current is weak (short black arrow in the middle of the Atlantic Ocean). Inversely, at a high latitude, conductivity is high; therefore the current is strong (long black arrow over Ireland). Meanwhile, the

conductivity', *Lenntech*. www.lenntech.com/applications/ultrapure/conductivity/water-conductivity.htm

531 About 0.2 ohm per meter.

532 Peabody, A. W., 'Considerations of telluric current effects on pipeline, in Solar System', *Plasma Physics*, pp. 349–352

533 Stephen, H., *Encyclopedia of Climate and Weather*, p. 67. This point is confirmed by two observations. First, lighting is more frequent around the equator. In this region, the atmosphere is less conductive, so the electric field can reach higher values and trigger massive discharges like lightning bolts. Second, auroras are more frequent near the poles. The huge quantity of charged particles carried by the solar wind during a solar storm enter the Earth's atmosphere more easily at high latitudes (where the atmosphere is more conductive) and partly ionize the air above the poles, hence the auroras.

electromagnetic force, represented by the vector B (grey arrows), is constant. Its intensity doesn't vary with latitude. The same applies in the southern hemisphere with stronger vertical atmospheric current at high latitude.

F (white arrows) represent the Lorentz force. Since it is proportional to the electric current that generates it, the **Lorentz force is stronger at high latitudes than it is at lower latitudes**. That's why the white arrow starting near Ireland is longer than the white arrow near the equator. The eastward Lorentz force exerted on the Atlantic waters is stronger at high latitudes and tends therefore to reinforce the Coriolis-induced clockwise water movement.

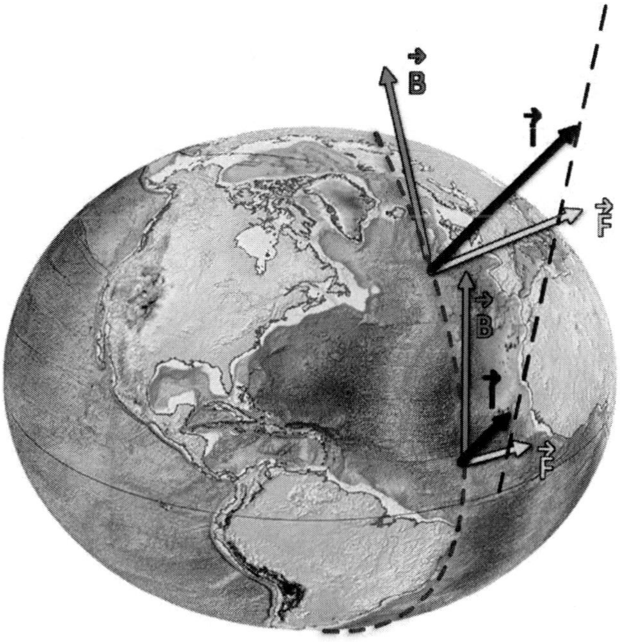

Figure 146: Variations in vertical atmospheric current and Lorentz force relative to the latitude. (© Sott.net)

Reciprocally, in the southern hemisphere, the atmospheric electric current is stronger at higher latitudes, therefore the Lorentz force is stronger near Antarctica than near the equator. Here again, the Lorentz force tends to reinforce the Coriolis-induced anti-clockwise water movement observed in oceans of the southern hemisphere.

The composite illustration in figure 147 also shows how solar activity affects the atmospheric vertical current, which in turn affects the Lorentz force, which has a direct influence on the Gulf Stream. On the left, solar activity is strong, inducing an intense atmospheric electric current, which in turn generates a strong Lorentz force that tends to increase the power of the Gulf Stream. On the right side, solar activity is weaker; consequently, the generated Lorentz force and the Gulf Stream are weaker.

The slowdown of the Gulf Stream plays a major role in terms of weather, because it provides an important amount of heat to the west coast of Europe and to the upper east coast of North America. The heat carried by the Gulf Stream is estimated to be 1.4 petawatts.[534] For comparison, in 2008 the total worldwide energy consumption was equivalent to an average power use of 15 terawatts.[535] So, at any instant, the **Gulf Stream carries approximately 100 times more energy than what the world population consumes in that year**.

Although affecting both continents, the effect of the Gulf Stream is more pronounced in Europe

---

[534] $1.4^{x}10^{15}$ W. See: 'Climate number: 1.3 petawatts', *Earth Gauge.* www.earthgauge.net/2012/climate-number-1-3-petawatts

[535] $1.504^{x}10^{13}$ W. See: 'World energy consumption', *Wikipedia.* en.wikipedia.org/wiki/World_energy_consumption

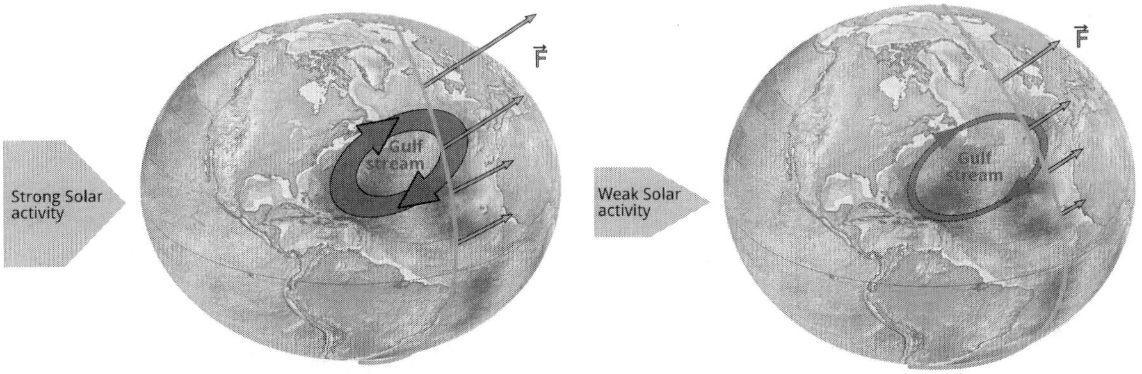

Figure 147: Effects of the latitude-dependent Lorentz force (F – grey arrows) on Gulf Stream intensity. (© Sott.net)

than in North America. When we compare regions of similar latitude, it appears that ocean heat storage makes winters in Western Europe 15 to 20° Celsius warmer than those in eastern North America (at equivalent latitudes).[536] For instance, New York and Madrid are located on the same latitude (40° north).[537] In February, the average temperature in New York is -1° Celsius,[538] while in Madrid it is 11° Celsius.[539]

The effects of the Gulf Stream are stronger in Western Europe for at least two reasons: First, the southward cold Labrador Current[540] squeezes between the eastern coast of North America and the Gulf Stream, pushing them apart. Therefore, the latter can only brush along the East Coast of North America, while it impacts fully on the whole Western European coast (figure 148).

Second, the dominant winds, driven by the northern jet stream (which we will examine soon[541]) flow eastward. Thus, the air heated up by the Gulf Stream rises above the Atlantic Ocean and is blown predominantly towards Europe.

The mainstream media have occasionally mentioned the 'erratic' behavior of the Gulf Stream. This was the case in 2004 when the Gulf Stream stalled for ten consecutive days,[542] and in January 2010 when it connected with the West Greenland Current after fluctuating for several weeks.[543]

But it was only in January 2013 that enduring weakening of the Gulf Stream was acknowledged by an international team of oceanographers led by Princeton University professor Tal Ezer, who demonstrated with 99.99% statistical confidence that the Gulf Stream had been continuously weakening since 2004:

---

[536] Seager, R., 'Climate mythology: The Gulf Stream, European climate and Abrupt Change', *Lamont-Doherty Earth Observatory of Columbia University*. www.ldeo.columbia.edu/res/div/ocp/gs/index.shtml

[537] '107 world cities by latitude', *Price of Travel* (19 April 2011). See: www.priceoftravel.com/1063/107-world-cities-by-latitude-things-line-up-in-surprising-ways/

[538] 'Climate of New York', *Wikipedia*. See: en.wikipedia.org/wiki/Climate_of_New_York

[539] 'Madrid time, date and weather', *Go Madrid*. See: www.gomadrid.com/madrid-weather.html

[540] See figure 144.

[541] See Chapter 28: 'Jet streams'

[542] Randerson, J., 'Sea change: why global warming could leave Britain feeling the cold', *The Guardian* (27 October 2004). See: www.theguardian.com/environment/2006/oct/27/science.climatechange

[543] 'Shutdown of thermohaline circulation', *Wikipedia*. See: en.wikipedia.org/wiki/Shutdown_of_thermohaline_circulation

Figure 148: Surface temperature in the North Atlantic. (© NASA)

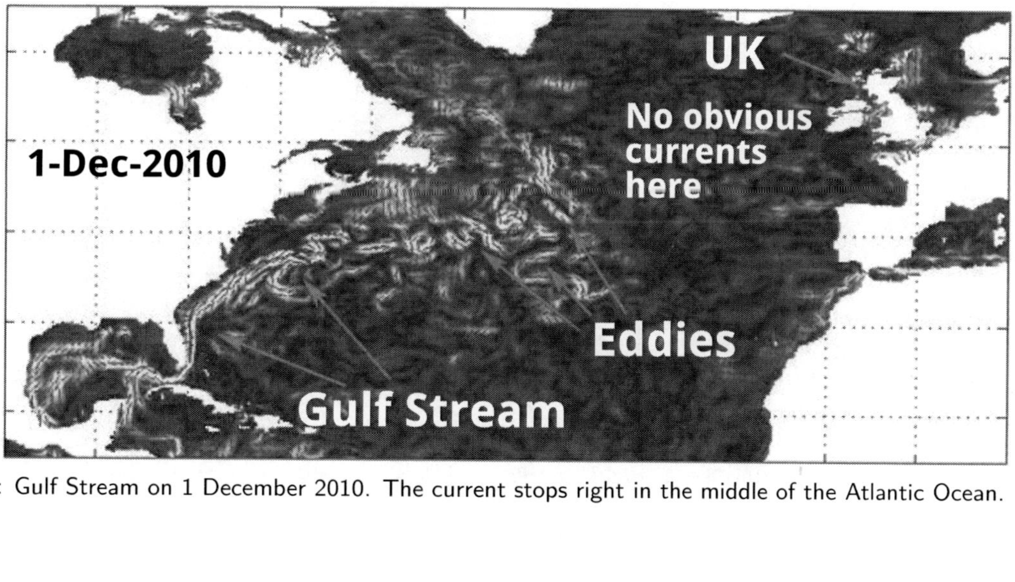

Figure 149: Gulf Stream on 1 December 2010. The current stops right in the middle of the Atlantic Ocean. (© DEOS)

> Recent studies indicate that the rates of sea level rise (SLR) along the U.S. mid-Atlantic coast have accelerated in recent decades, possibly due to a slowdown of the Atlantic Meridional Overturning Circulation (AMOC) and its upper branch, the Gulf Stream (GS). The coastal sea level variations were found to be strongly influenced by variations in the GS on time scales ranging from a few months to decades. It appears that the **GS [Gulf Stream] has shifted from a 6–8 year oscillation cycle to a continuous weakening trend since about 2004**, and that this trend may be responsible for recent acceleration in local SLR.[544]

For the scientific community, the causes of this weakening are unknown, as stated by a NOAA[545] representative:

> 'Why did the Gulf Stream slow down? Why did the fall wind pattern appear earlier?' NOAA's Edwing said. **'We don't have those answers.'**[546]

From what we've seen above, both the Lorentz force and the Coriolis effect are drivers of the Gulf Stream. Decreased solar activity reduces the Coriolis effect (by reducing the Earth's spin rate) and also weakens the Lorentz force (by reducing the atmospheric vertical current). These are probable causes for the recently acknowledged weakening of the Gulf Stream.

Below temperate latitudes, global cooling will be worsened by the weakening of ocean currents. The most affected regions will be the ones fed with warm ocean currents coming from the intertropical region: west coasts in the northern hemisphere (Western Europe, for example) or east coasts in the southern hemisphere (Argentina, for example).

---

[544] Ezer, T., *et al.*, 'Gulf Stream's induced sea level rise and variability along the U.S. mid-Atlantic coast', *Journal of Geophysical Research: Oceans* 118

[545] U.S. National Oceans and Atmosphere Administration

[546] Handwerk, B., 'Sea Levels Rose Two Feet This Summer in U.S. East', *National Geographic* (10 September 2009). See: news.nationalgeographic.com/news/2009/09/090 910-sea-levels-rise.html

Chapter 28.

Jet streams

Introduction

Now that we have covered oceanic currents we will focus on certain air 'currents', also known as 'jet streams'. There are, in fact, five jet streams of air blowing around the globe: two polar, two subtropical and one equatorial. Jet streams are fast-flowing (between 60 and 250 mph),[547] narrow (a few hundred miles wide and less than three miles tall) and high-altitude (about 7 miles) winds.[548] In this chapter, we're going to focus exclusively on the northern polar jet stream, also called the 'Arctic front' jet stream, an important driver of weather in North America and Europe.

As shown in figure 150, the northern polar jet stream (black arrow) whirls around the North Pole from west to east and defines a boundary between cold air (up north) and warm air (down south). If you are south of the Arctic jet stream you will enjoy a temperate climate; if you are north of the jet you will experience polar weather conditions.

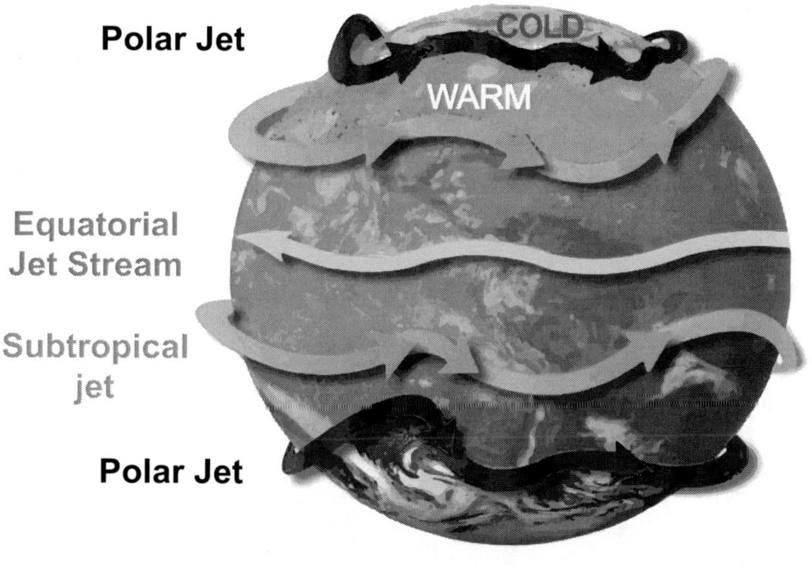

Figure 150: Schematic depiction of the five jet streams circling Earth. (© Sott.net)

As depicted in figure 151, the location, thickness, path and speed of the jet stream are all variable.[549] Sometimes, the jet stream blows at very high speeds, making the current fairly straight (left image). At other times, it slows down and charts a twisting path, making deep incursions towards the equator and far north towards the North Pole (right image). In fact, the jet stream's behavior is quite

[547] Between 90 and 400 km/h. See: 'Glossary: Jet Stream', *American Meteorological Society.* glossary.ametsoc.org/wiki/Jet_stream

[548] 'FAQ about the Jet Stream', *NOVA Online.* See: www.pbs.org/wgbh/nova/balloon/science/jetstream.html

[549] Newton, *Atmospheric circulation systems: Their structure and physical interpretation*, p. 216

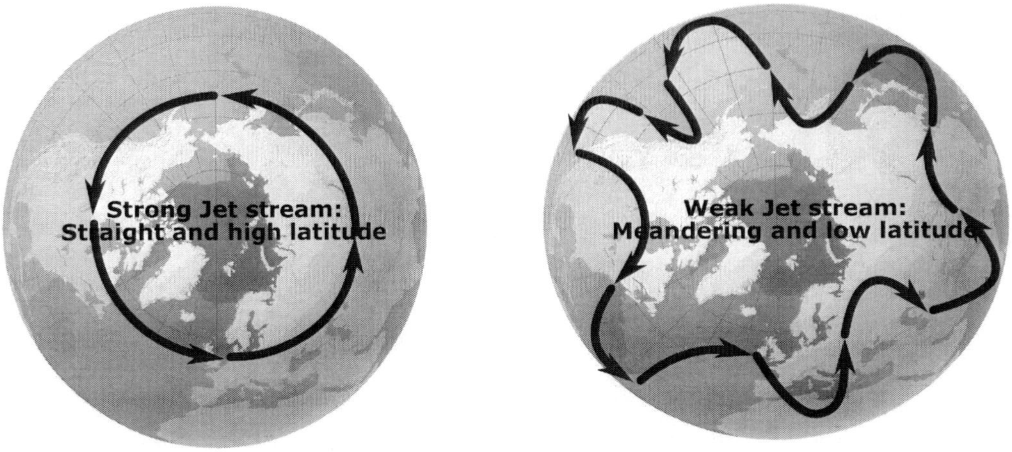

Figure 151: Two different jet stream configurations. (© Sott.net)

similar to that of a river: when its current is weak it starts to hesitate, meander and sometimes even stop (also known as 'jet stream blocking').[550] When the stream is strong, it circulates at high speed in a fairly straight direction.

Electric jet stream

For years, mainstream science claimed that the jet stream played a minor role in weather and was simply the result of 'frontogenesis', a complex process describing how convection (air movements) and temperature differences lead to jet stream circulation.[551]

Recently however, things have changed. Winter 2013–2014 was very harsh in North America[552] [553] with a persistent 'polar vortex' that lasted for weeks and whose shape followed almost exactly the 'cold drop' formed by the meandering Arctic jet stream pictured above. The evidence was quite obvious, so the narrative had to evolve:

> Scientists are trying to understand if the unusual weather in the northern hemisphere this winter ... is linked to climate change. One thing seems clear: **Shifts in the Jet Stream play a key role** and could become even more disruptive... Everyone is blaming the **jet stream, which drives most weather in mid-latitudes**. That would be a significant development. For what happens to the Jet Stream in the coming decades looks likely to be the key link between the abstractions of climate change and real weather we all experience.[554]

[550] Master, J., 'Arctic sea ice loss tied to unusual jet stream patterns', *Weather Underground* (2 April 2012). See: www.wunderground.com/blog/JeffMasters/arctic-sea-ice-loss-tied-to-unusual-jet-stream-patterns

[551] 'Air pressure and wind', *Eastern Illinois University*. See: www.ux1.eiu.edu/~cfjps/1400/pressure_wind.html

[552] For example on January 6th, 2014, the average daily temperature for the United States was 17.9 °F (-7.8 °C). The last time the average for the country was this low was January 13th, 1997. This 17-year gap is the longest on record. See: Borenstein, S., 'Weather wimps?', *Salisbury Post* (January 10, 2014)

[553] Dozens of cold-weather records were broken since 1870, when data began to be collected by the U.S. Weather Bureau. See: Livingston, I., 'Polar vortex delivering D.C.'s coldest day in decades, and we're not alone', *Washington Post* (7 January 2014)

[554] Pearce, F., 'Is weird winter weather related to climate change?' *Yale Environment 360* (24 February 2014).

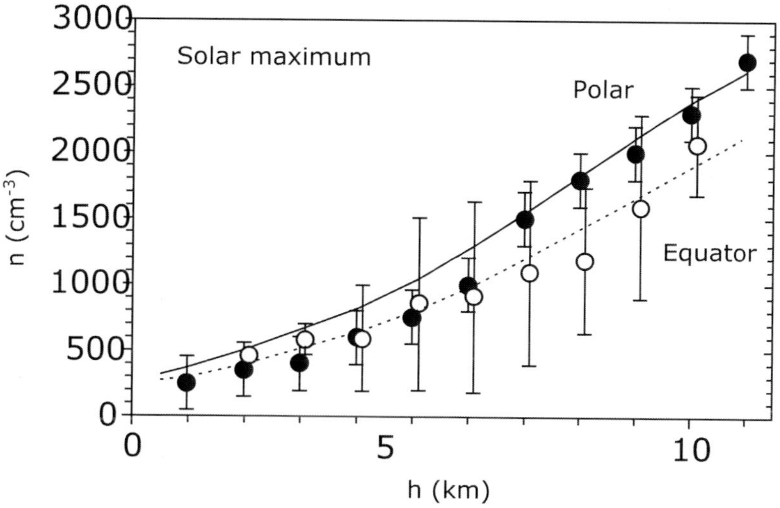

Figure 152: Ion concentration relative to altitude. Equator: white circles, Pole: black circles. (© Lehmacher and Offermann)

This analysis might indeed be closer to the true state of affairs. Evidence suggests that the jet streams are not powered by heat from the lower atmosphere but by electricity. At the altitude of the jet stream (about 10 km high), the atmosphere contains about 2,000 ions per cubic centimeter.[555] Air at this altitude is about 10 times more ionized than at ground level (figure 152).

In many respects the jet stream is similar to the Gulf Stream. However, the Gulf Stream is spread over a wide latitudinal range (equator to pole) and located in an enclosed space, so the effects of the Lorentz force and the Coriolis effect trigger a looping movement. **The jet stream is not limited by coasts and its latitudinal range is very narrow (a few kilometers), so no Coriolis effect applies to it and the Lorentz force freely propels it eastwards all around the globe.**

From this perspective, the **jet stream is not a consequence of lower atmospheric phenomena but, on the contrary, is a *driver* of lower atmospheric phenomena**, i.e. the weather we experience day to day. Physicist James McCanney describes a positively charged northern jet stream and its easterly flow as follows:

> The giant gas 'planets' and the Sun had many counter-rotating rings of electrical currents that are just like the three 'jet streams' [polar front, subtropical and equatorial] that circle Earth. The jet stream electrical currents snake around the high and low pressure cells as they wind their way eastward or westward. These reacted in kind to solar electrical storms and in turn drove the lower level weather systems.
>
> As the solar wind blows past Earth, the solar wind particles interact with the small permanent magnetic field. The electrons are forced to bend outwards and around Earth in one direction and the protons move in the other direction. **This flow causes the Earth's ionosphere to**

See: e360.yale.edu/feature/is_weird_winter_weather_related_to_climate_change/2742/

[555] Usoskin, I. G., *et al.*, 'Cosmic ray-induced ionization in the atmosphere: spatial and temporal changes', *Journal of Atmospheric and Solar-Terrestrial Physics* (December 2004) 66(18): 1791–1796

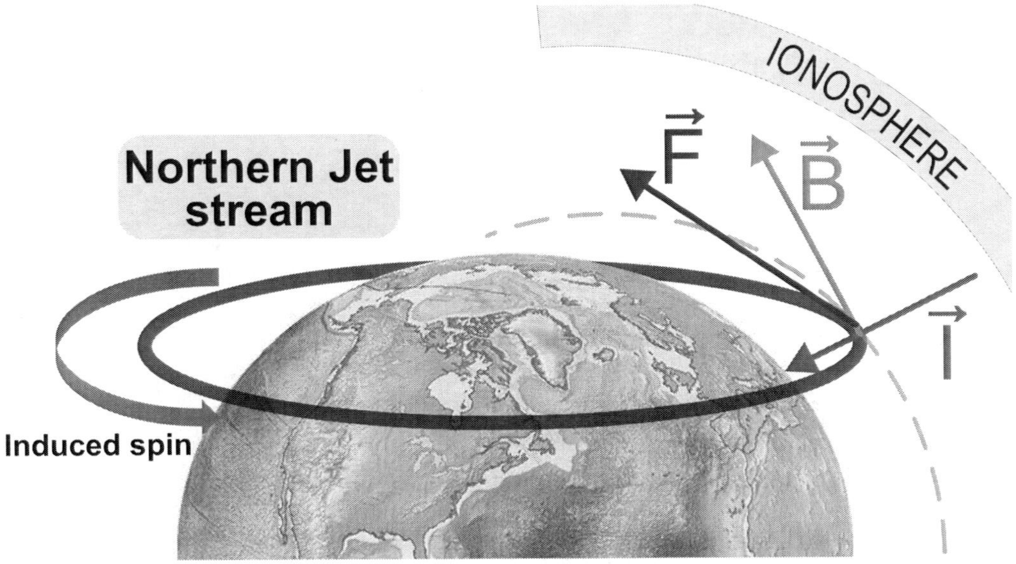

Figure 153: The Lorentz force (F) is the electromotive force that propels the jet stream. (© Sott.net)

> **set up three distinct current flows that make the major jet streams of our upper atmosphere.**[556]

As depicted in figure 153, the conjunction of the Earth's magnetic field (B – light grey arrow) and the atmospheric vertical current (I – dark grey vector) exerted on the charged molecules of the troposphere generate a Lorentz force (F – black arrow) that acts as an electromotive force powering the rotation of the Arctic jet stream (black torus).

Notice in figure 150, which depicts the five jet streams circling our planet, that the equatorial stream flows westwards (unlike the 4 other streams). This, posits McCanney, is because **the equatorial stream carries a negative charge, whereas the other streams carry an overall positive electric charge**:

> I further realized that Earth's jet streams were powered by the solar electrical currents, and that there were three belts of electrical current circling the globe. Two 'ion' current sheets traveled from west to east at higher latitudes (in the northern and southern hemispheres), while an equatorial 'electron' current sheet ran east to west.[557]

Similarly to the Gulf Stream, when solar activity is strong, the electrically driven jet stream blows at high speed and follows a straightforward path (light grey arrow in figure 154). If solar activity is weak, the ionosphere's electric charge decreases, and so do the atmospheric electric current and the subsequent Lorentz force. As a consequence, the jet stream slows down and starts meandering (dark grey arrow).

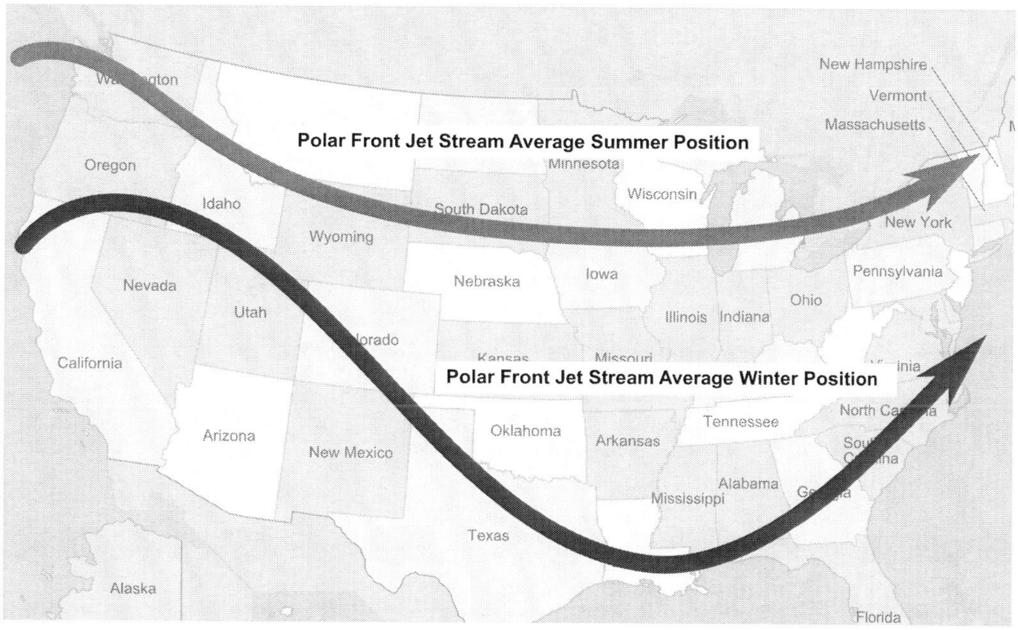

Figure 154: Typical winter/summer jet stream patterns over North America. (© Sott.net)

Latitude shift

Figure 154 shows that, in addition to meandering (dark grey arrow), a 'weak' jet stream also experiences an overall shift to the south. Typically, the jet stream blows roughly along the 45° north latitude in winter (light grey above) and around 60° north in summer (dark grey arrow), but in recent years, we've seen more and more anomalies, with a weaker-than-usual jet stream meandering and drifting further south (and subsequently exposing temperate latitude to 'polar vortices').[558]

The charts in figure 155 show recent historical trends in the jet streams. The curves on the left show latitude anomalies. The curves on the right show speed anomalies. The two black curves titled 'NH' relate to the northern hemisphere jet stream (Arctic stream). On the left chart, you can see that after 1999 (vertical light grey line), the Arctic jet stream moved closer to the equator (black curve). On the right chart, around 1998 (vertical light grey line), the Artic jet stream showed a drop in speed (black curve). The Antarctic jet stream (SHP – light grey curve) and the southern tropical jet stream (SHT – medium grey curve) show a similar reduction in speed after 1999 (figure 155).[559]

So since about 1998, *when the solar activity started to drop*, the Arctic jet stream has shown signs of weakness (lower speed and more southerly location).

Jet stream latitudinal oscillations have been acknowledged by mainstream science for years. They

556 McCanney, J., *Planet X, Comets and Earth Changes*, p. 72
557 *Ibid.*, p. 71
558 Archer, C. L. & Caldeira, K., 'Historical trends in the jet streams', *Geophysical Research Letters* 35
559 *Ibid.*

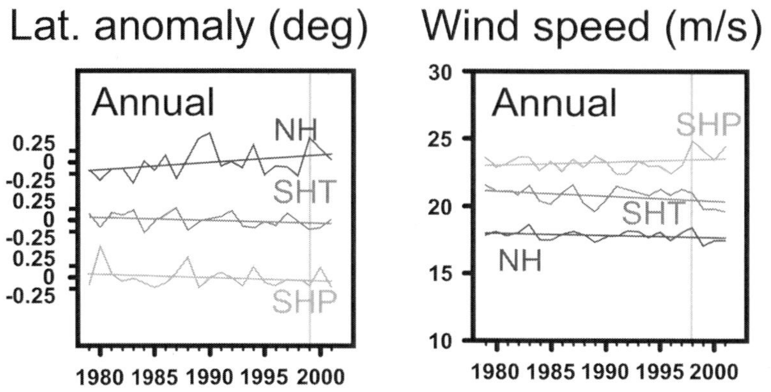

Figure 155: Historical trend of the jet streams (1975–2006). (Adapted from Archer et al.)

are allegedly due to changes in the Arctic oscillation.[560] So far, no convincing explanation has been provided for the causes of this 'oscillation'. However, if the electric nature of our solar system is taken into account, shifts in the jet stream begin to make sense. McCanney has suggested an explanation for the correlation between southerly shift of the jet stream and solar activity in the following terms:

> ... **these magnetic fields became compressed as 'solar storms' hit**, causing the transverse electric field to compress, which in turn compressed our ionosphere.[561]

Christopher T. Russell, a geophysics researcher at UCLA, has described a similar phenomenon.[562] Russell's model is depicted in figure 156. It involves planet Venus, but the fundamentals are similar for Earth, as demonstrated by geophysicist H.G. Zhuang.[563]

Russell's diagram (figure 156) shows a deformation of the ionosphere under pressure from the solar wind (grey arrow). However, this deformation is not uniform: the ionosphere is more compressed where the solar winds 'hit' (close to the equator) than above the polar regions.

Thus, solar activity induces a compression in the ionosphere, which is more prominent at the level of the equator. This equatorial compression of the positively charged ionosphere 'pushes' the positively charged polar jet stream towards the north. Conversely, **weak solar activity 'decompresses' the ionosphere at the level of the equator and allows the Arctic jet stream to move south.**

This explanation could also account for the latitude difference of the jet stream in winter and summer. In summer, the northern hemisphere faces the Sun, so the northern ionosphere is more compressed by solar winds and the jet stream is pushed northwards. In winter, when the northern

[560] 'Arctic oscillation' stands for a natural seesawing of atmospheric pressure between the Arctic and mid-latitudes of the North Pacific and North Atlantic Oceans. See: Mitchell, T., 'Arctic Oscillation time series', *JSAO*. jisao.washington.edu/data/aots/

[561] McCanney, *op. cit.*, p. 70

[562] Russell, C. T. & Vaisberg, O., 'The interaction of the solar wind with Venus', *Venus* (1983), pp. 873–940

[563] Zuang, H. C., *et al.*, 'The influence of the interplanetary magnetic field and thermal pressure on the position and shape of the magnetopause', *Journal of Geophysical Research* (1981) 86(A12): 10 009–10 021

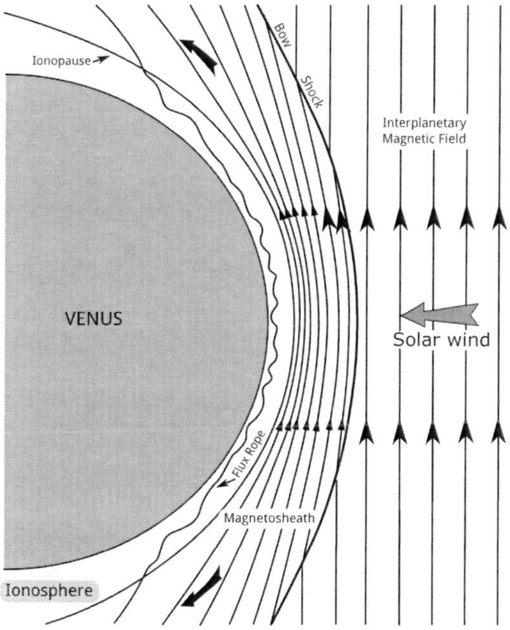

Figure 156: Solar winds impacting Venus's ionosphere. (Adapted from Russell)

hemisphere receives less solar wind, the ionosphere above the northern hemisphere relaxes and enables the jet stream to move back southward. (See figure 157.)

Therefore, if solar activity is weak, the jet stream should be observed at abnormally low latitudes. This is what has happened in recent years, particularly over Europe, with the jet stream as low as 15° north in winter (above North Africa) when it should be around 60° north (above Scotland).[564]

A low-latitude jet stream means that the Arctic air located to the north of the jet stream can reach lower latitudes, in particular the 'temperate' regions. This overall shift to the south exposes low-latitude regions (southern U.S., southern Europe) to very cold conditions. In this way, *a lasting decrease in solar activity would induce an overall cooling* of the 'temperate' latitudes that would be increasingly less separated from Arctic air by a more frequently and abnormally south-shifting polar jet stream. This could be an aggravating factor in the quick onset of an ice age.

Meandering jet stream

Low solar activity pushes the jet stream south, but it also makes it slow and meandering. The jet stream can even 'stall'. This is also called a 'blockage'. The correlation between weak solar activity and jet stream blocking was demonstrated in 2008.[565] A blockage prevents the moist and relatively warm air above the ocean from reaching Western Europe and heating that part of the continent.

[564] "Since 2007, and particularly in 2012 and early 2013, the jet stream has been at an abnormally low latitude across the UK, lying closer to the English Channel, around 50°N rather than its more usual north of Scotland latitude of around 60°N." 'Jet Stream, Long-Term climatic changes', *Wikipedia*. See: en.wikipedia.org/wiki/Jet_stream

[565] Barriopedro, D., Garcıa-Herrera, R., & Huth, R., 'Solar modulation of Northern Hemisphere winter blocking', *J. Geophys. Res.*

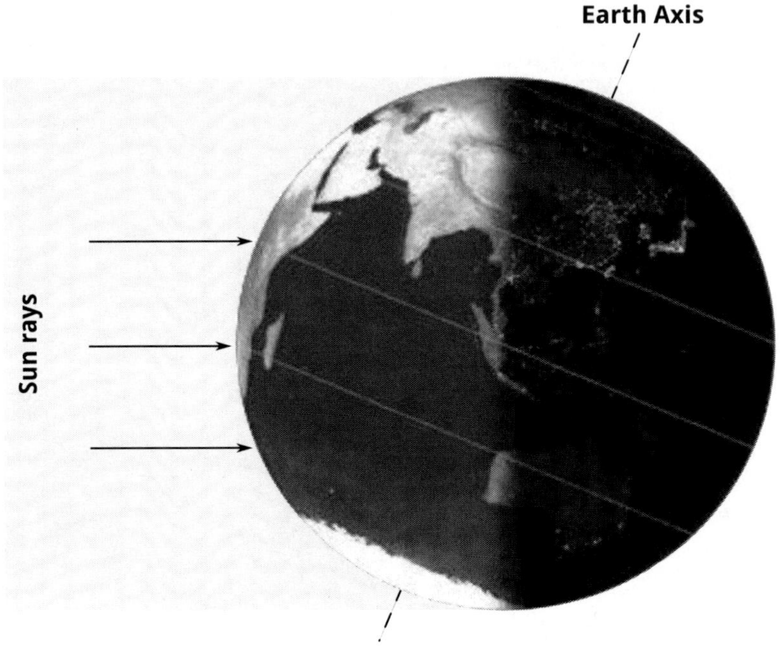

Figure 157: In winter, the northern hemisphere is tilted away from the Sun. The reduced solar 'pressure' decompresses the ionosphere. (© Polartrec.com)

It also stops separating the relatively warm air of the temperate latitudes from the cold air of the Arctic latitudes. The encounter of those masses of air exhibiting very different temperatures can lead to extreme weather phenomena like massive rainfall, snowfall and violent winds.

Instead of neatly separating cold regions from hot regions and providing fairly stable and predictable weather patterns, the meandering jet stream creates heterogeneous air regions that manifest locally in a succession of hot and cold air pockets. This succession of warm, moist air and dry, cold air is an important cause of weather perturbations (anticyclones, warm fronts, depressions, cold fronts). It generates unstable situations such as 'cold drops' surrounded by hot air and, conversely, 'hot drops' surrounded by cold air. In 2009, 2010, 2011 and 2012, Europe experienced 'abnormal' cold waves while North America experienced 'abnormal' cold waves in 2013 and in 2014.[566] During winter 2013–2014 the UK experienced flooding not seen for more than 250 years[567] and North America experienced the previously mentioned record cold temperatures.

A similar weak meandering jet stream configuration occurred in **February 2011, when more than half the northern hemisphere was covered in snow** (figure 158).

During this anomaly, the jet stream was exhibiting relatively low speeds, a meandering path and a quasi-'blockage', as shown in figure 159 depicting the jet stream over North America on February 2nd, 2011. The path of the jet stream is colored in gray. You can see that it meanders a lot,

[566] 'Cold Waves – Contemporary cold waves', *Wikipedia.* See: en.wikipedia.org/wiki/Cold_wave

[567] UK experienced its wettest winter since records began in 1766. See: Vaughan, A., 'England and Wales hit by wettest winter in nearly 250 years', *The Guardian* (27 February 2014). theguardian.com/uk-news/2014/feb/27/england-and-wales-hit-by-wettest-winter-in-nearly-250-years

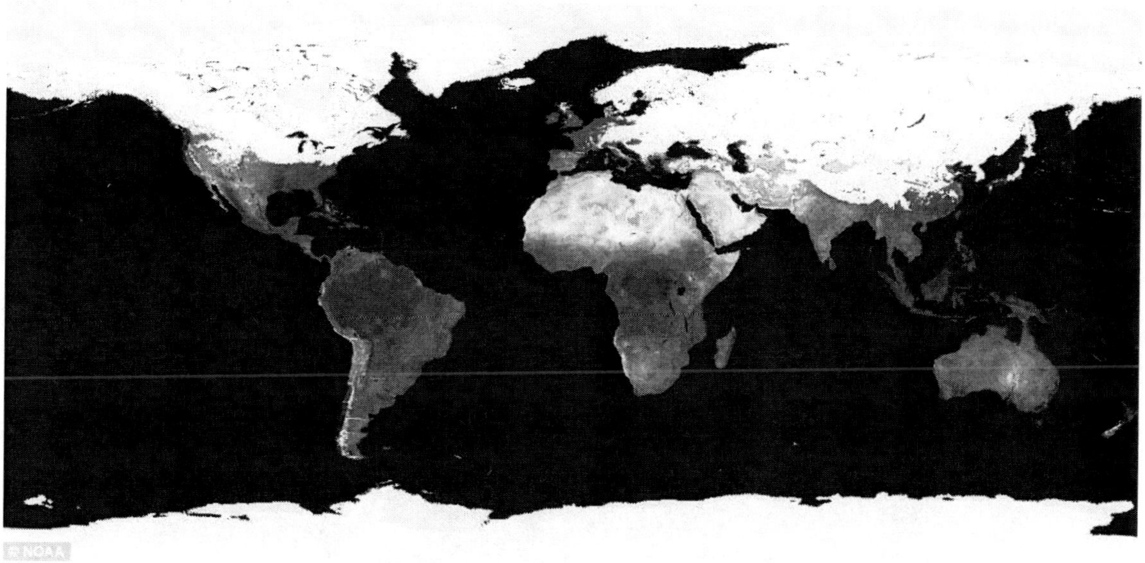

Figure 158: Satellite image showing the extent of snow cover (February 2nd, 2011). (© NOAA)

with its latitude varying between 70° and 20° north, while the section flowing above central U.S. literally moves 'backwards' (a pronounced westward direction instead of the usual strong eastward direction). The lower latitude reached by the jet stream is abnormally low (down to almost 20° north over Mexico).

The shades of gray (figure 159) are proportional to the speed of the jet stream (see scale at the bottom of the image). Most of the jet stream is medium gray, which corresponds to a speed of approximately 70 knots (80 mph). This figure is relatively low compared to the typical winter speed of the jet stream, which is about 110 knots (126 mph).[568] Above Alaska, it is experiencing a quasi-'blockage', where it is narrow and its color dark gray (low speed). This means that in this area the jet stream circulation is much reduced; it is almost 'stalling'.

The area south of the jet stream is in dark grey to represent the mass of warm air, while the area north is white to represent the cold Arctic air. A region exposed to a meandering jet stream will experience alternating warm moist air (dark grey) and cold dry air (white). Typically, when warm moist air follows or encounters a mass of cold air, it cools down and humidity condenses, leading to rainfall or even snowfalls. The higher the temperature difference and the faster the temperature change, the more dramatic the rainfalls and snowfalls. As a result, on February 1st and 2nd, 2011, more than half of the northern hemisphere was covered in snow, with some areas – Illinois for example[569] – witnessing two feet of snow.

Notice the arrows representing wind speed and direction in figure 159. In the white area, the arrows have an overall southerly direction, indicating that half the U.S. experienced a massive draft

[568] Swanson, B., 'Wind and Jet Stream', *USA Today* (27 June 2007).
See: usatoday30.usatoday.com/weather/resources/askjack/wawind.htm

[569] 'February 2nd, 2011 North American blizzard', *Wikipedia*. See: wikipedia.org/wiki/January_31_%E2%80%93_February_2,_2011_North_American_blizzard

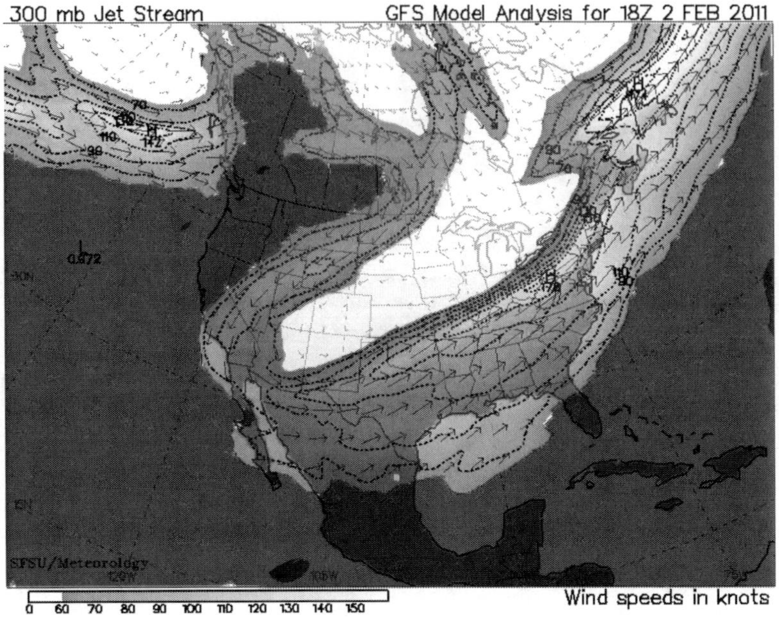

Figure 159: Jet stream over North America (February 2nd, 2011). (© SFSU)

of freezing air coming directly from the Arctic (polar vortex).

In regions like Europe and North America, global cooling will be worsened by the weakening of the jet stream and, as a consequence, it will shift towards a more southerly path, exposing usually temperate latitudes to Arctic air.

Chapter 29.

Strange sounds

Over the last few years, a number of 'strange sky sounds' (variously described as trumpet-like, metallic, hissing, cracking, buzzing and swishing noises) have been reported all over the planet.[570] Their origin is unclear, but eyewitnesses invariably describe them as emanating 'from the sky' or 'from the ground'. They are brief, lasting just a few minutes at a time, and they are localized, though they appear to occur in diverse locations simultaneously or in clusters.

The first 'strange sound' seems to have been reported in 2009,[571] but the phenomenon only really grabbed people's attention in 2011 when a strange sound was recorded in Kiev, Ukraine,[572] on August 11th.[573] Since then, hundreds of videos of these sounds have been published on the Internet. An amateur researcher has listed 187 strange sound reports on his personal website[574] (see figure 160).

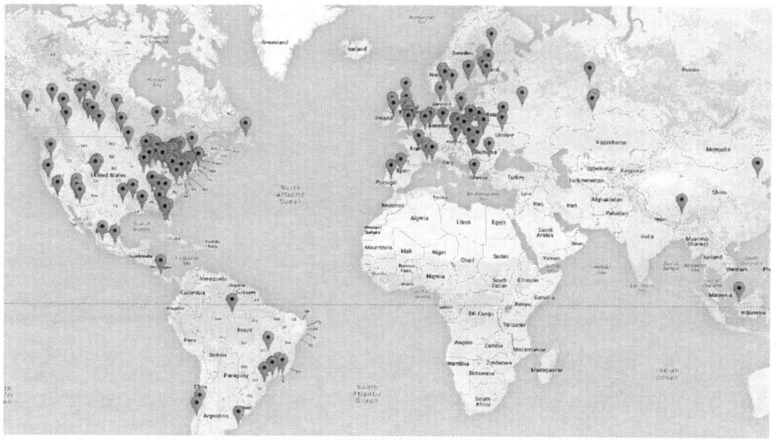

Figure 160: The 'strange sounds map'. (© seektress.com)

One of the more notable instances of these strange sound phenomena happened on August 14th, 2012, during a televised baseball game at Tropicana Field in Tampa, Florida.[575] Hundreds of thousands of supporters and TV viewers directly witnessed a ghostly noise that lasted for more than a minute.

[570] Quinn, J., 'Strange Noises in the Sky: Trumpets of the Apocalypse?', *Sott.net* (29 January 2012). See: www.sott.net/article/240 826-New-Sott-Report-Strange-Noises-in-the-Sky-Trumpets-of-the-Apocalypse

[571] *Ibid.*

[572] Ortega, X. 'More Unexplained Sounds, Tremors & Lights From The Skies', *ghosttheory.com* (20 August 2011). See: www.sott.net/article/233 902-More-Unexplained-Sounds-Tremors-Lights-From-The-Skies

[573] Interestingly, this sound occurred right at the peak of the Perseid meteor showers (July 17–Aug. 24). See: 'How to see the best meteor shower of the year', *Asteroid Watch – NASA* (31 December 2011). www.jpl.nasa.gov/asteroidwatch/meteor.cfm

[574] 187 events between September 9, 2008, and September 3, 2013. See: Kramer, L., 'Strange Sounds Maps', *Seektress* (21 January 2012). seektress.com/ssounds.htm

[575] Koehli, H., 'The Cs Hit List 06: Let's Do the Planetary Twist to the Tune of the Brothers Heliopolis', *Sott.net* (29 February 2012). See: www.sott.net/article/242 280-The-Cs-Hit-List-06-Lets-Do-the-Planetary-Twist-to-the-Tune-of-the-Brothers-Heliopolis

Figure 161: The Tropicana Field where a lasting eerie sound was heard on August 14th, 2012. (© Wikimedia Commons)

The mainstream and alternative media have variously labeled this phenomenon a hoax,[576] 'unexplained', the result of government secret weapons, the activation of HAARP or the by-product of construction work on underground bases.

Fake videos, with simulated or copied audio recordings of some legitimate events, have also been doing the rounds, injecting a lot of 'noise' into analyses of the phenomenon.[577] One of the few scientists quoted on the topic in the media is Elchin Khalilov.[578]

Khalikov provided the following explanation for the strange sounds that have been reported all over the globe:

> In our opinion, the source of such powerful and immense manifestation of acoustic-gravity waves must be very large-scale energy processes. These processes include powerful solar flares and huge energy flows generated by them, rushing towards Earth's surface and destabilizing the magnetosphere, ionosphere and upper atmosphere. Thus, the effects of powerful solar flares: the impact of shock waves in the solar wind, streams of corpuscles and bursts of electromagnetic radiation are the main causes of generation of acoustic-gravitation waves following increased solar activity.
>
> **Given the surge in solar activity as manifested in the higher number and energy of solar flares since mid-2011**, we can assume that there is a high probability of impact of

[576] Some debunkers cite the inspiration for the 'hoax' being the release of Kevin Smith's film, *Red State*, which premiered in September 2011, just weeks after the Kiev video emerged, and which features similar apocalypse-inspired trumpet sounds.

[577] Boyett, J., 'Are the 'Strange Sounds' Videos a Hoax?', *Huffington Post* (8 February 2012). See: www.huffingtonpost.com/jason-boyett/are-the-strange-sounds-vi_b_1258 049.html

[578] Azerbaijani geophysicist who has published over 250 scientific works and more than 300 popular science articles. Khalikov's analysis has been covered in international news magazines and encyclopedias, including the website of NATO International Program.

Figure 162: Azerbaijani geophysicist Elchin Khalikov.

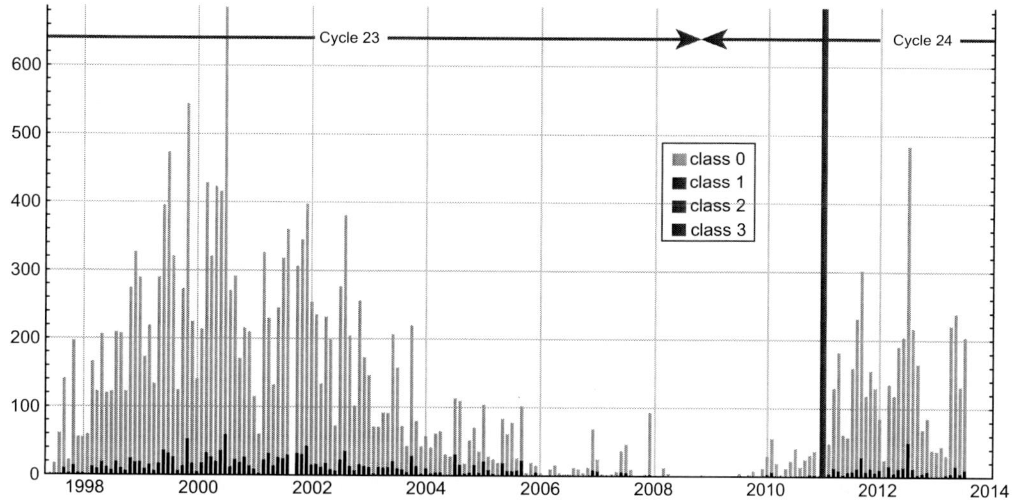

Figure 163: Monthly count of solar flares, 1998–2013. (© Solar Influences Data Analysis Center)

> the substantial increase in solar activity on the generation of the unusual humming coming from the sky.[579]

While Khalikov is certainly onto something, he's also leading the reader astray by blaming these strange sounds on "the surge in solar activity as manifested in the higher number and energy of solar flares since mid-2011."

Note that Khalikov refers to solar flares, not sunspots. So let's check solar-flare activity in recent years.

As shown in figure 163, solar flare activity has (timidly and erratically) increased since 2011 (black vertical line), although, as we've already seen,[580] we're in an unusually weak solar cycle (SC24) that was preceded by another weak cycle before it (SC23).

---

[579] 'Strange Sounds in Sky Explained by Scientists', *Sott.net* (28 January 2012). See: www.sott.net/article/240 863-Strange-Sounds-in-Sky-Explained-by-Scientists

[580] See Chapter 14: 'Variations in solar activity'

Strange sounds have been reported since 2009. Since then, there have been between 0 and 450 solar flares per month. If, as stated by Khalikov, the increase in flare activity is the cause of strange sounds, why then were they not reported in 2000 when the monthly solar flare activity ranged between 400 and 800?

In addition to pointing to the wrong cause, Khalikov does a major disservice to people by implying that the Sun is unusually active at this time when the truth is just the opposite.

However, "[the destabilization of] the magnetosphere, ionosphere and upper atmosphere" mentioned by Khalikov seems to be a valid and useful observation. These 'strange sounds' do appear to be some form of **electrophonic phenomenon**.

These are not acoustically propagated booms and rumbles, such as those heard seconds or minutes after the sighting of a fireball. Also, electrophonic sounds shouldn't be confused with electrophonic hearing, which is due to the passage of an electric current (of a particular frequency and intensity) through the human body.[581]

Electrophonic sounds were first described by astronomer Edmund Halley[582] who collected accounts of the large meteor fireball that was observed over England in 1719. Numerous witnesses reported hissing sounds as the fireball passed, as if it had been very close by.[583] But according to Halley's triangulation calculations, the fireball flew 60 miles overhead. At this altitude, **sound takes about five minutes to reach the Earth's surface**. So how was it possible to explain the fact that witnesses **simultaneously** heard and saw this meteor? Halley dismissed this as sheer fantasy on the part of the eyewitnesses', due to 'an affrighted imagination', and his conclusion became scientific consensus for centuries, despite recurring observations of electrophonic sounds accompanying meteor fireballs.

During the 1940s, some scientists started reconsidering the problem in terms of physics, and by the 1980s, Australian physicist Colin Keay demonstrated that fireballs can indeed produce electrophonic sounds. Keay found that, in addition to generating light, fireballs also emit very low frequencies (VLF),[584] **which travel at light speed**,[585] hence the simultaneous visual and auditory perceptions reported above. In 1988, Watanabe *et al.* achieved the first detection of a VLF signal from a meteor fireball.[586]

These VLFs are due to turbulences generated in the geomagnetic field (i.e. Earth's magnetic field) by the magnetic influence of the fireball. This is not a mechanical phenomenon, but an electromagnetic one: the fireball not only creates turbulence in the surrounding air (i.e. a sonic boom), but also electromagnetic radiation, which creates turbulence in the surrounding geomagnetic field (VLF).[587] VLF *radio* waves cannot be heard by humans, of course. However, VLF *sound* waves

---

[581] Gordon, F., 'Effect of Different Types of Electrodes in Electrophonic Hearing', *J. Acoust. Soc. Am.* 24(4): 447–448

[582] 1656–1742. English astronomer, geophysicist, mathematician, meteorologist and physicist who is best known for having calculated the orbit of the eponymous Halley's Comet.

[583] 'Listening to Leonids', *NASA* (26 November 2001).
See: science1.nasa.gov/science-news/science-at-nasa/2001/ast26nov_1/

[584] Radio frequencies in the 3 kHz to 30 kHz range, with wavelengths between 10 and 100 kilometers.

[585] 299,792 km/s.

[586] Christou *et al.*, 'Comparative studies of meteoroid-planet interaction in the inner solar system', *Planetary and Space Science* (2007) 55(14): 2049–2062

[587] Keay, C., 'Progress in Explaining the Mysterious Sounds Produced by Very Large Meteor Fireballs', *Journal of Scientific Exploration* (1993) 7(4): 337–354

match part of the normal human auditory range.[588]

The electrophonic sounds produced by meteors are usually reported as a hissing or crackling sound, but they can cover a whole variety of sounds and frequencies.[589] Keay demonstrated in lab experiments how meteors could generate VLFs and also how these **VLFs could then be transduced into audible sounds.**[590] **The human body itself can act as a transducer, but external transduction enabled by a nearby object** (like a pair of glasses or an antenna, for example) is more effective than transduction occurring within the ear.[591]

Colin Keay, Australian physicist and astronomer. Keay is the father of geophysical electrophonics. (© vicskeptics.com)

Also, the previously mentioned 'opening up' of the Earth[592] can be the source of electrophonics. Most of the *Earth's crust can become highly conductive* if subjected to mechanical stress/shock, and this high conductivity might very well produce turbulence in the electromagnetic field of the Earth, both underground and in the atmosphere. Maybe when rocks 'wake up', they don't just sparkle and glow as indicated by Freund[593] but, under the right circumstances, they also 'sing'?

This would explain why numerous testimonies state that the sound 'came from ground level', or from several locations, or from a nearby location, or from the sky, or 'everywhere and nowhere'. Keay also mentions that people's sensitivity to electrophonic sounds is variable,[594] perhaps explaining why some people report hearing an electrophonic sound and others don't, despite being present in the same location.

Some fundamentalist Christians identify these sounds with the 'Trumpet of Jericho'.[595] Actually, this last explanation may hold a grain of truth, particularly if the story of Jericho records a mythicized meteoric event. In such a scenario, meteoric electrophonic sounds might have eventually come to be remembered as 'trumpet-like', and the subsequent destruction of the walls of the city due to a cometary bombardment–induced earthquake and fire might have later been transformed in the biblical narrative into an attack by the Israelite army.

The Old Testament's Jericho is not an isolated example. From John's 'Revelations'[596] to Ovid's

[588] The human ear picks up sound that is between 20 and 20,000 Hertz.

[589] Keay, C., 'Continued progress in electrophonic fireball investigations', *Earth, Moon, and Planets* (1995) 68(1): 361–368

[590] Keay, C., & Ostwald's Laboratory, 'Test of the Production of Electrophonic Sounds', *Journal of the Acoustical Society of America* (1991) 89(4) Pt(1): 1823–1824

[591] Keay, C., 'The 1978 New South Wales Fireball', *Nature* (1980) 285: 464–466

[592] See Chapter 23: 'Earth opening up: earthquakes and volcanic eruptions'

[593] Freund, T. F., 2003. 'Rocks That Crackle and Sparkle and Glow: Strange Pre-Earthquake Phenomena', *Journal of Scientific Exploration* 17(1): 37

[594] Keay, C., 'Continued progress in electrophonic fireball investigations', *Earth, Moon, and Planets*, (1995) 68(1): 364

[595] Joshua 6:20: "When the trumpets sounded, the army shouted, and at the sound of the trumpet, when the men gave a loud shout, the wall collapsed."

[596] The Apocalypse of John, Revelation 8: "And I saw the seven angels which stood before God; and to them were given seven trumpets ... The first angel sounded, and there followed hail and fire mingled with blood, and they were cast upon the earth: and the third part of trees was burnt up, and all green grass was burnt up ... And the second angel sounded, and as it were a great mountain burning with fire was cast into the sea: and the third part of the sea became blood. And the third part of the creatures which were in the sea, and had life, died; and the third part of the ships

Metamorphosis,[597] the New Testament's 'Acts',[598] Homer's *Iliad*,[599] chronicles of the Ancient Roman Empire, and numerous myths and legends,[600] ancient and historical accounts have linked blasting horns and trumpets to mass destruction.

Figure 165: Jean Fouqet, 'The taking of Jericho'. (Oil on canvas, c. 1450)

Often associated with such sounds were images of 'fire-breathing' dragons and gods bringing death and destruction with their thunderbolts, rocks, fire and brimstone. As shown by Clube and Napier,[601] among others, these are most likely depictions and memories of ancient encounters with comets, from a time when such heavenly occurrences were much more frequent and dramatic.

In addition to eerie noises, electrophonics may also trigger mutations within living creatures, including humans, even in the absence of mutagenic radiation, as explained in the following excerpt:

> Some **genetic anomalies were reported in the plants, insects and people of the Tunguska region**. Remarkably, the increased rate of biological mutations was found not only within the epicenter area, but also **along the trajectory of the Tunguska Space Body** (TSB). At that **no traces of radioactivity were found**, which could be reliably associated with the Tunguska event. The main hypotheses about the nature of the TSB, a stony asteroid, a comet nucleus or a carbonaceous chondrite, readily explain the absence of radioactivity but give no clues how to deal with the genetic anomaly. A choice between these hypotheses, as far as the genetic anomaly is concerned,

were destroyed ... And the third angel sounded, and there fell a great star from heaven, burning as it were a lamp, and it fell upon the third part of the rivers, and upon the fountains of waters. And the name of the star is called Wormwood: and the third part of the waters became wormwood; and many men died of the waters, because they were made bitter. And the fourth angel sounded, and the third part of the sun was smitten, and the third part of the moon, and the third part of the stars; so as the third part of them was darkened, and the day shone not for a third part of it, and the night likewise."

[597] Ovid, *Metamorphosis*, Book 15 (8 AD): "[Gods send signs, earthquakes, darkness] ... still they gave no uncertain portents of the woe that was at hand. They say that the clashing of arms amid the dark storm-clouds and fear-inspiring trumpets and horns heard in the sky forewarned men of the crime; also the darkened face of the sun shone with lurid light upon the troubled lands. Often firebrands were seen to flash amidst the stars; often drops of blood fell down from the clouds; the morning-star was of dusky hue and his face was blotched with dark red spots, and Luna's chariot was stained with blood."

[598] Acts of the Apostles 2:2–3: "Suddenly a sound like the blowing of a violent wind came from heaven and filled the whole house where they were sitting. They saw what seemed to be tongues of fire that separated and came to rest on each of them."

[599] Homer, *Iliad, Book XXI*: "They fell on one another with a mighty uproar – Earth groaned, and the spacious firmament rang out as with a blare of trumpets."

[600] For further myths associating horns, comets and destruction, like Islam's Israfel, the Norse god Heimdall, the Roman Triton and the Arthur legend, see Mike Baillie's books, *Exodus to Arthur* and *The Celtic Gods*, and Knight-Jadczyk, L., *Secret History of the World, Vol. II: Comets and the Horns of Moses*

[601] Clube, V. & Napier, W., *Cosmic Winter*

is like the choice between 'blue devil, green devil and speckled devil,' to quote late Academician N.V. Vasilyev. However, if another mysterious phenomenon, **electrophonic** meteors, is evoked, the origin of the Tunguska genetic anomaly becomes less obscure.[602]

In conclusion, alongside atmospheric dust, surface impacts, overhead explosions, electromagnetic pulses and airborne viruses, electrophonic sounds are another documented effect of meteors.

Figure 166: A tree from Tunguska exhibiting ring enlargement right after 1908. (© University of Bologna)

As noted, meteors are not the only phenomena that can generate electrophonic sounds. Earthquakes,[603] lightning[604] and Northern Lights (*aurora borealis*)[605] have been shown to be sources of electrophonics also. They display a strong electromagnetic activity that explains the disruption of the geomagnetic field and the emission of VLF. More globally, any substantial change in the electromagnetic environment of the planet might induce turbulence in the geomagnetic field.

The association between electrophonic sound and disasters in general (and cometary activity in particular) has been repeatedly mentioned in ancient texts. It seems that what we are currently experiencing is nothing new, just another wave of cosmic changes and subsequent earthly destruction.

And, once again the elite rulers are trying to depict this phenomenon as manmade, harmless and unrelated to anything else of any import, whereas these sounds are in fact intimately related to the ongoing cosmic changes and to very destructive natural events.

---

[602] Silagadze, Z. K., 'Tunguska genetic anomaly and electrophonic meteors', *Astronomy & Astrophysics* (2008)

[603] Keay, C., 'Audible fireballs and geophysical electrophonics', *Astronomical Society of Australia* 11(1): 12–15

[604] Vinkovic, D., *et al.*, 'Global Electrophonic Fireball Survey: a review of witness reports, *WGN, the Journal of the IMO* (11 November 2002)

[605] Laine U. K., Turunen E., Manninen J., & Nevanlinna H., 'Measurements and analysis of sounds during active aurorae in Finland 2000–2001', *URSI General Assembly*, Maastricht, 17–24 August 2002

Chapter 30.

Geographic tilt and magnetic reversal

The Earth has two kinds of poles: geographic and magnetic. The geographic poles define the axis around which the planet spins. The magnetic poles are the locations where the lines of Earth's magnetic field are vertical.

As you can see in figure 167, the magnetic poles and the geographic poles are set along two distinct axes. **Neither of them is perpendicular to the plane of the ecliptic** – the plane of the solar system (medium grey). Instead of being vertical (black dashed line), the geographic axis (dark grey line) has a 23.5° tilt relative to the ecliptic, and the magnetic axis (black line) has an 11° tilt relative to the geographic axis.

If our planet had experienced a uniformitarian, uneventful developmental history, the geographic and magnetic poles should both be perfectly vertical (figure 168).

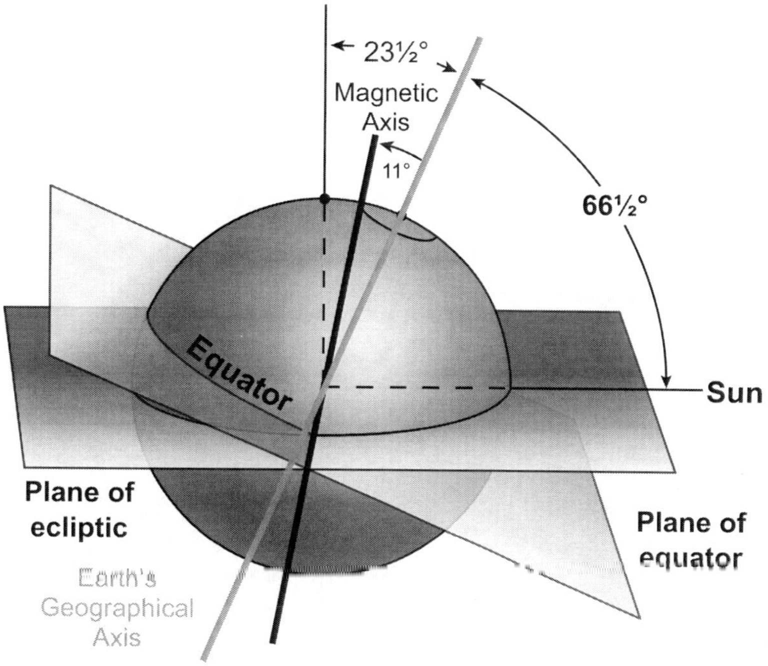

Figure 167: Magnetic poles and geographic poles relative to the ecliptic. (© electricYouniversE.com)

Tilt of the geographic poles

Let's first focus on the geographic pole and its 23.5° tilt. As mentioned above, if our planet had been accreting matter to itself, rotating quietly within its solar system cradle with its fellow nursery mates for billions of years, which is the official theory, then its rotation axis should be vertical. Even classical mechanics – which doesn't take into account the electric nature of our universe – shows that an external factor is necessary for the Earth's rotation axis to change and not be vertical. The following was stated as early as 1878:

> When a mass of matter is in rotation about an axis, it cannot be made to rotate about a new one except by external forces.[606]

Later on, scientists discovered the existence of an electric field in which our planet is embedded, which led to a better explanation for the tilt of the Earth's geographic axis:

> ... the Earth's field is tied up in some ways with the rotation of the planet. And this leads to a remarkable finding about the Earth's rotation itself ... The Earth's axis of rotation has also changed. In other words, **the Earth has rolled about, changing the location of its geographical poles** ...[607]

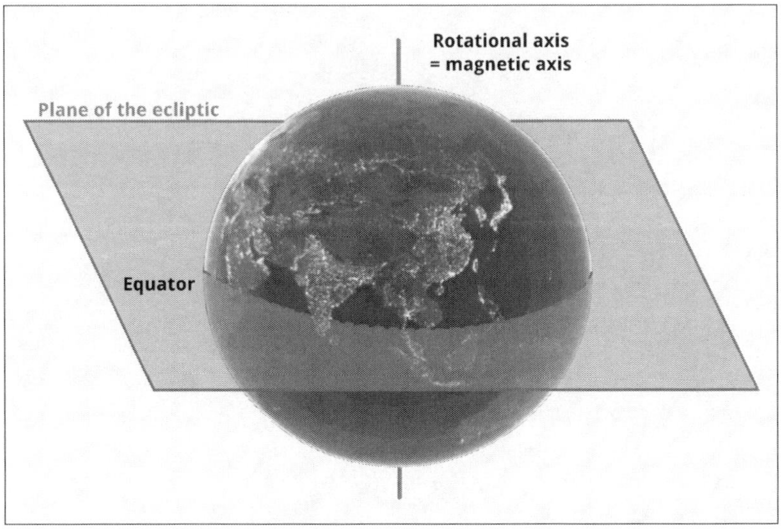

Figure 168: Aligned magnetic and geographic poles. (© Sott.net)

Zoologist François de Sarre confirms that the rotation axis of the planet should be vertical and explains how the electric force generated by a highly charged comet passing close to Earth could easily interact with Earth and shift its rotation axis:

> Armin Naudiet[608] noted that the attraction exerted by the Sun on the Earth gyroscope would cause, in a long run, its straightening [rotation axis perpendicular to the plane of the solar system], because our planet is not a perfect sphere. Geometrically speaking, it is an ellipsoid. Due to its own rotation and the existence of a higher centrifugal force at the equator, the Earth deforms: it is not spherical but slightly flattened at the poles. In other words, **if, as astronomers admit, our planet rotates 'quietly' (that is to say without having been 'shaken') for billions of years around the Sun, there should be no axis inclination**... and no seasons, since they result from the inclination (23°26') of the Earth's axis relative to the plane of the solar system.
>
> In any case, the recent discovery near the poles of a fossil fauna (dinosaurs!) rather used to heat seem to support the thesis of an Earth without seasons during the Secondary era.[609] It was hot all over the globe! ...
>
> How to explain that there are now seasons (in our latitudes) and that the Earth 'nods' its head, oscillating around its tilted axis? **Presumably it received a 'big shock'. The asteroid that ended the saga of dinosaurs is now widely accepted in science – in terms of the**

[606] Hill, E., 'On the possibility of changes in the Earth's axis', *Geol. Mag.* 5: 262–266

[607] Runcorn, S., 'The Earth's magnetism', *Scientific American* (1955): 152–162

[608] Naudiet, A., 'Das Geheimnis der Präzession', *Efodon Synesis* (1995) 9: 17–23

[609] Among other sources, the presence of mid-latitude and tropical flora in the circumpolar region is confirmed by the findings described in: Kropotkin *et al.*, 'Baron Von Toll on New Siberia and the Circumpolar Tertiary Flora', *Geogr. J.* 16: 95–98

> **impact and its immediate consequences. But not a word about the disruption that could result in the Earth's rotation.**
>
> In his book,[610] German engineer Hans-Joachim Zillmer thinks **the current tilt of the Earth is due to a celestial body carrying a strong electric potential (a comet?), which, passing close to Earth, has tilted [it] ...**
>
> Suppose a **close-passing comet has recurred here for a few thousand years. Exposed to the sudden presence of electrokinetic forces, the inclination of the axis of the Earth relative to the ecliptic could have increased sharply, up to 30 or 35 degrees...** No, no danger that Earth would tip over; let us recall the image of the toy top!
>
> **Then the Earth began to recover slowly. This would explain the phenomenon of the precession of the equinoxes. Contrary to what astronomers usually think, it would not be constant, but be steadily declining, since the Earth returns to its original position, perpendicular to the plane of the solar ecliptic** (despite the inertia of the system).[611]

As emphasized in the above excerpt, an external agent has to be factored in to explain the current tilt of the Earth's rotation axis. A highly charged comet passing close enough to Earth could very well be the responsible agent, exerting massive electro-gravitational forces on the planet and 'pulling' the geographical axis away from its default vertical position. If this scenario is accurate, the Earth has been slowly returning to its default vertical position (because of the inertia of the gyroscope that is planet Earth).

However, for mainstream science, Earth's obliquity (i.e. its axial tilt angle) has always been there and is gently oscillating between 22.1° and 24.5° every 40,000 years, on average.[612] There are even pre-made formulae, like the Newcomb formula,[613] that calculate the Earth's tilt one million years in the future or in the past. But all these figures about the past or future tilt of our planet are based on theoretical models that don't even fit past observations.

George Dodwell[614] conducted an extensive study of ancient observations of the obliquity of Earth. He gathered 120 measurements[615] spanning the last four millennia, from 1100 BC up to the 20th century.[616]

The graph in figure 169 covers the period 2450 BC to 1960 AD. It shows the evolution of Earth obliquity (in degrees) according to the Newcomb theoretical model (grey line) and the observations compiled by Dodwell (grey curve).

The observations compiled by Dodwell not only diverge from the theoretical model held by mainstream science, but also define an almost perfect logarithmic curve. This logarithmic curve – showing a sudden and pronounced change (at least several degrees) in obliquity c. 2345 BC – strongly suggests the occurrence of a major cosmic event that rolled the Earth over.

---

610 Zillmer, R., *Irrtümer der Erdgeschichte*
611 De Sarre, F., *Mais où est donc passé le Moyen Âge*, p. 39
612 Berger, A., 'Obliquity and Precession for the Last 5,000,000 Years', *Astronomy and Astrophysics*, pp. 127–135
613 Newcomb, S., *Tables of the Four Inner Planets*
614 George F. Dodwell (1879–1963), leading Australian astronomer.
615 'Dodwell main chart', *Setterfield*. See: www.setterfield.org/Dodwell%20main%20chart.html
616 Sources included Thales, about 558 BC; Eratosthenes, about 230 BC; Hipparchus, 135 BC; Ptolemy, 126 AD; and several medieval astronomers up to the time of Tycho Brahe, 1587 AD, and Wendelin, 1616 AD.

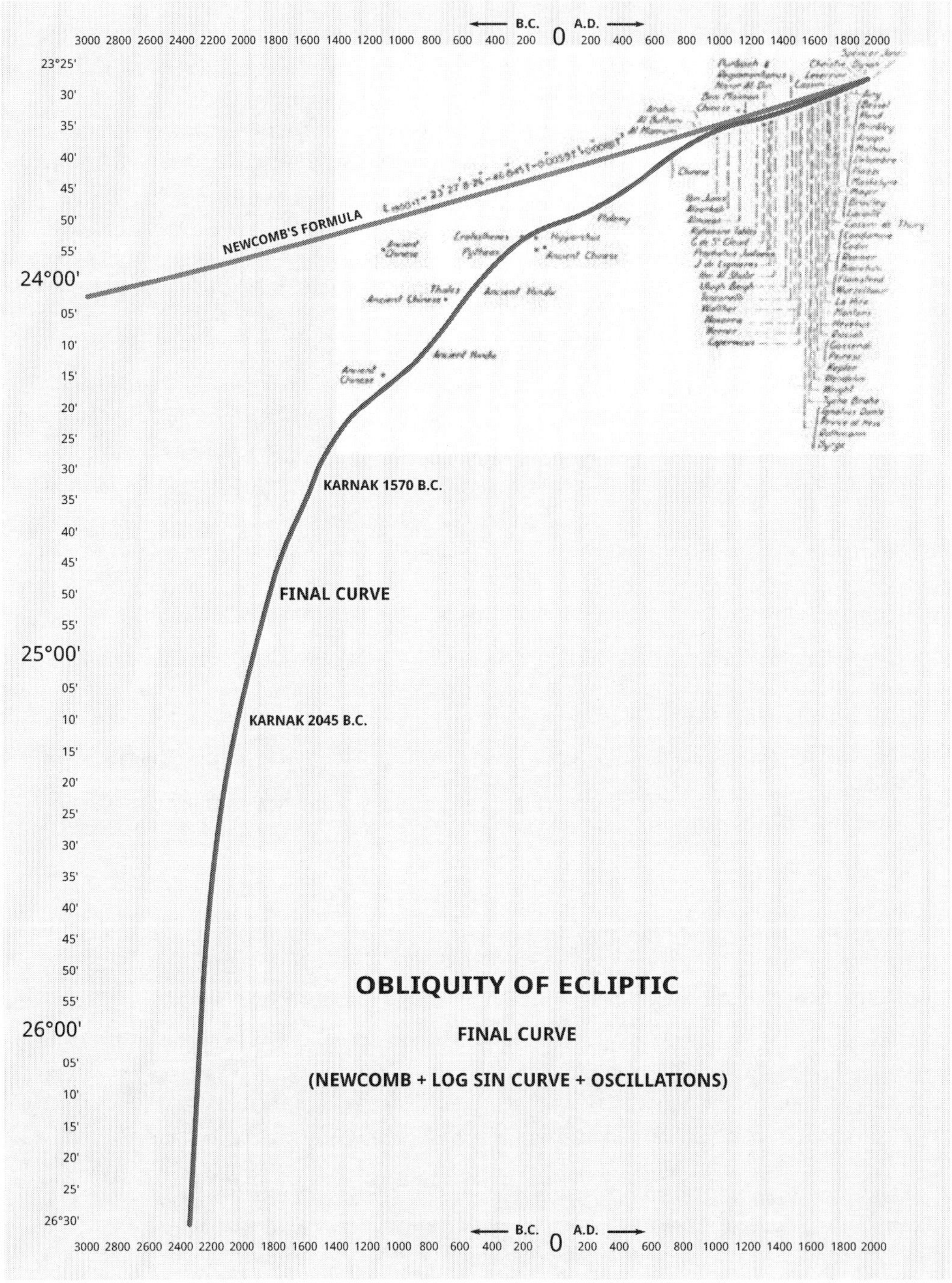

Figure 169: Obliquity of the ecliptic over the period 2345 BC–2000 AD. Newcomb theoretical formula (grey line) vs. Dodwell's compiled observations (grey curve). (© Dodwell)

At this time, severe climatic, geological and archaeological changes occurred in which the great Bronze Age civilizations in Egypt, Mesopotamia and Greece were wiped out. For example, out of the 350 Early Bronze Age sites in ancient Greece, more than 300 were destroyed and many others were abandoned.[617]

The scientific community generally agrees that a major disaster occurred at that time.[618] The discovery of half a dozen craters[619] that were formed within a century of 2350 BC, including a massive one (3.4 km diameter) that was discovered in Iraq,[620] confirmed the asteroid hypothesis promoted for years by several scientists.[621]

Dodwell supposed that a very strong cometary impact had been necessary to jolt the Earth from its default vertical position to its new tilt. However, only a very large body could account for a substantial change in tilt.[622] But, as described by François De Sarre, cometary bodies don't necessarily have to impact Earth's surface in order to alter its rotation. A highly charged comet passing close enough to our planet could even exert greater electro-gravitational forces, sufficient to tilt Earth's rotation axis, than the mechanical consequences resulting from a direct impact.

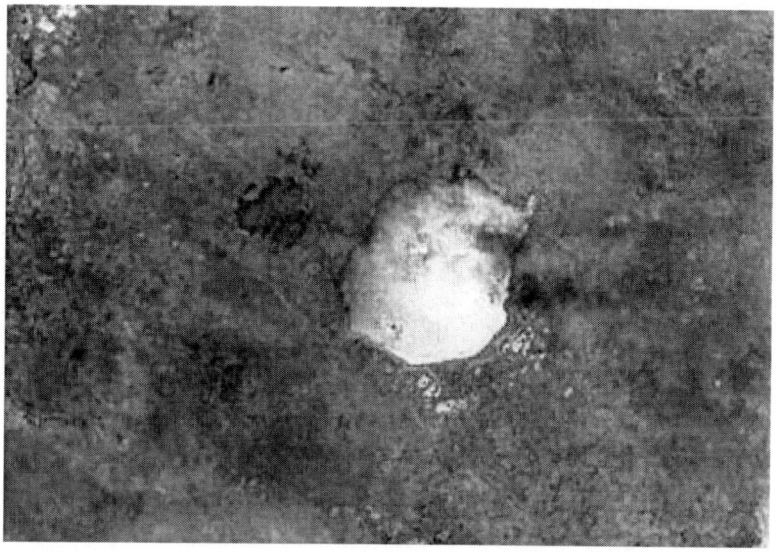

Figure 170: The Umm Al Binni Lake in the Al'Amarah marshes, Iraq. (© NASA)

Several researchers[623] have shown that meteors do not have to hit the Earth directly in order to have a devastating effect, including the creation of craters.[624] [625] In 1908, an object exploded three miles above Tunguska, in Siberia, with the force of thousands of Hiroshima bombs, laying waste to over 1,250 square miles of land

[617] Peiser, B. J., 'Collapse of early bronze age civilizations: has the smoking gun been found?', *Cambridge Conference Correspondence*

[618] *Ibid.*

[619] Syal, R., 'Meteor showers blotted out man's first civilisations', *The Sunday Times* (14 December 1997). See: www.times-archive.co.uk/cgi-bin/BackIssue

[620] Master, S., 'A possible Holocene impact structure in the Al 'Amarah marshes' - *Meteoritics & Planetary Science* (2001) 36(supplement): A124

[621] Courty 1997, 1998; Peiser 1997; Napier 1997; Bjorkman 1973; Weiss *et al.* 1993; Master 2001, 2002

[622] The impact of a body 640 km in diameter and travelling at 72 km/s would induce a 4°15 change in Earth's tilt. See: Dachille, 'Axis Changes in the Earth from Large Meteorite Collisions', *Nature* (April 1963) 198(176)

[623] Peiser, B. J., 'Catch a falling star', *Jewish Chronicle* (6 March 1998). See: www.jchron.co.uk.index.htm

[624] Firestone *et al.* consider the Carolina Bays (more than 500,000 elliptical craters concentrated along the Atlantic seaboard) as remnants of overhead cometary explosions. See: Firestone, R., West, A. & and Warwick, S., *The Cycle of Cosmic Catastrophes: How a Stone Age Comet Changed the Course of World Culture*, 2006

[625] "In a near miss encounter between two bodies of different net charge, a discharge between them would be likely. The discharge if concentrated on the surface would leave a crater." See: White, J., *Pole Shift*, p. 177

below.[626] The Tunguska overhead explosion might even have produced a crater. In 2007 Italian geologist Gasperini investigated Lake Cheko:

> We report results from the investigation of Lake Cheko, located 8 km NNW of the inferred explosion epicenter. Its funnel-like bottom morphology and the structure of its sedimentary deposits, revealed by acoustic imagery and direct sampling, **all suggest that the lake fills an impact crater.**[627]

Figure 171: Lake Cheko in the Siberian region of Tunguska. (© University of Bologna)

The 'asteroid-induced geographic tilt of the Earth' hypothesis is also strongly supported by the numerous earthquakes our planet experienced around this time. Only a substantial and global crustal slippage could explain the disruption that affected the whole planet:

> ... the most significant aspect of the geological evidence is the **crustal movements**[628] **that apparently began at about the same time around 2300 BC at many regions of the Earth.**[629]

As shown by Dodwell's research, since 2345 BC the Earth's tilt has decreased, quickly at first (since it was far away from its normal rotation axis) and then slower and slower, hence the logarithmic curve. Eventually the Earth's rotation axis might reach its default vertical setting – assuming nothing happens in the meantime to interfere with this process.

The Earth behaves like a spinning top, rotating around its vertical axis, which is then jolted out of its default movement. The rotation axis of the top is tilted, and it wobbles because of this external disturbance. Then, due to its gyroscopic properties, which push it to keep its axis pointed always in the same direction, the top returns to its initial configuration.

This wobble of the Earth is also called 'Chandler's wobble' or 'nutation'. Such an instability is symptomatic of a flawed equilibrium: the Earth's axis of rotation does not coincide with its center of inertia.[630] If the Earth's current state was solely the result of an eventless linear evolution, its axis would be vertical and the wobble would be non-existent.

[626] Bevan, A., & De Laeter, J., *Meteorites: A Journey Through Space and Time*, p. 183

[627] Gasperini, L., *et al.*, 'Possible impact crater for the 1908 Tunguska Event', *Terra Nova* (August 2007) 19(4): 245–251

[628] Sea-level changes, land uplift and land subsidence, volcanic eruptions and earthquakes happened concomitantly in many places around the world.

[629] Mandelkehr, M. M., 'An Integrated Model for an Earthwide Event at 2300 BC', *Chronology & Catastrophism* (1988) 10

[630] Allan, D. & Delair, J., *Cataclysm!*, p. 190

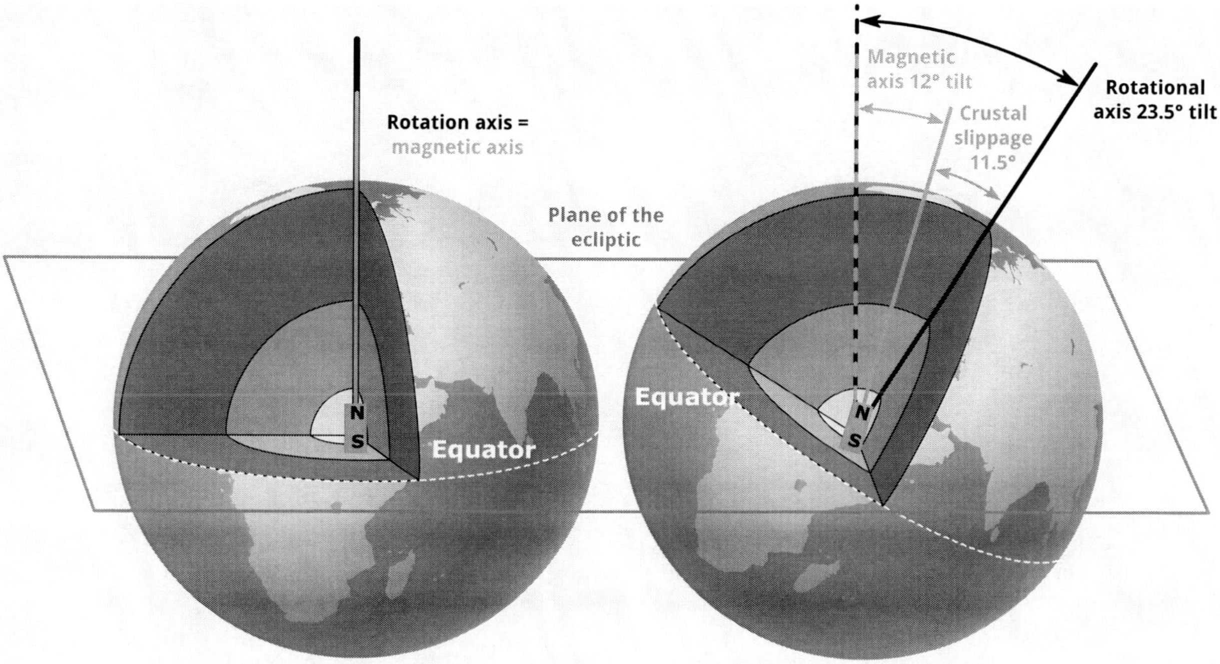

Figure 172: Geographic and magnetic axis before and after a cometary disturbance. (© Sott.net)

One of the few plausible explanations for such major disruption is a change in the orientation of the Earth's rotation axis induced by electromagnetic forces exerted by a nearby comet. In such a scenario:

1. The change in rotation axis would move the equatorial bulge from the previous equatorial region to the new one. That would amount to massive amounts of compression around the former equatorial region (decreased centrifugal forces) and massive extension forces around the new equatorial region (increased centrifugal forces).

2. A sudden change in rotation axis can induce 'crustal slippage'. Because of the viscous properties of the mantle and core of our planet, only part of the torque exerted by the close fly-by of a comet can be transmitted to the inner region of the Earth. The solid crust rotates more than the more liquid mantle. The difference in rotation between the core and the crust is equal to the crustal slippage.

On the left of the drawing in figure 172, the Earth is shown with its two vertical and aligned axes (magnetic and rotational). On the right, the effects of a nearby comet tilting the geographic axis 23.5^{o} relative to the initial vertical axis and tilting the core by 12°, hence an 11.5^{o} crustal slippage.

Notice that the effects of such a sudden slippage would be much more dramatic than the effects of the limited slippage due to the Earth's minute slowdown described previously.[631]

[631] See Chapter 22: 'The slowdown of the Earth'

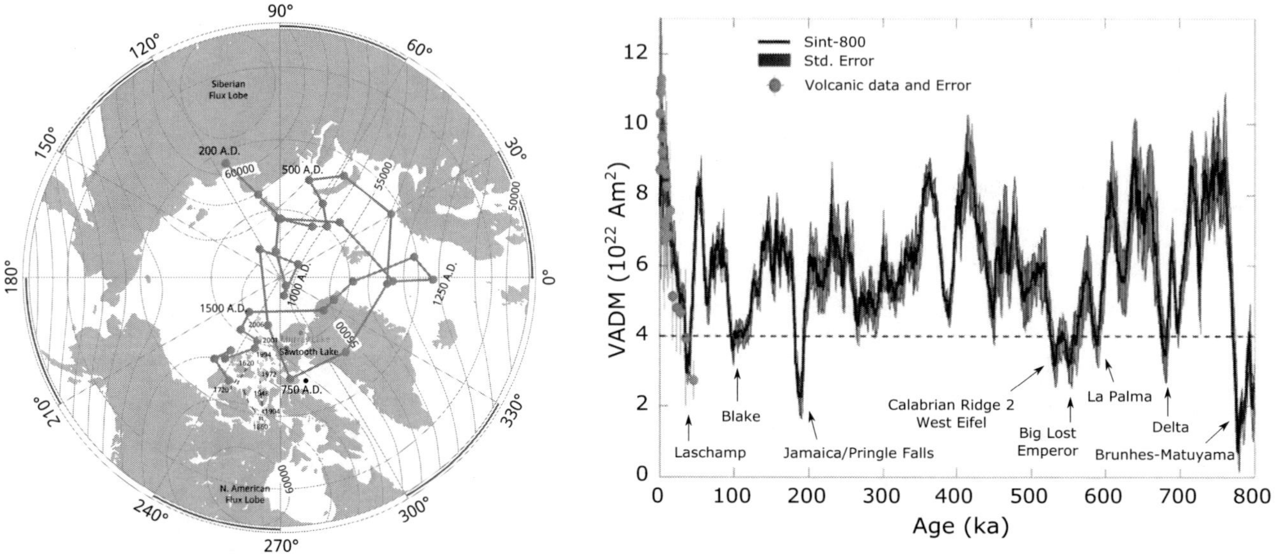

Figure 173: On the left, variation in the location of the magnetic pole (200 AD – now). On the right, variation in the intensity of the magnetic field over the last 800,000 years. (© COAS / NASA)

Magnetic poles

Let's now consider the Earth's magnetic field, also known as the geomagnetic field. We might believe it is immutable when thinking about sailors who use their compass to find their way on the open ocean. In reality, the **geomagnetic field keeps changing in intensity and in direction** (figure 173).

The change in location of the magnetic poles is the reason why sailing maps are provided with a date of print and a declination (angle between the geographic pole and the magnetic pole). Depending on the date when the sailor calculates his route, he will adjust this declination because of the ever-changing location of the magnetic pole.

Actual measurements of the intensity of the total geomagnetic field began in Gauss's observatories in 1840.[632] Since then, it has been *continually decreasing* at a rate of 6.3% per century. At this rate, the intensity of the field will reach 0 in 1,600 years.[633] This trend has been confirmed by studies of the magnetization in ancient clay pots, showing that the *Earth's magnetic field was approximately twice as strong in Roman times.*[634]

The geomagnetic field comes from several sources. One of them seems to be linked to solar activity. Several scientists have demonstrated a positive correlation between solar activity and the intensity of Earth's magnetic field,[635] as depicted by figure 174.

This variable magnetic field might be induced, at least partly, by the jet stream (see figure 175).

[632] Hulot, G., *et al.* 'The Magnetic Field of Planet Earth', *Space Sci. Rev.*, p. 5

[633] Courtillot, V., *et al.*, 'Time Variations of the Earth's Magnetic Field: From Daily to Secular', *Annual Review of Earth and Planetary Science* (1988) 16: 435

[634] 'Reversals: Magnetic Flip', *British Geological Survey.* See: www.geomag.bgs.ac.uk/education/reversals.html

[635] Courtillot, V. & Le Mouel, J., 'Time variations of the earth's magnetic field - From daily to secular', *Annual review of earth and planetary sciences*(1988) 16: 389–398

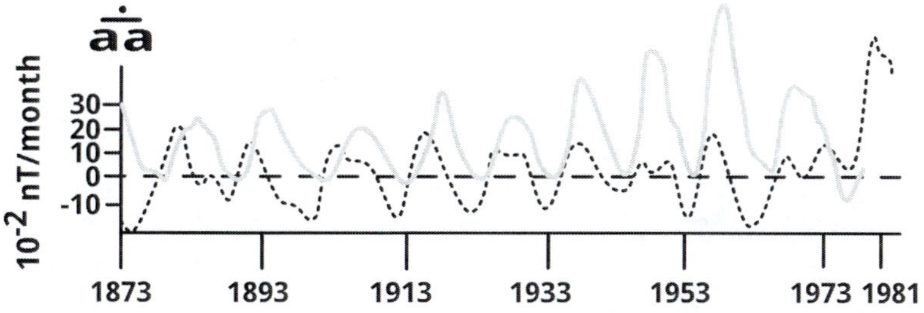

Figure 174: Earth's geomagnetic activity (dashed black curve) vs. solar activity (light grey curve) over the period 1873–1981. (© Courtillot & Le Mouel)

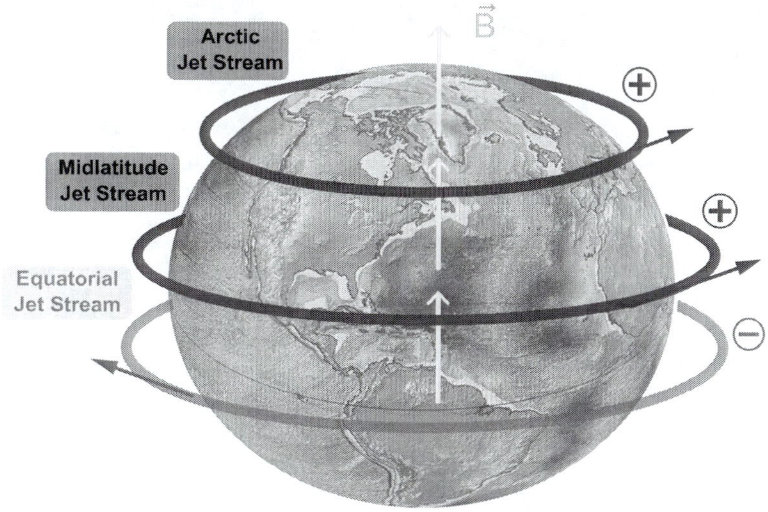

Figure 175: Magnetic fields (B) induced by the jet streams. (© Sott.net)

Indeed, if a positive charge is moved in a circle eastwards (black arrows), it generates a magnetic field exhibiting the same polarity as the geomagnetic field (light grey vertical arrows). This is the case with the polar and the mid-latitude jet streams. The induced magnetic field will have the same direction if a negative charge is moved westwards, which is the case with the equatorial jet stream (dark grey arrow).

This might explain why when solar activity is low, the geomagnetic field decreases in intensity. Notice, however, that **those variations are extremely small** and amount to less than 0.01% of the geomagnetic field intensity.[636]

Another non-mutually exclusive possibility is that solar activity and the geomagnetic field are driven by the same cause.

A second source of the geomagnetic field is the Earth's crust. Iron, a magnetic element, is a major constituent of our planet's crust, accounting for 32% of its composition.[637] We will focus on the

[636] Monthly variations are about 30 nT (nanoTesla), while the geomagnetic field is about 30,000 nT. See: Reynolds, J., *An Introduction to Applied and Environmental Geophysics*, p. 92

[637] Sharp, T., 'What is Earth Made Of?', *Space* (26 September 2012). See: www.space.com/17777-what-is-earth-made-of.html

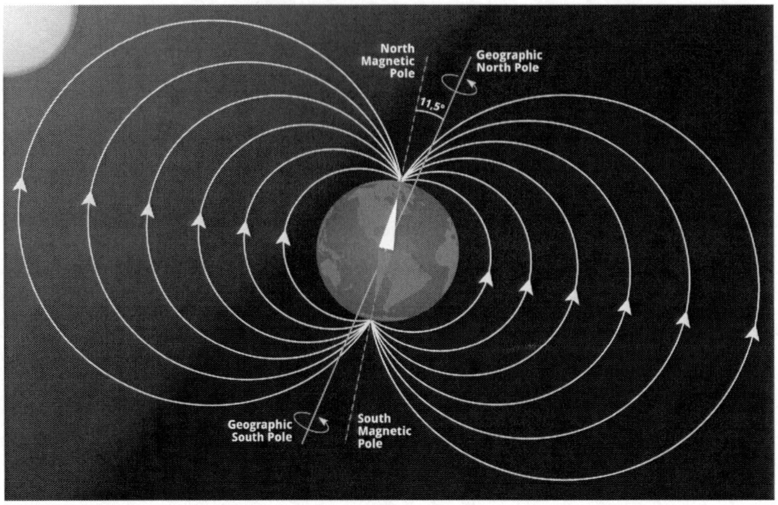

Figure 176: The Earth's magnetic field. (© SBG System)

crust because the mantle and core exhibit temperatures higher[638] than the Curie temperature,[639] where magnetic elements lose their magnetism.

Notice that magnetized elements like iron can be almost instantly demagnetized or re-magnetized. For example, an iron rod can be magnetized simply by rubbing a magnet on it, i.e. subjecting the rod to the magnetic field of the magnet.

Cometary bodies carry a high electric charge and therefore, when moving, emit a strong magnetic field. Thus, the magnetic field of a nearby comet may be capable of demagnetizing crustal iron. Similarly, the electric discharge between a nearby comet and Earth can generate a magnetic field capable of magnetizing crustal iron. Examples of crustal demagnetization induced by a cometary body have been found on Mars, where a whole region located within and around a crater exhibits no magnetism, unlike the rest of the Red Planet.[640]

So far we have identified two factors, the jet streams and the remnant magnetism held by the iron-rich crust, contributing to the Earth's magnetic field. However, as suggested by its overall shape and confirmed by a number of scientists, 90% of the geomagnetic field comes from the core of the Earth.[641]

The geomagnetic field is evidence of a powerful energy source at the center of the Earth.[642] Traditional geophysical models of energy sources for the geomagnetic field invoke the decay energy of naturally occurring long-lived radionuclides[643] or the movement of iron in the core that produces electricity and subsequently magnetism (the 'geodynamo effect'[644]) or the assumed ongoing growth

---

638 Mantle temperatures range between 500° and 4,000° Celsius (930° and 7,200° Fahrenheit). See: Louie, J., *Earth's Interior*, University of Nevada, Reno, 1996

639 The Curie temperature for iron is about 1,400° Fahrenheit or 770° Celsius.

640 Louzada, K., Stewart, S. & Weiss, B., 'Shock Demagnetization of Pyrrhotite', *Lunar and Planetary Science* (2005) 36

641 Langel, R. & Hinze, W. J., *The Magnetic Field of the Earth's Lithosphere: The Satellite Perspective*, p. 24

642 Hollenbach, D. F., *et al.*, 'Deep-Earth reactor: Nuclear fission, helium, and the geomagnetic field', *PNAS* 98(20)

643 Mainly uranium and thorium.

644 Fearn, D. R., *The Geodynamo* (19 August 2004)

of the inner core by cooling, which is supposed to release gravitational potential energy and latent heat of crystallization.[645]

However, those energy sources *can only change gradually and in only one direction* over time. The Earth's magnetic field, by contrast, exhibits massive intensity fluctuations in both directions.[646]

In order to explain those two-way variations, some scientists have proposed the hypothesis of a nuclear fission reactor at the core of the Earth,[647] but deficient neutrino counts tend to falsify this theory.[648]

In any case, the Earth seems to be able to provide a variable and endless supply of energy. The Earth manages to 1) refill itself and keep providing electrons into the Earth/ionosphere capacitor, 2) provide a massive amount of heat to the mantle, 3) maintain a magnetic field that has been in existence for millions of years.

This is actually very similar to the Sun's situation, which provides a seemingly endless supply of energy and for which a deficient neutrino count falsifies the nuclear reaction hypothesis.[649]

Like the Sun and other stars, the Earth might be tapping into an external energy source that powers its core, which in turn heats the mantle, provides electrons for the atmospheric capacitor and generates most of the geomagnetic field.

Geomagnetic reversal

At high temperatures, rocks do not exhibit any magnetic properties: they are nonmagnetic. But when they are cooled down below the Curie Point,[650] they acquire and keep a magnetic orientation equal to that of the Earth at the time when the cooling occurred. This is an interesting property since it can provide information about the orientation of the geomagnetic field at the time of major volcanic activity (the eruption of rock in liquid form: magma).

In particular, evidence of magnetic pole reversals can be seen at ocean ridges, where tectonic plates move apart and the seabed is filled in with magma. As the magma seeps out of the mantle, the magnetic particles contained within it are oriented in the direction of the magnetic field at the time the magma cools and solidifies.

Paleo-magnetism, the scientific study of past orientations of Earth's magnetic field, has shown that old magma frequently exhibits reversed magnetic polarity, indicating that at the time it solidified, the magnetic field was reversed. Thanks to the study of ocean ridges,[651] scientists have managed to reconstruct the geomagnetic past of our planet and have discovered that *magnetic pole reversal*

645 Gubbins, D., *et al.*, 'Can the Earth's dynamo run on heat alone?', *Geophys. J. Int.* (2003) 155: 609–622

646 Tauxe, L., *Paleomagnetic Principles and Practice*, Modern Approaches in Geophysics, Volume 18, 1998

647 Hollenbach, *op. cit.*

648 'Rethinking Nuclear Fission: A fundamental and natural reaction', *The Bad Science Blog* (13 December 2011). See: depletedcranium.com/rethinking-nuclear-fission-a-fundimental-and-natural-reaction/

649 See Chapter 9: 'External power source of celestial objects'

650 Around 600° Celsius or 1,100° Fahrenheit.

651 In ocean ridges, the seafloor spreads, so the magma wells up from the mantle and solidifies, forming new basaltic rocks. During the cooling, the basalt records the direction of the Earth's field. This new basalt forms on both sides of the ridge and moves away from it. When the Earth's field reverses, new basalt records the reversed direction. See: 'Earth's inconstant magnetic field, *NASA Science News* (29 December 2003). science.nasa.gov/science-news/science-at-nasa/2003/29dec_magneticfield/

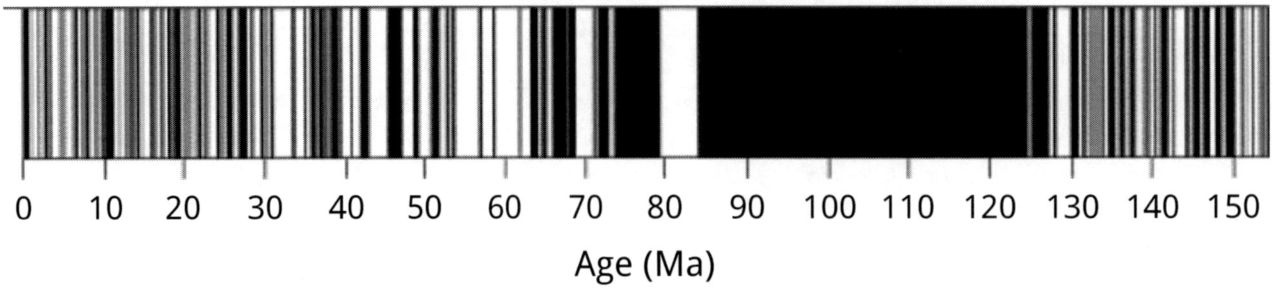

Figure 177: Orientation of the geomagnetic field over the last 150 million years. (© all-geo.com)

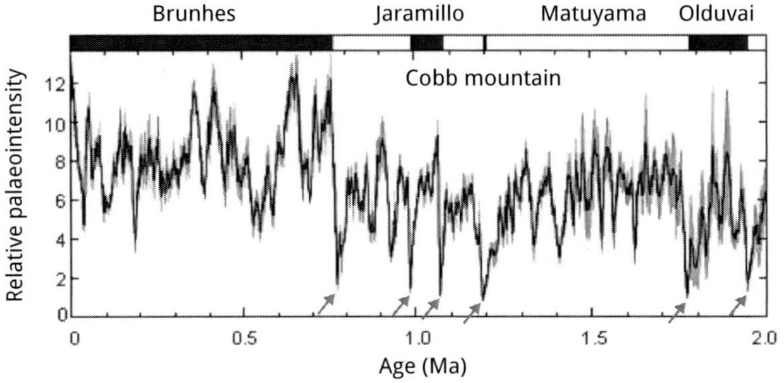

Figure 178: Earth's magnetic field intensity and magnetic pole reversal over the last two million years. (© Jean Pierre Valet/CNRS)

was actually a fairly common phenomenon.

In figure 177, the black areas show periods of 'normal' geomagnetism, similar to the magnetic polarity we are experiencing today. The white areas show periods of reversed magnetic polarity.

The intensity of the Earth's magnetic field has been reconstructed over the last two million years,[652] and it shows that magnetic reversals occur frequently – about once every 100,000 years – and that most of the time the reversal follows a very sudden drop in the intensity of the geomagnetic field. Figure 178 shows six reversals (light grey arrows) preceded by a steep decline in the intensity of the geomagnetic field.

It was previously believed that the timescale for pole reversal was between 10,000 and 100 years.[653] However it appears that **magnetic pole reversals can happen much quicker**,[654] with orientation changes up to 6° per day.[655] This rate is exceptionally high, being 10,000 times faster than the usual rate of change in magnetic orientation[656] and leading to reversals that could take place

[652] Valet, J.-P., Meynadier, L. & Guyodo, Y. 'Geomagnetic field strength and reversal rate over the past two million years', *Nature* (2005) 435: 802–805

[653] White, *op. cit.*, p. 138

[654] The speed of reversal is calculated by comparing magnetic field direction in different parts of the same lava flow. See: Felix, R. W., *Magnetic reversals and evolutionary leaps*, p.95

[655] Coe, R., et al. 'New evidence for extraordinarily rapid change of the geomagnetic field during a reversal', *Nature* (20 April 1994) 374: 687–692

[656] Langel, R., *Geomagnetism* Vol. 1, pp. 249–512

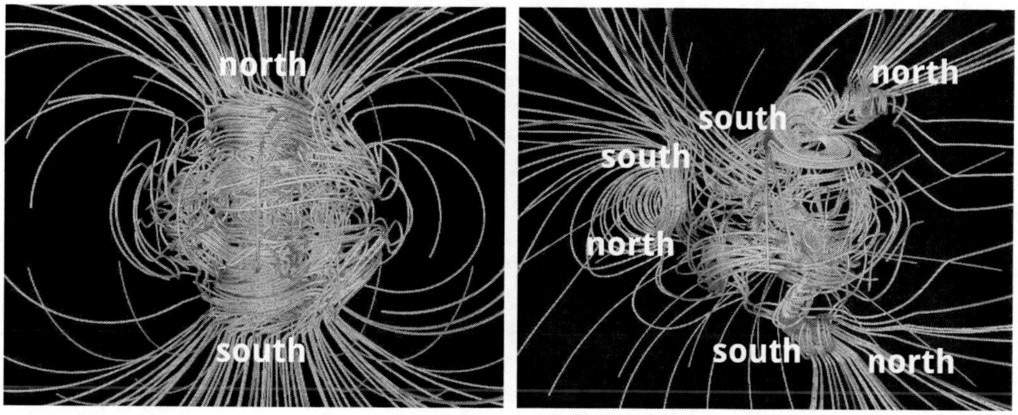

Figure 179: The geomagnetic field: between reversals on the left and during reversal on the right. (© Wikimedia Commons)

in a matter of days.[657]

In addition, it appears that rocks exhibiting reversed magnetic polarity show magnetic charges up to 100 times what could have been generated by the Earth's magnetic field.

The above suggests an external agent that could dramatically increase the magnetic field measured on Earth. Comets and their high electrical activity could, if close enough, trigger a discharge exchange with the Earth.

Such a discharge could generate massive magnetic fields (hence the high magnetic charges measured in rocks exhibiting reverse polarity and spikes in the geomagnetic field before reversals), destabilize the existing geomagnetic field, and be the trigger for the 're-set' that sees 'north' become 'south' and *vice versa.*

Interestingly, such geomagnetic 'spikes' seem to be also linked to global cooling episodes:

> Four potential geomagnetic events ('archeomagnetic jerks'), marked by strong intensity increases, are observed and appear to be synchronous with cooling episodes in the North Atlantic. This temporal coincidence strengthens the recent suggestion that the geomagnetic field influences climate change over multi-decadal time scales.[658]

As a matter of fact, **comets are a prime source of dust and magnetism**. Thus, a cometary event can be the cause of both geomagnetic jerks (induced by cometary discharges) and cooling events (induced by cometary dust).

However, notice that **most, but not at all**, geomagnetic reversals are preceded by magnetic surges, accompanied with global cooling episodes and *followed by mass extinction.*[659]

Therefore, cometary discharges can't be the only cause of geomagnetic reversal. The Sun experiences a magnetic reversal every 22 years, although no cometary activity could cause such a phenomenon. The magnetic fields of the Earth and the Sun might be modulated by a factor that science has not yet discovered.

[657] Warlow P., 'Does pole-flipping account for earth magnetism?', *New Scientist* (9 November 1978): 1224

[658] Gallet, Y., 'Possible impact of the Earth's magnetic field on the history of ancient civilizations', *Earth and Planetary Science Letters* (15 June 2006) 246(1–2): 17–26

[659] Felix, *op. cit.*, p. 101

Chapter 31.

One cosmic cause, numerous earthly effects

Most of the phenomena analyzed above seem to have started or accelerated around the turn of the 21st century. Figure 180 recapitulates them. However, keep in mind that most of these observations are based on publicly available data. I suppose only a very paranoid mind would suspect that the data has been 'cooked'. However, it wouldn't be the first time. For example, researchers of the University of East Anglia did just that to global temperature data. The Climate Research Unit incident soon became known as 'Climategate', quite an apt name when we reflect on what happened:

> ... AGW[660] proponents are under fire resulting from release of more emails in the 'Climategate' scandals. These emails show, as in the previous release two years ago, that a **small cabal of influential scientists promoted the AGW line, schemed to suppress dissenting views, bullied journal editors, hid the decline, and in private expressed much more uncertainty than their public statements would connote**. While we are waiting to see the fallout from recent revelations, I offer for your consideration an essay by British scientist and engineer Dr. John Brignell titled: 'How we know they know they are lying.' Among other things, he discusses the difference between real science and bureaucratic science (BS).[661]

As we will see in the final part, the elite rulers of our world, as well as their servants, a largely corrupt scientific community (there are numerous exceptions) and the bought-and-paid-for (no exceptions) mainstream media, have several good reasons to hide what is going on, or at least to minimize it. So it's not outside the realm of probability that the available data underestimates the enormity of 'Earth changes'. Anyhow, even though we only have publicly available data to go on, the current situation, and probable developments, are pretty clear based on personal observations and the global sharing of information via the internet.

The overall trend is confirmed in figure 181. Since 2003, the increase in natural catastrophes has intensified (light grey curve – ascending linear regression), with 2012 and 2013 seeing countless records being broken[662] (floods, earthquakes, cold, meteor fireballs and drought). Apart from 2000, the last eight years (2006–2013) have been the most chaotic since the beginning of the records in 1980.

---

660 Anthropogenic global warming

661 Brignell, J., 'How we know they know they are lying', *Numberwatch* (January 2009). See: www.numberwatch.co.uk/lying.htm

662 See, for example: Lyall, S., 'Heat, Flood or Icy Cold, Extreme Weather Rages Worldwide', *The New York Times* (10 January 2013). www.nytimes.com/2013/01/11/science/earth/extreme-weather-grows-in-frequency-and-intensity-around-world.html?pagewanted=all&_r=0

Phenomenon	Beginning / Pick-up	Note	Source
Drop in Solar activity	1997-1998	Weak Solar cycles 23 and 24	SOHO
Earth's slowdown	About 2003	Acceleration of day lengthening	IERS
Global cooling	About 2002	Observed global temperatures	IPCC - HADCRUT3
Increase in fireballs	2002	AMS data only start in 2005. Increase starter around 2002	AMS - Fireballs & Meteors blog
Increased volcanism	1998	Yearly number of volcanic eruptions	GVP
Increase in earthquakes	1995	Magnitude 6.0+ earthquakes	USGS
Sinkholes	2004	Earliest entry on SOTT database	SOTT
Gulf stream	2004	Constant weakening since 2004	Princeton University
Jet Stream	1998-1999	Weakening speed and shift towards the Equator	NCAA
Hurricanes	1997	Increase in hurricane frequency	Dr. Maue
Tornadoes	2002	Increase in tornadoes frequency	NOAA
Global dimming	About 2000	Global dimming increased around 2000 after dropping during the 90's	
Nocticulent clouds	1980-1990	First observed in 1885 but 28% increase per decade since 1980	NASA
Contrails	1995-1997	'Chemtrail' propaganda begins to cover up lingering jet contrails	Various sites about 'chemtrails'
Strange sounds	2008	First modern report	Seekstress.com

Figure 180: Concomitant symptoms of the approaching Nemesis and its swarm (year it picked up, notes and sources). (© Sott.net)

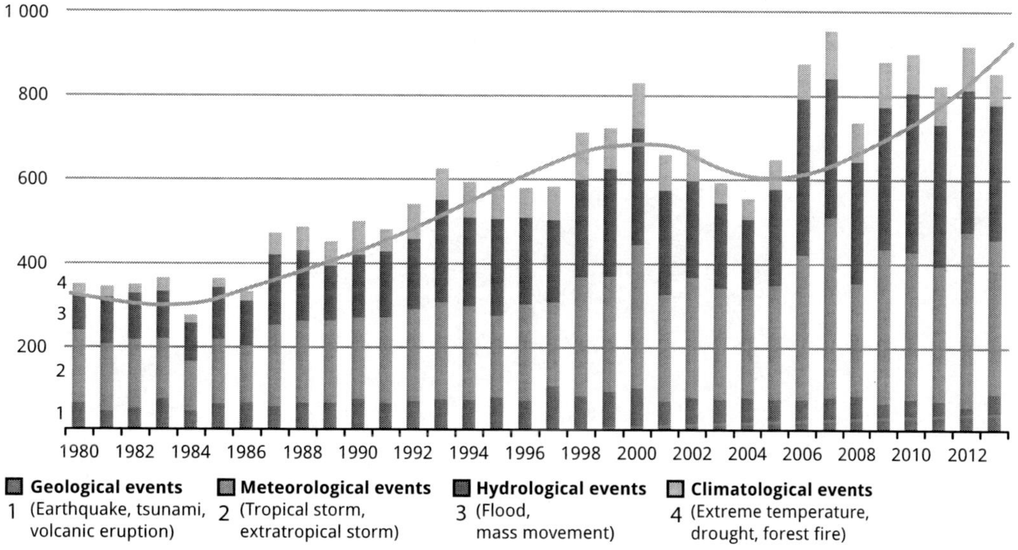

Figure 181: Natural catastrophes over the period 1980–2013. (© Munich RE)

The fact that natural catastrophes and the 'symptoms' described above are becoming more frequent and intense suggest that **Nemesis and its accompanying cometary swarm have not yet reached perihelion** (closest point to the Sun) and may still be approaching somewhere in the far reaches of the solar system. Until Nemesis passes perihelion, we should, in the coming years, continue witnessing an increase in both the frequency and intensity of its effects. And no, it is not likely that such a body would ever come into the inner solar system[663] itself. All of the phenomena described in this book can manifest as a result of the electrical disruptions to the Sun by a relatively close approach, and the concomitant increase in cometary flux that acts as Nemesis' calling cards.

All of the mentioned phenomena – the quiescence of the Sun, the increase in cometary activity and cataclysms, global cooling, the increase in sinkholes, the disruption of oceanic streams and jet streams, etc. – are typically considered to be 'anomalies', to be explained away, where possible, as the results of human activity (e.g. anthropogenic climate change, man-made 'chemtrails', 'missile tests' or space junk concealing fireballs, pipes breaking explaining sinkholes, et cetera). While their reality is sometimes publicly acknowledged, mainstream sources systematically underreport these 'anomalies' and downplay their consequences. The constituted authorities consider them harmless, meaningless, stable (in terms of their frequency) and, therefore, ultimately unrelated.

But this is simply not true. These 'anomalies' are not due to random chance or human activity; they are the *result of fundamental cosmic changes*. They are not harmless; they are becoming *increasingly destructive*, and they are responsible for a growing number of casualties and natural disasters. Their frequency is not stable; they have been increasing since the turn of the century. And ultimately, it is extremely unlikely that they are unrelated; rather, they all are likely to *stem from the same fundamental cause*: the possible approach of a solar companion and its cometary

[663] Region encompassing the Earth and its neighbors (Mercury, Venus and Mars).

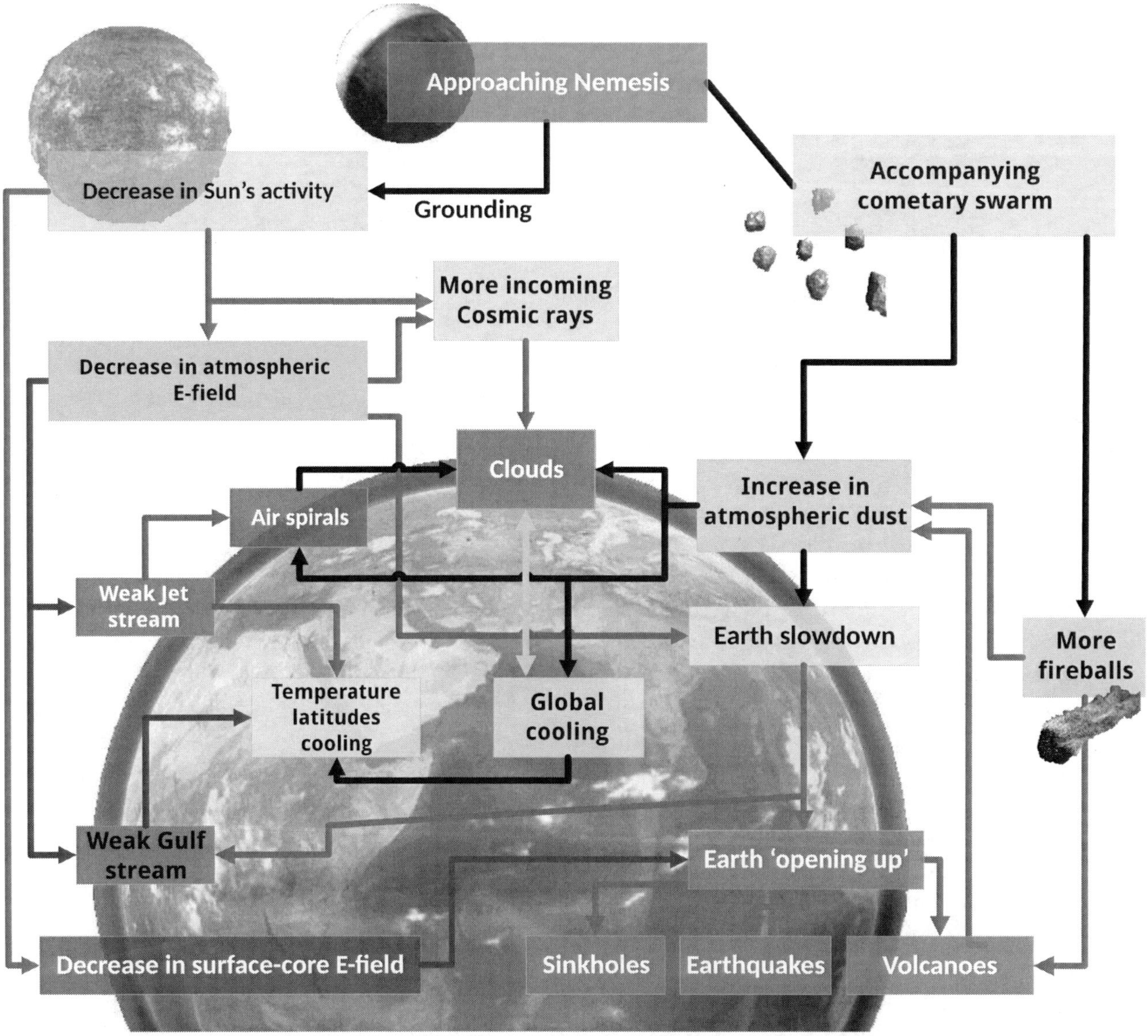

Figure 182: Global effects of the approaching Nemesis and its accompanying cometary swarm. (© Sott.net)

swarm. In that sense, they are all related to a *systemic, growing and global phenomenon*, and they affect all regions of the planet.

For decades, some 'heretical' scientists have studied the electric nature of the universe, cometary cycles, the impact of solar activity on Earth, and **cyclical catastrophes**. Their work presents fundamental challenges to the prevailing orthodoxy, which **assumes that we live in a predictable and stable universe**.

The problem for the high priests of official science is that, if they were to recognize the role played by the Sun, its companion and a cometary swarm in these phenomena, they would ultimately be faced with the following fundamental questions: 'What factor(s) modulate the Sun's activity?' 'Do we have any power to alter these phenomena?' 'What future awaits us?' 'What is the cause of the ongoing changes?'

The vast disinformation campaign led by the authorities regarding these 'anomalies' is betrayed by their dubious theories, which obfuscate the common cause connecting all these phenomena, the true gravity of the situation, and the probable results in our near future. *If people knew the truth, they would realize what our future has in store. They would realize the lies they've been fed. They would realize that the elites do not have their best interests in mind or at heart, and that so much more could be done for humanity at large if the rulers would stop the destruction, the raping and pillaging of the Earth, and started fulfilling their responsibilities.*

And they might even realize – as populations have done in the past – that those very elites who are supposed to protect them might very well be the cause of this threatening situation and our gloomy future. This is what we will focus on in the final part of this book.

Part IV.

Role of the elites, the human-cosmic connection

Chapter 32.

The mandate of Heaven

The legitimacy of the ruling class – whatever political form it may take – is based on the *illusion* that it can protect the people, whether from war, famine, economic hardship or any other kind of disaster that disrupts the everyday routine of their lives and livelihood. The following excerpt illustrates this point:

> I want this country to realize that we stand on the edge of oblivion. I want every man, woman and child to understand how close we are to chaos. **I want everyone to remember why they need us**![664]

If the earthly disruptions described in Part III were, as these very authorities claim, man-made and/or negligible, the elites would indeed be able to control them. As frightening as the problems really are, this is a comforting thought, and it is used expertly on the public in this craze to control CO2 (see figure 185).

In fact, the anthropogenic global warming scheme was essentially a propaganda campaign – not to mention a vast financial scam and guilt trip imposed by those who consume the most – designed to convey the message that the cause of the so-called climate change is known and, therefore, ultimately fixable. Indeed, the authorities even designed workable solutions, which were then implemented, such as banning CFCs,[665] promoting clean energy[666] and reducing CO2 emissions.[667]

When people begin to become aware of atmospheric, geological and climatic disruptions and all the ills they bring upon a society, they individually and collectively look to their leaders to do something about it – to 'fix things'. This is the origin of the concept of the Divine King.[668] Traditionally, the king is supposed to be able to intercede with the gods for his people. If the king is unsuccessful with his intercession, a solution has to be found. Sacrifices are made, rituals performed and, of course, if all that doesn't work – if the gods remain angry – then the king has to die.

---

[664] Wachowski, A. & Wachowski, L., *V for Vendetta* (film)

[665] Van Dyke, D., 'The CFC Ban: Global Warming's Pilot Episode', *americanthinker.com* (4 February 2010). See: www.americanthinker.com/2010/02/the_cfc_ban_global_warmings_pi.html

[666] See, for example: 'Clean Energy Solutions', *Toowarm.org.* www.toowarm.org/globalwarming/cleanenergy/

[667] See, for example: 'Overview of Greenhouse Gases', *United States Environmental protection Agency.* www.epa.gov/climatechange/ghgemissions/gases/co2.html

[668] Sommerville, J., 'The Divine Right of Kings', *Wisconsin University.* See: faculty.history.wisc.edu/sommerville/367/367-04.htm

Figure 183: Chancellor Adam Sutler urging his lieutenants to hystericize the population. Snapshot from the film 'V for Vendetta'. (© Warner Bros)

Figure 184: The high priest places the sins on the scapegoat. (© Davis Collection)

This impulse may stem from the same or a similar 'brain switch' that drives people to seek whatever relieves stress, no matter how irrational the means. In other words, if the gods are angry, find a scapegoat. When it is the nation that is threatened, the most obvious guilty person or persons are those in charge: the king and his elite. What's more, rulers instinctively know their vulnerability to this reaction and take steps to avoid it.

Looked at from a different angle, given that human history appears to be defined by a succession of more or less corrupt ruling elites, and if we assume that such corruption (and its spread throughout society) is the mechanism by which a civilization attracts cosmic catastrophes (as we will argue below), then blaming and deposing the ruling elite *may actually be an effective solution.* The problem, however, is that the underlying mechanism is not understood by the people, which means that they lack the knowledge that, if they are to avoid further destruction, they must, at all costs, first prevent the establishment of any future corrupt elite;[669] they must avoid the 'meet the new boss, same as the old boss' trap.

In the end, the people and the elite both seek a paradigm that downplays cyclical catastrophes, but they do it for different reasons. The people want to relieve the enormous stress of a certain but unpredictable major catastrophe – they wish to believe in the overall stability and safety of the cosmos – while the elite wants to remain in power. The compromise that serves both objectives is the illusion of an elite that is able to protect the people from any disaster. This illusion can take various forms: rituals to appease the gods, revision of history to show a uniformitarian evolution of humanity, the manipulation of science, lots of propaganda, etc.

This lie works especially well during periods of calm between major catastrophes. When the skies

[669] Le Bon, G., *The Crowd: A Study of the Popular Mind*

are quiet, it's easier to believe that they have always been and will always be so. However, history shows that whenever a situation arises over which the ruling elites have no control – when famines, earthquakes and plagues have struck and taken a heavy toll, when volcanoes erupt or comets blaze across the sky, or when meteor storms and weather anomalies increase – the illusion falls apart, the *raison d'être* of the elites (i.e. protecting the people) collapses and the target has always, and ever will be, ultimately, the ruling classes. And they know it.[670]

That is why the symptoms of increased cometary activity are systematically covered up by the elites as man-made phenomena. The contrails due to higher concentrations of atmospheric cometary dust are depicted as 'chemtrails' sprayed by government agencies, the ever more frequent overhead cometary explosions are presented as 'missile tests' or sonic booms from military jets, the cosmically-induced global cooling is labeled 'anthropogenic global warming', the increase in atmospheric dust of cometary origin is considered a consequence of industrial pollution, increasing weather extremes are blamed on HAARP and other 'top secret' government-military programs, sinkholes are blamed on defective piping, etc.

By attributing the cause of these cosmically-induced events to humans, the elites maintain the illusion that they are, at least to some degree, in control; if they are causing it, then theoretically, at least, they could stop it.

'Chemtrails' can be stopped if enough people petition the world's governments to cease the 'sprayings'; missile tests can be stopped if military budgets are reigned in; global warming can be reduced by controlling man-made greenhouse gas emissions; burst pipes can be repaired; global dimming will vanish if we stop polluting the atmosphere: 'Man is in control and all is right with the world'.

However, contrails triggered by increased cometary activity, overhead cometary explosions, cosmically-induced weather disruptions and the accumulation of cometary dust in the atmosphere cannot be changed by the elites, even if just theoretically. The intimation that the cosmos is threatening the survival of humanity is a stress too intense to be borne by most people.

Even if 'chemtrails', global warming and missile tests are not actually stopped in practice, the very idea that they *can be* stopped is enough to invite people to continue supporting the illusion of control, wishfully thinking that all it will take is a public awareness campaign, new elections, political advocacy, protests, the scapegoating of 'guilty parties', etc. But it's all a lie, and if the public realizes this, the only thing their stressed brains can do for relief is to think that 'the gods are angry' and, collectively, to try to find the real reason. Again and again throughout history, this has resulted in the masses coming to the conclusion that the 'gods' are angry because of the corruption and violence perpetrated by the elites in their efforts to attain and maintain greater power. Based on the evidence we will discuss below, it is entirely possible that this is an accurate assessment.

If the masses of humanity were to become aware of the real causes of such phenomena, that knowledge would not only entail recognizing the ultimate powerlessness of the elites, but also the end of their 'Mandate of Heaven', a concept that dates back to the Zhou dynasty[671] and which is still prevalent in contemporary China:

> The Chinese believed that an emperor could reign only while he enjoyed the Mandate of Heaven,

[670] Lippman, W., *Public Opinion*

[671] The Zhou dynasty ruled China during the 11th century BC.

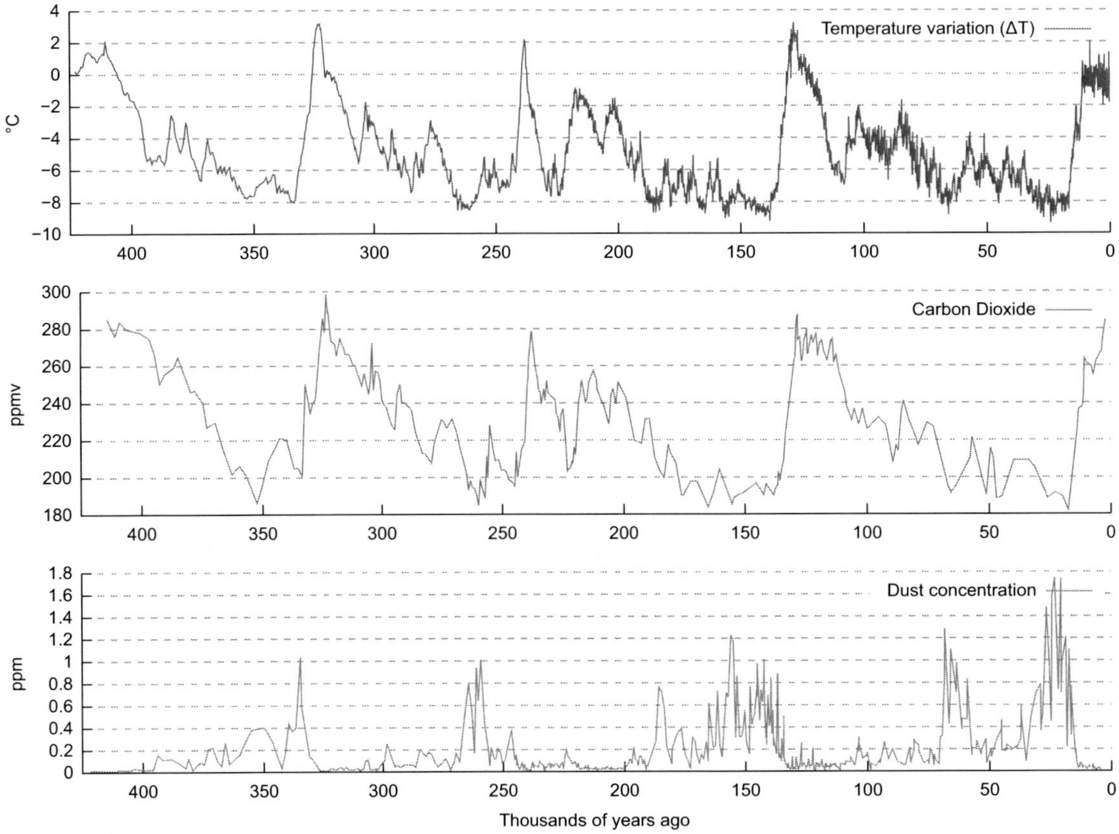

Figure 185: For the past 425,000 years, much earlier than the industrial revolution, the Vostok ice cores show a strong correlation between high dust concentration (bottom curve), CO2 reduction (middle curve) and temperature drop (top curve).

> that is, while he 'looked after his people'; **if for any reason he failed to look after their well-being, Heaven would withdraw its Mandate and the emperor and probably his ruling dynasty would be deposed.** ... Heaven would have been seen to withdraw its Mandate when the sky darkened, the crops failed and famine ensued bringing death to large numbers of people. The emperor, guilty or not, gets the blame for failing his people. In the aftermath of a calamitous dust-veil event the political upset could easily lead to the deposing of the ruling regime.[672]

Figure 186: The Mandate of Heaven, 'Tianming' in Chinese, is the conditional right to rule (Ming) bestowed by Heaven (Tian). (©Sott.net)

Of course, when the first signs of trouble appear, people want to believe that their government (their kings and ruling elite) is powerful enough – or pure enough – to control nature such that the floods, wildfires, earthquakes and volcanic eruptions cease, and all returns to normal. The ruling elites take advantage of this state of affairs by quickly identifying likely scapegoats in other nations, some minority group in their own country, their personal enemies or whomever else they deem to be suitable targets – including those who are pointing out that it just might be the elite's own corruption that is bringing on all the evils – so as to distract the larger populace from noticing the elite's own possible sins of commission or omission.

This means that such a period can include protracted wars and the accompanying necessary burden of taxes, persecutions of this or that group, generally increasing social hysteria and unrest, until finally, one day, the people as a whole wake up and see that their rulers have behaved very badly, place the blame squarely on their shoulders and clean house. Complete destruction of worn-out civilizations has historically been accomplished by masses of people under the rule of crowd psychology as stated by Gustave Le Bon:

> A crowd may easily enact the part of an executioner, but not less easily that of a martyr. It is crowds that have furnished the torrents of blood requisite for the triumph of every belief.[673]

[672] Baillie, M., *Exodus to Arthur*, p. 62
[673] Le Bon, *op. cit.*, p. 18

Chapter 33.

The cover-up

It is very unlikely that the elites believe the lies they are spreading via mainstream science and the media. This is particularly true for the man-made global warming scam. In the following paragraph, we will see that at least some members of the elite obviously know what is happening and what is going to happen (and which they hopelessly try to hide). But we will also see that they seem to lack some essential pieces of information.

As early as 2003, while the mainstream media were hammering home the global warming scam, the elites were starting to prepare for the opposite scenario. In the midst of patronizing and fear-inducing images such as a poor polar bear desperately hanging on to an almost-melted raft of ice (see figure 187), oceans rising up to engulf the Statue of Liberty, glaciers inexorably melting and CO2-generating cow flatulency, the Pentagon was seriously considering the possibility of global cooling:

> The Pentagon released a controversial report titled *An Abrupt Climate Change Scenario and Its Implications for United States National Security.*[674] The report explained how global warming might lead to **rapid and catastrophic global cooling** through mechanisms such as the slowing of North Atlantic deep-water circulation.[675]

Figure 187: Fine piece of propaganda evoking the myth of the last polar bear hanging to the last chunk of ice.

Obviously, this report omits the actual causes of global cooling and only mentions it in the context of 'man-made global warming'![676] This omission might be an attempt to save face by preserving the myth of man-made global warming, or it might be a case of deliberately hiding the real cause. In any case, the implications of this report seem to have been taken seriously by the Pentagon itself.

[674] Schwartz, P. & Randall, D., *An Abrupt Climate Change Scenario and Its Implications for United States National Security*, October 2003

[675] Fleming, J. R., 'The-Climate-Engineers', *The Wilson Quarterly*

[676] See Chapter 24: 'Global cooling'

Figure 188: Luxury Hotel Dolce Sitges, south of Barcelona, venue for the 2010 Bilderberg meeting.

In addition, in 2010, while the elites were implementing carbon taxes[677] and raking in billions from carbon credits,[678] they were also – privately – addressing global cooling:

> The 58th Bilderberg Meeting will be held in Sitges, Spain 3–6 June 2010. The Conference will deal mainly with Financial Reform, Security, Cyber Technology, Energy, Pakistan, Afghanistan, World Food Problem, **Global Cooling**, Social Networking, Medical Science, EU-US relations.[679]

Actually, as early as 1974, the CIA was preparing for another ice age.[680] But the elites didn't simply limit themselves to theoretical work and academic discussions about a potential global cooling: **they acted upon this knowledge**. One of the most interesting features of their response is the recent construction of so-called seed banks. Such a structure was recently built in the Himalayas (figure 189).

> Inside a stone and wood-paneled building on a frozen mountain top in Ladakh, on the road from Leh to Pangong Lake, seeds of vegetables and other plants sealed in moisture-proof packets sit on steel racks. '**This is Noah's Ark type of activity**,' said William Selvamurthy, a senior scientist who heads the life sciences division at India's Defence Research and Development Organization, which has funded the Rs 2 crore Chang-La seed vault.[681]

[677] According to the Australian Climate Commission, by 2013 the carbon tax should be paid in 33 countries, amounting to around 850 million people, around 30 percent of the global economy and around 20 percent of global emissions. See: Flannery, T., Beale, R. & Hueston, G., *The critical decade: International action on climate change*, Climate Commission, 2012

[678] The carbon credits market is expected to become the largest financial commodities market in the world, worth some $1 trillion by 2017. See: Kanter, J., 'Carbon trading: Where greed is green', *New York Times* (20 June 2007). See: www.nytimes.com/2007/06/20/business/worldbusiness/20iht-money.4.6234 700.html?_r=2&

[679] Delingpole, J., 'Global Cooling and the New World Order', *The Telegraph* (26 September 2010). See: http://blogs.telegraph.co.uk/news/jamesdelingpole/100055 500/global-cooling-and-the-new-world-order/

[680] Office of Research and Development of the CIA, *A study of Climatological Research as it Pertains to Intelligence Problems*

[681] Mudur, G., 'Seed safety vault in Ladakh', *The Telegraph India* (16 May 2010). See: http://www.telegraphindia.com/1100 517/jsp/nation/story_12457 330.jsp

Besides the Ladakh 'Noah's Ark',[682] there is the giant underground facility built in Svalbard, Norway (figure 190), and funded by 'insiders' like David Rockefeller, Bill Gates and Monsanto, who have invested millions in this project. The 'seeds' will be very well protected indeed:

Figure 189: Above-ground buildings of the Chang-la seed vault. (© thehindu.com)

> In 2006 when most people in such a situation might think of retiring to a quiet Pacific island, Bill Gates decided to devote his energies to his *Bill and Melinda Gates Foundation*, the world's largest 'transparent' private foundation as it says, with a whopping $34.6 billion. No project is more interesting at the moment than a curious project in one of the world's most remote spots, Svalbard. **Bill Gates is investing millions in a seed bank on the Barents Sea** near the Arctic Ocean, some 1,100 kilometers from the North Pole. On this God-forsaken island **Bill Gates is investing tens of his millions along with the Rockefeller Foundation, Monsanto Corporation, Syngenta Foundation and the Government of Norway, among others**, in what is called the 'doomsday seed bank'. The bank will have **dual blast-proof doors with motion sensors, two airlocks, and walls of steel-reinforced concrete one meter thick**. It will contain up to three million different varieties of seeds from the entire world, 'so that crop diversity can be conserved for the future,' according to the Norwegian government. Seeds will be specially wrapped to exclude moisture. There will be no full-time staff, but the vault's relative inaccessibility will facilitate monitoring any possible human activity.[683]

Did we miss something here? The press release stated: "so that crop diversity can be conserved for the future." What future do the seed bank's sponsors foresee that would threaten the global availability of current seeds? Why create seed vaults when there are already numerous existing seed banks all over the world?[684]

One can legitimately wonder: why is Monsanto investing millions of dollars in a 'seed vault' while at the same time destroying existing seeds because of cross-contamination with their GMO seeds?[685] Why did Bill Gates quit his prominent professional career and invest millions in a 'seed

[682] The Chang-La seed vault is directly protected by the Indian Army. The vault itself is within defense-protected land. See: *The Times Of India*, Bangalore (22 February 2010), p. 12

[683] 'Doomsday Seed Vault in the Arctic', *globalresearch.com* (23 May 2013). See: http://www.globalresearch.ca/doomsday-seed-vault-in-the-arctic-2/23 503

[684] There are already about *1,300* seed banks in the world. See: Wei, J., 'A one world seed bank vault', *Jerilee Wei Blog*. http://jerileewei.hubpages.com/hub/A-One-Bank-World

[685] Sharratt, L., '15 years of GMO contamination in Canada', *Common Ground* (7 February 2011). See: http://www.sott.net/article/223 340-15-years-of-GMO-contamination-in-Canada

bank'? When did the Rockefellers start caring enough for posterity to preserve humanity's heritage flora? One can wonder if the elites are actually *using seed vaults as a pretext or 'cover' to hide the construction of massive underground facilities.* Are the elites aware of the gloomy near future and preparing for their own survival while lying to the rest of the population and making them even more vulnerable to coming events?

Figure 190: Above-ground section of the Svalbard 'Seedbank'.

Along with seed vaults, there seems to be a strong increase in *massive land purchases by rich individuals, companies and countries.* One can only wonder if these two moves are motivated by the same factor.

> The World Bank is playing a leading role in a global land grab, say farmers' movements and international allies. The World Bank's policies for land privatization and concentration, have paved the way for corporations from Wall Street to Singapore to take land from rural communities across the world in the past few years.[686]

The phenomenon of 'land grabbing' is particularly aimed at lands in the southern hemisphere (Latin America and Africa).[687] Depending on sources, land grabbing has amounted to 50 to 80 million hectares over the last few years.[688]

Some interesting examples illustrate this land grabbing dynamic. The **Bush family** purchased a 45,000-hectare (100,000 acres) ranch in Paraguay in 2006.[689] In 1999, **Sun Myung Moon** purchased 600,000 hectares (about 1.4 million acres) of land in the very same area of Paraguay, above one of the biggest aquifers in Latin America.[690] [691]

---

686 'World Bank And Wall Street Support Land Grabbing In Developing Countries', *globalresearch.com* (20 April 2012). See: www.globalresearch.ca/world-bank-and-wall-street-support-land-grabbing-in-developing-countries/30423

687 Hall, R., 'Land Grabbing in Africa and the New Politics of Food', *Future Agricultures* (June 2011)

688 That's about 200,000 square miles, more than California's total surface area. See: 'Land Grabbing', *Wikipedia.* http://en.wikipedia.org/wiki/Land_grabbing

689 'Paraguay in a spin about Bush's alleged 100,000 acre hideaway', *The Guardian* (23 October 2006). See: http://www.guardian.co.uk/world/2006/oct/23/mainsection.tomphillips

690 'Moonshadow', *Rigorous Intuition Blog* (16 October 2006).
See: web.archive.org/web/20110728003314/http://rigint.blogspot.com/2006/10/moonshadow.html

691 'News of the Puerto Casado Railway', *The International Steam Pages.*
See: www.internationalsteam.co.uk/trains/para09.htm

Figure 191: Puerto Casado, one of the villages included in the land purchased by Korean tycoon Sun Myung Moon. (© ea.cpm.py)

In 2009, Prince Willem and Princess Maxima of the Netherlands purchased 1,500 hectares (about 4,000 acres) in Patagonia (Southern Argentina).[692] In 2007, the New Zealand Overseas Investment Office investigated David de Rothschild over the work planned for the Banks Peninsula farm, 442 hectares (about 1,000 acres) of land located in New Zealand that Rothschild bought in 2002.[693] In South Sudan, one of the most prominent land-grabbers is former AIG partner Philippe Heilberg. He garnered attention in *Rolling Stone* for his aggressive pursuit of land in conflict-ridden regions, not least Sub-Saharan Africa, where Heilberg is planning to invest in 800,000 hectares of land in partnership with many of South Sudan's top generals and civilian officials.

This land-grabbing business is growing so much that several companies have sprung up, dedicated to promoting the activity. This is the case with two French corporations: Charles Beigbedor's Agrogeneration Group, which has invested heavily by purchasing land in Ukraine, and Olivier Combastet's Pergam Finance, which is based in South America and gives international clients the opportunity to buy land in Argentina, Chile and Uruguay.[694] Some observers suspect that these massive investments in remote places are not driven by profit-seeking, but *fear*:

> It may seem a little odd that in 2011 anyone's thinking of putting money into assets that would have seemed attractive in 1911, but there's something in the air – namely, fear. The hedge

[692] 'Royal couple buy land in Argentina', *Dutchnews.nl* (25 November 2009). See: www.dutchnews.nl/news/archives/2009/11/royal_couple_buy_land_in_argen.php

[693] 'Rich-lister in a bit of bother down south', *New Zealand Herald* (20 May 2007). See: www.nzherald.co.nz/nz/news/article.cfm?c_id=1&objectid=10440 712

[694] Petric, Boris, 'The land rush - Transnational strategies for land grabbing', *Revues transcontinentales* (10 November 2011). See: www.transcontinentales.revues.org/1303

> fund manager and others like him *envision a doomsday scenario* catalyzed by a weak dollar, higher-than-you-think inflation and an uncertain political climate here and abroad.[695]

Here is a glimpse at the elites' state of mind, how they envision the future and how they plan to face it:

> Al Corbi's residence in the Hollywood Hills has the requisite white walls covered in artwork and picture windows offering breathtaking views of downtown Los Angeles, but it has more in common with NSA headquarters than with the other contemporary homes on the block. The Corbi family doesn't need keys (thanks to biometric recognition software), *doesn't fear earthquakes* (thanks to steel-reinforced concrete caissons that burrow 30 feet into the private hilltop) and sleeps easily inside a 2,500-square-foot home within a home: a *ballistics-proof* panic suite that Corbi refers to as a 'safe core.'
>
> Paranoid? Perhaps. But also increasingly commonplace. Futuristic security technologies – many developed for the military but sounding as though they came straight from James Bond's Q – have made their way into the home, available to deep-pocketed owners.[...]
>
> Chris Pollack, president of *Pollack&Partners*, a design and construction advisor based in Purchase, N.Y., says that while security has always been a given in building homes for his ultra-high-net-worth clients, the spending for home defense has increased markedly over the last five years. ...
>
> The **panic room** has also undergone a high-tech evolution that makes the old Jodie Foster movie look quaint. A prominent author who declined to be identified has had his South Florida home jerry-rigged with a **physical perimeter alarm, motion sensors throughout rooms and stairways, and a heat sensor** that detects room temperature differentials caused by a sudden change in body heat. If worse comes to worst: The third-floor master suite is outfitted as a **2,500-square-foot safe haven**. Switches installed throughout the house will encapsulate the space, locking down its three entrances with **reinforced doors** while alerting local authorities. Taking it further, the space's bathroom doubles as an inner panic room, protected by 'a **silent home defense system with sufficient armament**.' Fortunately, it has never come into use. ...
>
> Luxury home builders have discovered that buyers will pay top dollar for such extensively fortified mansions. One compound on hyperexclusive Indian Creek Island in Miami, completed in early 2012, includes a **fully stocked safe room that operates on its own generator and has a separate ventilation system, a wine cellar wired with facial recognition software and an infrared surveillance system**. The property traded last year for $47 million – the highest price ever paid for a house in Miami-Dade County. ...
>
> Corbi says his SAFE has projects catering to such anxieties under way on four continents. These bunkers can span tens of thousands of square feet and tunnel as much as 30 stories into the ground. They are typically hidden in plain sight, accessed by secret passageways like rotating fireplaces and floors that lift up to expose an underground staircase.
>
> Corbi's own home has such a facility in the basement – a space that doubles as a wine cellar. If the 'safe core' is compromised or the rooftop helipad (complete with evacuation plans signed into a contract with a local helicopter company) isn't accessible, **the family can retreat to**

[695] Kamer, F. 'Hedge Farm! The Doomsday Food Price Scenario Turning Hedgies into Survivalists', *observer.com* (18 May 2011). See: www.observer.com/2011/05/hedge-farm-the-doomsday-food-price-scenario-turning-hedgies-into-survivalists/

Figure 192: In billionaire Corbi's home the wine cellar doubles as an underground bunker, stocked with provisions of food and water. (© Forbes)

> **this stronghold, thwarting natural disaster, nuclear holocaust or electromagnetic pulses emitted by the sun.** [696]

While billionaires get ready for the worst, they claim everything is all right, or worse, they push people in the wrong direction, priming them to buy air conditioners to prevent global warming while they should be buying wood-burning stoves to face global cooling. Of course, if ordinary people get clued in and also begin to 'prepare', they are labeled 'dangerous, anarchical preppers'.

[696] Brennan, M., 'Billionaire Bunkers: Beyond the Panic Room, Home Security Goes Sci-Fi', *Forbes Magazine* (16 December 2013). See: www.forbes.com/sites/morganbrennan/2013/11/27/billionaire-bunkers-beyond-the-panic-room-home-security-goes-sci-fi/

Chapter 34.

Historic evidence of a human-cosmic connection

As noted above, our ancestors believed deeply in the connection between human behavior and natural catastrophes. If humans behaved well, Nature would remain calm, but if humans behaved badly, this would trigger the wrath of the 'gods' and Nature would react fiercely.

It was not only the 'common' layers of society that held a belief in this human-cosmic connection. Scholars, oracles and leaders subscribed to a similar belief. Chroniclers (the historians of the ancient times) wrote endless texts about cosmic calamities (the gods' wrath) triggered by abuses perpetrated by the elites. This was actually the core of their writing.

Chronicler Michael the Syrian[697] went even further. He structured each page of his chronicles in two columns. On the right, he listed the political affairs and conduct of the elites, while the left was dedicated to the subsequent natural catastrophes. Of course, he also abundantly commented on the links between the two because of the difficulties discerning exactly which political scandal caused which environmental calamity.[698]

To give just one example of the correlation between cosmic disasters and leaders' misconduct, we will focus on the last decades of the Roman Empire (530–580 AD). The following is an excerpt from Procopius's *Secret History*, entitled 'How Justinian Killed a Trillion People'. Keep in mind that Procopius was probably the most prominent historian of the 6th century. Most scholars consider him the last major historian of the ancient world.[699] Procopius was not only an historian; for more than ten years he was also a counselor to the greatest general of the time, Belisarius.[700] As such, during this period of numerous travels, battles and negotiations, he was a *direct witness* of Roman political and military life. After his retirement from military affairs, he became a judge and, in that capacity, he had the ways and means to obtain inside information that made up his *Secret History*.[701]

Procopius's account is such a vivid description of a plutocratic regime, with all its violence, corruption, betrayal, cruelty and incompetency that it's well worth quoting at length. Of course, any resemblance to modern times is purely coincidental.

---

697 One of the most important Syrian Orthodox patriarchs who wrote a universal chronicle in Syriac, covering events from the Creation until 1195. See: Weltecke D., 'The World Chronicle by Patriarch Michael the Great (1126–1199): Some reflections', *Journal of Assyrian Academic Studies*

698 Michel le Syrien, *Chronique de Michel le Syrien*, books I, II, III and IV (not translated into English).

699 'Procopius of Caesarea', *About.com.* See: ancienthistory.about.com/od/historians/g/013 108Procopius.htm

700 'Medieval Sourcebook: Procopius of Caesarea: The Secret History', *Fordham university.* See: www.fordham.edu/halsall/basis/procop-anec.asp

701 See Anthony Kaldellis's excellent work: *Procopius of Caesarea: Tyranny, History, and Philosophy at the End of Antiquity*

Figure 193: Michael the Syrian, also known as Michael the Great.

Figure 194: Mosaic possibly representing General Belisarius (Ravenna Cathedral). (© The York Project)

That Justinian was not a man, but a demon, as I have said, in human form, one might prove by considering the enormity of the evils he brought upon mankind. For in the monstrousness of his actions the power of a fiend is manifest. Certainly an accurate reckoning of all those whom he destroyed would be impossible, I think, for anyone but God to make. Sooner could one number, I fancy, the sands of the sea than the men this Emperor murdered.

Examining **the countries that he made desolate of inhabitants**, I would say he slew a trillion people. For Libya, vast as it is, he so devastated that you would have to go a long way to find a single man, and he would be remarkable. Yet eighty thousand Vandals capable of bearing arms had dwelt there, and as for their wives and children and servants, who could guess their number? Yet still more numerous than these were the Mauretanians, who with their wives and children were all exterminated. And again, many Roman soldiers and those who followed them to Constantinople, the earth now covers; so that if one should venture to say that **five million men perished in Libya** alone, he would not, I imagine, be telling the half of it.

The reason for this was that after the Vandals were defeated, Justinian planned, not how he might best strengthen his hold on the country, nor how by safeguarding the interests of those who were loyal to him he might have the goodwill of his subjects: but instead he foolishly recalled Belisarius at once, on the charge that the latter intended to make himself King (an idea of which Belisarius was utterly incapable), and so that he might manage affairs there himself and be able to plunder the whole of Libya. Sending commissioners to value the province, he **imposed grievous taxes where before there had been none**. Whatever lands were most valuable, he seized, and prohibited the Arians from observing their religious ceremonies. Negligent toward sending necessary supplies to the soldiers, he was over-strict with them in other ways; wherefore mutinies arose resulting in the deaths of many. For he was **never able to abide by established customs, but naturally threw everything into confusion and disturbance**.

Italy, which is not less than thrice as large as Libya, was everywhere desolated of men, even worse than the other country; and from this the count of those who perished there may be imagined. The reason for what happened in Italy I have already made plain. All of his crimes in Libya were repeated here; sending his auditors to Italy, **he soon upset and ruined everything**.

The rule of the Goths, before this war, had extended from the land of the Gauls to the boundaries of Dacia, where the city of Sirmium is. The Germans held Cisalpine Gaul and most of the land of the Venetians, when the Roman army arrived in Italy. Sirmium and the neighboring country was in the hands of the Gepidae. **All of these he utterly depopulated. For those who did not die in battle perished of disease and famine,** which as usual followed in the train of war. Illyria and all of Thrace, that is, from the Ionian Gulf to the suburbs of Constantinople, including Greece and the Chersonese, were overrun by the Huns, Slavs and Antes, almost every year, from the time when Justinian took over the Roman Empire; and intolerable things they did to the inhabitants. For in each of these incursions, I should say, **more than two hundred thousand Romans were slain or enslaved, so that all this country became a desert** like that of Scythia. ...

For neither the Persians nor the Saracens, the Huns or the Slavs or the rest of the barbarians, were able to withdraw from Roman territory undamaged. In their inroads, and still more in their sieges of cities and in battles, where they prevailed over opposing forces, they shared in disastrous losses quite as much. Not only the Romans, but nearly all the barbarians thus felt Justinian's **bloodthirstiness**. For while Chosroes himself was bad enough, as I have duly shown elsewhere, Justinian was the one who each time gave him an occasion for the war. For **he took no heed to fit his policies to an appropriate time, but did everything at the wrong moment: in time of peace or truce he ever craftily contrived to find pretext for war**

with his neighbors; while in time of war, he unreasonably lost interest, and hesitated too long in preparing for the campaign, grudging the necessary expenses; and instead of putting his mind on the war, gave his attention to stargazing and research as to the nature of God. Yet he would not abandon hostilities, since he was so bloodthirsty and tyrannical, even when thus unable to conquer the enemy because of his negligence in meeting the situation.

So **while he was Emperor, the whole earth ran red with the blood** of nearly all the Romans and the barbarians. Such were the results of the wars throughout the whole Empire during this time. But the civil strife in Constantinople and in every other city, if the dead were reckoned, would total no smaller number of slain than those who perished in the wars, I believe. Since justice and impartial punishment were seldom directed against offenders, and each of the two factions tried to win the favor of the Emperor over the other, neither party kept the peace. Each, according to his smile or his frown, was now terrified, now encouraged. Sometimes they attacked each other in full strength, sometimes in smaller groups, or even lay in ambush against the first single man of the opposite party who came along. For thirty-two years, without ever ceasing, they performed outrages against each other, many of them being punished with death by the municipal Prefect. However, punishment for these offenses was mostly directed against the Greens.

Furthermore the **persecution of the Samaritans and the so-called heretics filled the Roman realm with blood**. Let this present recapitulation suffice to recall what I have described more fully a little while since. **Such were the things done to all mankind by the demon in flesh for which Justinian, as Emperor, was responsible.** But what evils he wrought against men by some hidden power and diabolic force I shall now relate.

During his rule over the Romans, many disasters of various kinds occurred: which some said were due to the presence and artifices of the Devil, and others considered were effected by the Divinity, Who, disgusted with the Roman Empire, had turned away from it and given the country up to the Old One. The **Scirtus River flooded Edessa**, creating countless sufferings among the inhabitants, as I have elsewhere written. The **Nile, rising as usual, but not subsiding in the customary season**, brought **terrible calamities** to the people there, as I have also previously recounted. The **Cydnus inundated Tarsus, covering almost the whole city for many days**, and did not subside until it had done irreparable damage.

Earthquakes destroyed Antioch, the leading city of the East; Seleucia, which is situated nearby; and Anazarbus, most renowned city in Cilicia. Who could number those that perished in these metropoles? Yet one must add also those who lived in Ibora; in Amasea, the chief city of Pontus; in Polybotus in Phrygia, called Polymede by the Pisidians; in Lychnidus in Epirus; and in Corinth: all thickly inhabited cities from of old. **All of these were destroyed by earthquakes during this time, with a loss of almost all their inhabitants**. And then came the **plague**, which I have previously mentioned, killing half at least of those who had survived the earthquakes. To so many men came their doom, when Justinian first came to direct the Roman state and later possessed the throne of autocracy.

Now I think I have shown sufficiently how **this destroyer devoured all the public moneys** and robbed each member of the Senate, publicly and privately, of all his estates; and how by bringing false charges he confiscated the properties of everybody else who was reputed to be wealthy, I imagine I have adequately told: as in the case of the soldiers, subordinate officers, and the palace guard; the farmers and landowners; those whose business is in words; merchants, ship owners and sailors; mechanics, artisans, and market dealers; those whose livelihood is in the theater; and indeed everyone else, who was affected in turn by the damage done to these. ...

One of the municipal aqueducts, which furnished not a small share of the city water, collapsed;

> but the rulers disregarded the matter and refused to repair it, though the constant crowds who had to use the wells were fairly stifling, and all the baths were shut down. On the other hand, he threw away great sums of money senselessly on buildings by the seashore and elsewhere, in all the suburbs, as if the palaces in which all the former emperors had been content to dwell were not enough for this pair. So **it was not to save money, but to destroy his subjects, that he refused to rebuild the aqueduct; for no one in all history had ever been born among men more eager than Justinian to get hold of money, and then to throw it immediately away again**. Through the two things left to them to drink and eat, water and bread, **this Emperor injured those who were in the last extremes of poverty**; making the one hard to procure at all, and the other too expensive to buy.
>
> The deeds of Justinian were such that all eternity would not be long enough in which to describe them adequately. So a few examples will have to suffice to illuminate his whole character to future generations: what a dissembler he was, how he disregarded God, the priests, the laws, and the people who showed themselves loyal to him. He had no shame at all, either when he brought destruction on the State or at any misdeed; he did not bother to try to excuse his actions, and his only care was how he might get sole possession of all the wealth of the world.
>
> Knowing countless other such acts of Justinian, I cannot include them, since the end of this book draws near. In any case, what I have told will be enough to show the nature of the man.[702]

Procopius clearly established a link between political abuses and calamities. In this regard, he was no different than other chroniclers. They all drew the parallel simply because they were direct witnesses of those two kinds of events, *experiencing them firsthand for years.*

Figure 195: Justinian (Mosaic of San Vitale cathedral in Ravenna). (© The York Project)

Some modern scholars try to explain away the chroniclers' remarkable accounts by saying they simply tended to *exaggerate*. This may be the case when it comes to such things as kill counts or military victories, but when it comes to obvious descriptions of cosmic events and catastrophes, how can we explain the fact that chroniclers who were sometimes hundreds of kilometers apart, and never met, give the same account of the same event? Embellishment is one thing; wholesale fabrication of calamitous events is quite another. When read with the possibility of such disasters in mind, their reports come across as objective and balanced as their writing about more mundane events.

Procopius' *Secret History* is undoubtedly one of the most bitterly passionate denunciations ever written. Procopius made his loathing of Justinian, Theodora (Justinian's wife) and everything they did, absolutely clear. He found no spark of humanity in them at all. It seems that he was even writing the text as a way to try to come to grips with what he was seeing and experiencing, because he had no classical examples in his training that could explain this to him. He asks the question at one point: "*How can anyone possibly explain in*

[702] Procopius, *The Secret History* (translated by Richard Atwater), chapter XVIII

words the character of Justinian?" The desperate urgency of the text reveals his frustration.

Figure 196: Empress Theodora. Mosaic detail from San Vitale.

Considering the fact that modern psychology studies reveal that individuals like Justinian and Theodora do, truly, exist, anyone who has studied psychopathology deeply, or who has experienced the type of individual Justinian was, in person, knows immediately and viscerally what he is talking about and why it affected him the way it did.

It is important to note that the Procopius' description of Justinian and Theodora and their reign has been confirmed by other sources dealing with the same events and validated by incontrovertible evidence. Anthony Kaldellis writes:

> A dispassionate analysis of the evidence points to the conclusion that Theodora was in fact petty, arrogant, wrathful, unforgiving, ruthless, and willing to sacrifice everything to her personal feuds and avarice. Such people do exist, and we should not dismiss the testimony of historians who try to tell us about them at great personal risk.[703]

Justinian reigned over the Byzantine Empire (i.e. the Eastern Roman Empire) from 527 until his death in 565.[704] The table in figure 197 shows a list of the catastrophes reported by the chroniclers[705] of the time between 535 and 591.

The frequency of calamities is high. This compilation is even more striking when we take into account the geographic scope of the recorded history (limited to the Mediterranean region), the time span (encompassing only 56 years) and the number of sources (limited to a dozen chroniclers).[706]

Comets are mentioned 16 times (medium grey), earthquakes 18 times (dark grey) and plague 19 times (medium grey).[707] What this table *doesn't* provide, however, is the magnitude of each event. Here are three excerpts from the previously mentioned chronicler Michael the Syrian that give an idea of what was going on at the time. Let's start with 'a sign in the Sun', which is quite the euphemism, as we'll see in the following description:

> A little earlier, in 536, there was a sign in the Sun. We had never seen such a sign before and it is nowhere written that it ever happened before in the world. If we hadn't found records in the most credible and proven writings and statements made by men worthy of faith, we would not have written it down because it is difficult to conceive. So it is said that the Sun was darkened,

[703] Kaldellis, *op. cit.*, p. 130

[704] Evans, J. A., 'Justinian', *De Imperatoribus Romanis*, University of British Columbia. See: http://www.roman-emperors.org/justinia.htm

[705] MtS = Michael the Syrian, Ps.D. = Pseudo Dyonisos, JoE = John of Ephesus, GoT = Gregory of Tours.

[706] Let's also keep in mind that for the most severe cataclysms there are no written accounts, simply because no witness survived. This may explain the numerous 'blank spells' of history, like the 'Dark Ages' from the 6th to 9th centuries AD.

[707] Interestingly, the Justinian plague appeared in 541, right after an extended period of cometary activity (538–541). See Chapter 21: 'Increase in cometary activity'

535	Double roll of "thunder" (Nan Shi) Krakatau eruption (Eye-witness)
536	Dim Sun almost a whole year. (Lydos, MtS, Zuqnin) Failure of bread 536 to 539 AD (Inisfallen, MtS) Pestilence (Procopius)
537	Drought, Mecca (Arab historians)
538	Pompeioupolis earthquake (Malalas, Mts, Zuqnin) Great and terrible comet (PZ, MtS) Great Beirut earthquake and tsunami (MtS)
539	Comet, famine, Vesuvius rumble (Procopius) Earthquake Antioch (Malalas, MtS)
540	Comet bombardment (Glidas) Flood (Sheba)
541	Plague Egypt (Malalas) Comet, Gaul (Roger of Wendover) Earthquake, Kyzikos (Malalas, Zuqnin) Comet, drought (Zuqnin)
542	Dim Sun, plague in the East (Boethius)
543	Plague, Mesopotamia (Zuqnin)
544	Plague, Italy, southern France, Spain
545	Plague in Persia (Procopius)
546	Famine, plague, Mesopotamia (Zuqnin)
547	Tremendous thunder and lightning (Malalas)
549	Cicilia, flood (Zuqnin) Plague British territories (Bishop of Llandaf)
551	Earthquake Beirut, tsunami, earthquake Middle East Sea retreats (Malalas, Ps.D., JoE) Plague (MtS)
553	Earthquake, terrible thunder and lightning (Theophanes)
554	Earthquake Constantinople (Malalas)
555	Earthquake Constantinople (Zuqnin) Plague (Zuqnin)
556	Famine + plague, Constantinople (Malalas) Ashes from the sky (Zachariah of Mitylene)
557	Plague (Zuqnin, MtS) Earthquake Constantinople (Malalas, Zuqnin)
558	Constantinople plague (Malalas)
559	Fire in Julian's harbor, plague+earthquake, Cicilia, Antioch (Theophanes)
562	Drought Constantinople (Malalas)
563	Shortage of water (Malalas)
565	Comet (Zuqnin, MtS)
566	Comet (MtS)
570	Earthquake, Edessa (MtS) Plague and famine, Yemen Bombardment of stones from the sky
574	Earthquake (MtS)
577	Famine, cometary bombardment (MtS) Comet (20 rays) (GoT)
580	Great Floods, comet, earthquake, epidemics (GoT)
581	Heavens aflame. Flood. Wind. Comet epidemic (GoT)
582	Flood. Comet. Sky in fire. Rained blood. Plague. Earthquake (GoT)
583	Great ball of fire fell from the sky, flood (GoT)
584	Locusts, plague, hailstorm, drought, earthquake (GoT)
585	Rays of light in the northern sky, clouds were blood-red (GoT)
586	Heavy rain, two islands were consumed by fire from the sky (GoT)
587	Rained heavily, a fall of snow buried everything, frost (GoT)
588	Plague (GoT)
589	Heavy rain and hail. Flood (GoT)
590	Comets, earthquake, eclipse of the Sun, plague, flood (GoT)
591	Eclipse of the Sun, violent earthquake, plague Constantinople (MtS)

Figure 197: Reported calamities (535–591 AD). (© Sott.net)

Figure 198: Veiled sun due to a high concentration in atmospheric particles. (© Leslie McCluskey)

> and *the eclipse lasted a year and a half, that is to say, eighteen months.* Every day he shone for about four hours and still this light was only a weak shadow. Everyone said the Sun would not return to its initial state. The fruits did not ripen, and the wine tasted like the one that comes from acid grapes.[708]

An explanation that comes to mind for such an extended 'sign in the Sun' is an atmospheric dust concentration of a magnitude such that the Sun was veiled for 18 months. The origin of this dust may have been strong cometary or volcanic activity, or a combination of the two, as we know they are often correlated.[709]

Now, here is an account of the plague that struck Constantinople in 543:

> When the plague fell upon the imperial city, it first occurred among the poor. There were days when we carried away 5,000 bodies, others 7,000, others 12,000, and 16,000 in one day. As it was in the beginning, men were staying on the ports, at the doors of the city, on the other bank, and reported their number. And if one wants to count, in fact, there had been more than 300,000 that were taken away from public places, when the number of individuals counting reached 1,000, they stopped counting. After the poor died, the devastation reached the powerful and the noble individuals of the empire, and the ones who had momentarily escaped death by avoiding the groin disease and the smallpox, which we call in our language 'tumors' and which they call bubons. We saw that this wound did not only occur amongst men, but also amongst animals, even wild animals, even reptiles of the earth. You could see the cattle and dogs, and even mice, with swollen groin, which succumbed and perished.
>
> ...The number of gravediggers was too small, and this big city became a gloomy sepulcher for its inhabitants. As we could not find any more tombs, they piled the corpses in boats and threw

[708] Michael the Syrian, *Chroniques Universelles* (translated by Sott.net), Book IX, Chapter XXVI

[709] See Chapter 23: 'Earth opening up'

> them like dung on the other bank. The Emperor saw what happened, ordered to make about six hundred stretchers, and appointed one of his clerks, whom he ordered to take and give as much gold as required to hire men to dig large pits in which corpses were to be piled up. It was done in the mountains, and *in each of these pits 70,000 cadavers were dropped.* He placed men around the pits who piled the corpses in rows, one above the other, in the same way hay is stacked. As a result, the town was freed from the cadavers. One of our deacons, engaged in this work, found a closed and stinky house. He entered and found about twenty people rotting and eaten by worms. He brought over a few men, and they carried the corpses away. There were women who had died and whose young children, still alive, held their mother's breast while crying.[710]

As you can see, we're not dealing with your usual epidemic here. The figures are staggering, even *if* we accept the possibility of some exaggeration. Before Justinian's plague, Constantinople's population was 500,000.[711] The plague killed nearly half its inhabitants. According to *modern day* estimates, it was said to have caused the deaths of at least one quarter of the human population in the Eastern Mediterranean;[712] and that was *just the first wave*; the plague returned again and again.

Here is one final excerpt describing the great earthquake that occurred in 559:

> When this terrible earthquake happened in Beirut and in other coastal cities of Phoenicia,[713] *the sea withdrew, by the order of God, over about two miles.* The bottom of the sea was uncovered, and we saw many things, wrecked ships full of goods. And instead [of being scared] by these horrors, those who [were on the edge of the sea] ran [eagerly] to capture these treasures hidden [in the depths], they entered [in the wrecks] and loaded [the treasures] to move them away. In turn, others rushed to get into the wrecks, and then the terrible power of the sea came back and swallowed them in its depths by the secret counsel of God. Those who were still on the coast, seeing the sea coming back, ran away to escape, but the quake toppled buildings on them, and buried them. This happened in all the coastal cities, and especially in Beirut, where the fire broke out after the destruction of the city. The fire lasted two months. *Even the stones were consumed and transformed into lime.*[714]

Here we find descriptions of a massive tsunami and a most peculiar fire that lasted two months and exhibited unusually high temperatures.[715]

The veracity of chroniclers' accounts of this time[716] was finally proved during the 1990s, when scientists discovered and explored several lines of evidence showing that a massive cometary body entered Earth's atmosphere in 536 and that this period was marked by exceptional climate stress. Several independent studies show a sequence of unprecedented cold winters starting in 536 (one of

[710] Michael the Syrian, *op. cit.*, Book IX, Chapter XXVIII

[711] Harl, W. H., 'Early Medieval and Byzantine Civilization: Constantine to Crusades', *Tulane University.* See: www.tulane.edu/~august/H303/handouts/Population.htm

[712] Mango, C., *Byzantium: The Empire of New Rome*

[713] Mediterranean coast of Lebanon.

[714] Michael the Syrian, *Chroniques Universelles* (translated by Sott.net), Book IX, Chapter XXIX

[715] Limestone transforms into lime at 900° Celsius (1,652° Fahrenheit).

[716] The severe disruptions experienced around 536 AD are independently reported by Procopius, Cassiodorus, John of Ephesus and Evagrius Scholasticus.

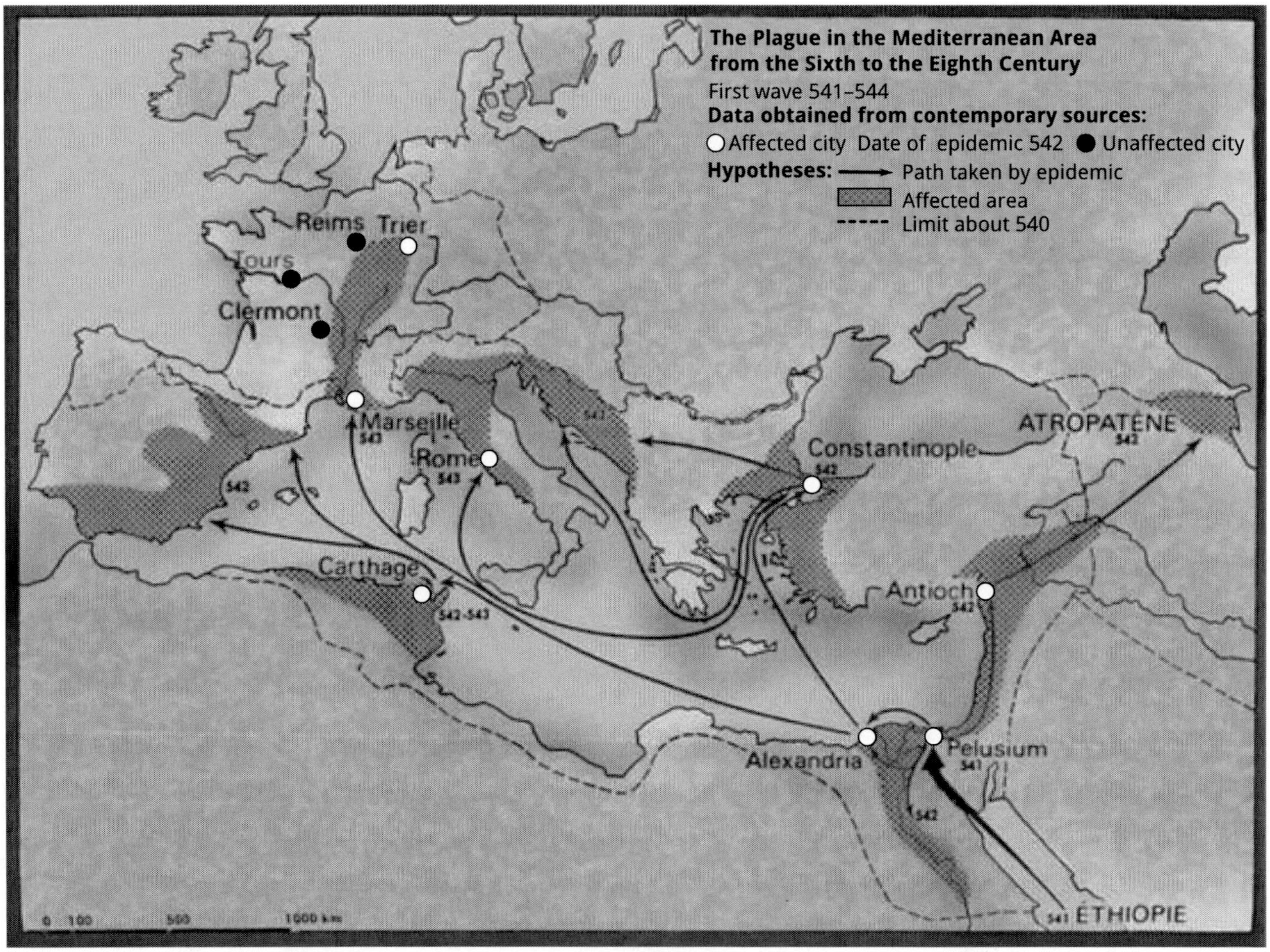

Figure 199: The first wave of the plague of Justinian (541–544 AD). (© University of California, Irvine)

the coldest years ever)[717] and ending around 545, as demonstrated by tree-ring analyses from wood samples found in Ireland,[718] Scandinavia,[719] continental Europe, North America,[720] Mongolia[721] and Argentina.[722] These studies all point to the same conclusion: the sudden cooling,[723] was due to an atmospheric dust veil event probably generated by the overhead explosion of massive cometary bodies[724] [725] – multiple Tunguska-like events or Shoemaker-Levy, if you prefer.

Bryan Ward-Perkins writes about the fall of the Roman Empire:

> The post-Roman world reverted to levels of economic simplicity, lower even that those of immediately pre-Roman times, with little movement of goods, poor housing, and only the most basic manufactured items. The sophistication of the Roman period, by spreading high-quality goods widely in society, had destroyed the local skills and local networks that, in pre-Roman times, had provided lower-level economic complexity. *It took centuries for people in the former empire to reacquire the skills and the regional networks* that would take them back to these pre-Roman levels of sophistication. Ironically, viewed from the perspective of fifth-century Britain and of most of the sixth-and seventh-century Mediterranean, the Roman experience had been highly damaging.[726]

Extensive studies of the available evidence show that during the 6th century AD, any kind of human activity beyond basic survival vanished. Between the 7th and the 10th centuries, human activity in the Roman Empire was virtually nonexistent:

- House building regressed from numerous elaborate stone houses to scattered wooden shacks.[727]
- The metal industry all but ceased to exist, and it was only during the 16th century that the level of industrial activity prior to the fall of Rome would be once more attained.[728]
- Widespread and elaborate potteries became very rare and progress stagnated so much that it's impossible to distinguish 7th-century ceramics from those produced 2 centuries later.[729]

---

717 Scuderi L., 'A 2000-Year Tree Ring Record of Annual Temperatures in the Sierra Nevada Mountains', *Science* (5 March 1993) 259(5100): 1433–1436

718 Baillie, M., *World Archaeology* (1991) 23; 233 43; *Exodus to Arthur*

719 Briffa, K., *et al.*, 'A 1,400-year tree-ring record of summer temperatures in Fennoscandia', *Nature* (2 August 1990) 346: 434–439

720 LaMarche, V., 'Paleoclimatic Inferences from Long Tree-Ring Records', *Science* (15 March 1974) 183(4129): 1043–1048; Scuderi, L., 'A 2000-Year Tree-Ring Record of Annual Temperatures in the Sierra Nevada Mountains', *Science* (5 March 1993) 259(5100): 1433–1436

721 D'Arigo, S., *et al.*, '1,738 years of Mongolian temperature variability inferred from a tree-ring width chronology of Siberian pine', *Geophys. Res. Lett.* (2001) 28: 543–46

722 Baillie, *op. cit.*

723 Briffa *et al.*, *op. cit.*

724 Clube, V. & Napier, W., 'The Microstructure Of Terrestrial Catastrophism', *Monthly Notices Of The Royal Astronomical Society* 211(4): 953–968

725 Rigby, E., *et al.*, 'A comet impact in AD 536?', *A&G* (2004) 45(1): 1.23–1.26

726 Ward-Perkins, B., *The Fall of Rome*, pp. 136–137, excerpts

727 *Ibid.*, pp. 95, 111

728 *Ibid.*, p. 96

729 Hans-Ulrich, N., *Did the early middle age really exist*, p.5

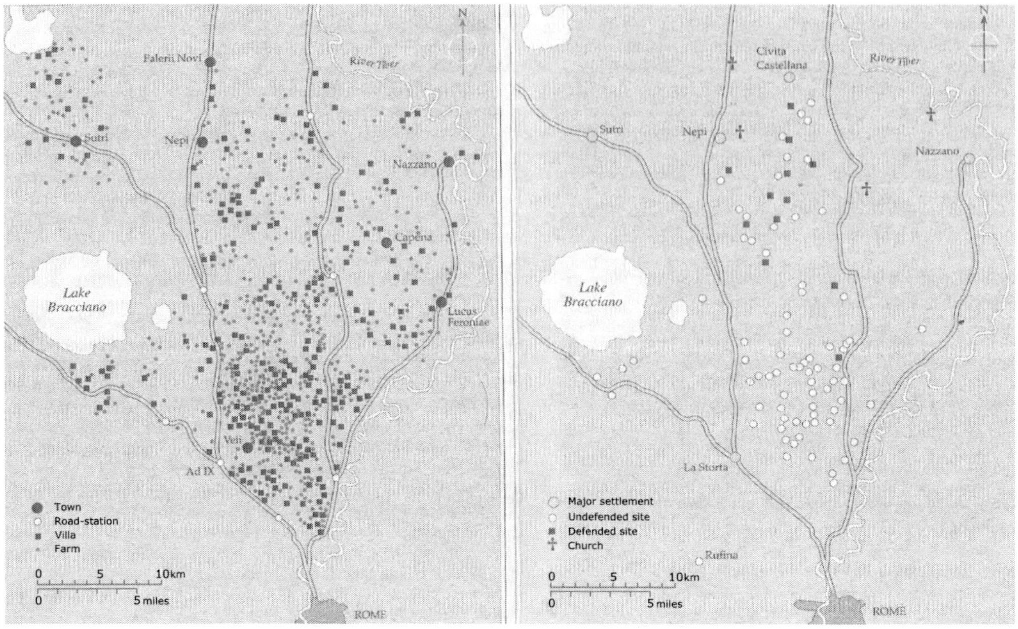

Figure 200: Settlements north of Rome, c. 100 AD (left) and 5th to 8th century (right). (© Ward-Perkins)

- The very few churches built during this period were more than 10 times smaller than the 4th-century St. Peter's Basilica of Rome.[730] Only the cathedrals of the 11th century would attain similar sizes.
- Farming and warfare techniques stagnated for 3 centuries.[731]
- The quality and quantity of coinage declined in most regions of the Empire.[732]
- Written documents were almost totally absent for 3 centuries.[733]

In figure 200, we see an archaeological picture of the problem we are dealing with. On the left we see rural settlements north of the city of Rome, in Roman times, as revealed by an archaeological field survey. The date is about 100 AD.

Now, let's look at the image on the right, the 'after' image. It represents sites revealed by pottery of the 5th to 8th centuries. That's 300 years. It's a long time to be covered in one image so you can imagine that, at any one time, the number of actual living sites was much, much smaller than what you see here where those *300 years are aggregated*!

The decline in population is not just startling, it is horrifying to imagine. From a landscape that was densely settled with modern buildings, extensive cultivation, and was intricately networked, we move rapidly to a landscape that is, for the most part, *empty space*. What's more, the evidence

[730] Ward-Perkins, *op. cit.*, pp. 148–149
[731] White, L. Jr., *Die mittelalterliche Technik und der Wandel der Gesellschaft*
[732] Ward-Perkins, *op. cit.*, pp. 110–117
[733] Hardouin, J., *The Prolegomena*, p. 64

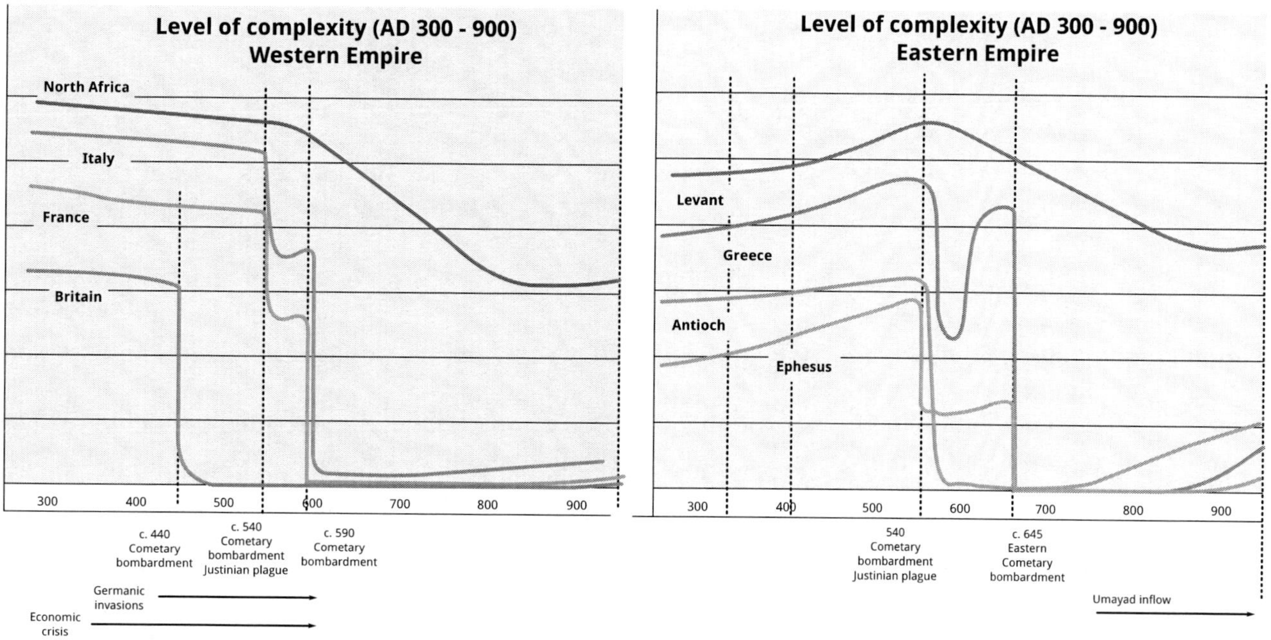

Figure 201: Collapse of level of development in the Roman Empire (300–900 AD). (© Sott.net, adapted from Ward-Perkins)

for occupation in towns and cities decreases dramatically at the same time.[734] That means people were not fleeing from the countryside to the cities. We are looking at something like 90% mortality rate here.

The collapse was not uniform. Figure 201 shows the collapse in level of development in various areas of the Roman Empire.

There is general agreement that things happened in Britain very early and very fast. Post-Roman Britain sank to a Pre-Roman Iron Age level of survival. Even before the coming of the Romans, Britain had had an active and productive economy of its own. None of this can be found in 5th- and 6th-century, post-Roman Britain. It took *300 years* for Britain to recover enough to even equal its own Iron Age developmental state *before* the Romans had come to begin with.

At the same time, the Eastern Empire continued to thrive in some respects *for a while.* New tableware was being produced in Cyprus and Phocaea (west coast of Turkey). New types of amphorae appear transporting oil and wine from different areas of the Levant.[735] Coins continued to be produced, buildings continued to be built in elegant style, at least in Constantinople.

The Levant to Egypt, and central Greece, appear to have experienced remarkable expansion. Settlement not only increased during this period, but so did prosperity overall. This increase left behind a mass of newly built houses and a whole raft of *monasteries and churches.* (This fact, too, is important.) But then, in the Aegean, all this came to a sudden end, as shown in figure 201, around 600 AD. The great cities of Athens, Corinth and Ephesus shrank to a fraction of their

734 Ward-Perkins, *op. cit.*

735 Eastern Mediterranean region.

former size just as had already happened in the Western Empire. It appears to have happened with the same suddenness as what took place in Britain sixty years earlier. Constantinople also shrank dramatically in population and wealth.

By 700 AD, the only areas of the former Roman Empire that had not 'fallen' were the Levant and Egypt. This may account for the sudden rise of the Arab empire, or Caliphate, beginning in the 7th century. Perhaps the Arabs didn't *conquer* the Roman Empire; they may have simply entered empty territories.

Thus, all around, the same decline can be measured, though it progressed somewhat more slowly. What this all seems to indicate is that there was no single moment of total collapse for the entire Empire. *It disappeared at different times in different places, and proceeded at different speeds while some places only declined moderately.*

But, to bring this discussion home, the facts outlined above show a positive correlation between an oppressive, cruel and corrupt regime and large-scale cataclysmic events. This is not the only example that could be brought forward. There have been a number of such periods – discontinuities in history – after which stability is established and things get back on track. (Maybe the Gallic peoples were not so crazy when they stated that their only fear was that the sky would fall on their heads!)[736]

Such periods are generally signaled by a tremendous effort to re-establish norms, writing, building on a large scale, and very often, imperialistic strivings in the form of wars of conquest and domination. The most recent serious Dark Age in our history is that which followed the fall of Rome. Rome itself came into being several hundred years after a previous Dark Age, which we have already discussed in the context of the transformation of observational science to myth.[737]

It is very important to understand Rome since it appears that major aspects of our present civilization – particularly in the legal and governing arenas – are largely based on what Rome began allegedly 2,500 years ago. It's a certainty that Christianity was shaped by the Roman Empire and could be said to be an extension of it. Christianity then shaped Western civilization because it was the tool, the ideology, with which the pathological leaders dazzled and engaged the growing masses of people that had survived the Dark Ages,[738] to do their bidding. Throughout the period when these phase transitions were manifested, there are many descriptions of comet/asteroid-related disasters. The bottom line is this: what happened to the Roman Empire *was no mere transformation*: it was the End of a Civilization that did not regain the ground lost for *over 800 years.*

[736] Celtic leaders, when asked by Alexander the Great what in the world they feared the most, answered that their worst fear was that the sky might fall on their heads. No doubt Alexander was disappointed not to hear himself named. See Arrian, *Campaigns of Alexander*, Book I:4.

[737] Knight-Jadczyk, L., *Comets and the Horns of Moses: Secret History of the World Vol. 2*

[738] See Chapter 35: 'The dynastic cycle'

Chapter 35.

The dynastic cycle

As depicted in figure 202, far from our linear, uniformitarian vision of the universe, the Chinese believed – and some continue to believe – that the behavior of rulers, the state of the world and natural catastrophes are intimately related. This was also the way our ancient ancestors understood the world.

It seems that when a large enough number of people exhibit high levels of ignorance and false beliefs, cosmic turmoil soon follows; and if you study history long enough, you'll see the pattern emerge again and again and again. A negative feedback loop may develop at this point (bottom right of the diagram). When natural catastrophes increase to the point that people finally realize something is 'up', it also dawns on them that their leaders are unable to provide for their safety and so they turn against them. Social unrest grows and the elites turn to increasingly repressive measures of violence and control, which, in turn, increase the cosmic 'reaction', which further fuels the people's anxiety and urge to revolt.[739]

The 500–1300 cycle

While the dynastic cycle can apply to any empire from the depths of antiquity up until today, let's continue with our example of the Roman Empire to explicate the steps shown in figure 202.

During Justinian's reign, the Empire treated such a great number of human beings so badly that a kind of critical mass was reached. During the 6th century AD, the Roman Empire collapsed like a row of dominoes,[740] obliterated by a *series* of cataclysmic events. The population was reduced to a few pockets of survivors living in an extremely harsh 'post-apocalyptic' environment. The people of the time experienced an almost total reset in terms of civilization; technologies were lost and knowledge forgotten.

A mini ice age set in, epidemics were rampant, resources were scarce and crops barely grew. The survivors were literally reduced to Stone Age conditions (see figure 202 bottom left).

The cultural inactivity during this period is so pronounced that mathematician and historian Anatoly Fomenko,[741] among others, has developed a theory that the centuries of the Dark Age never really existed and were simply added into later histories. But what seems to have happened is that those centuries really did 'happen'; they were simply blank.

[739] See Le Bon, G., *The Crowd*

[740] See Chapter 34: 'Historic evidence of a human-cosmic connection'

[741] Fomenko, A., Nosovskiy, G., Tamdhu, F. & Zinoviev, P., *History: Fiction or Science?*

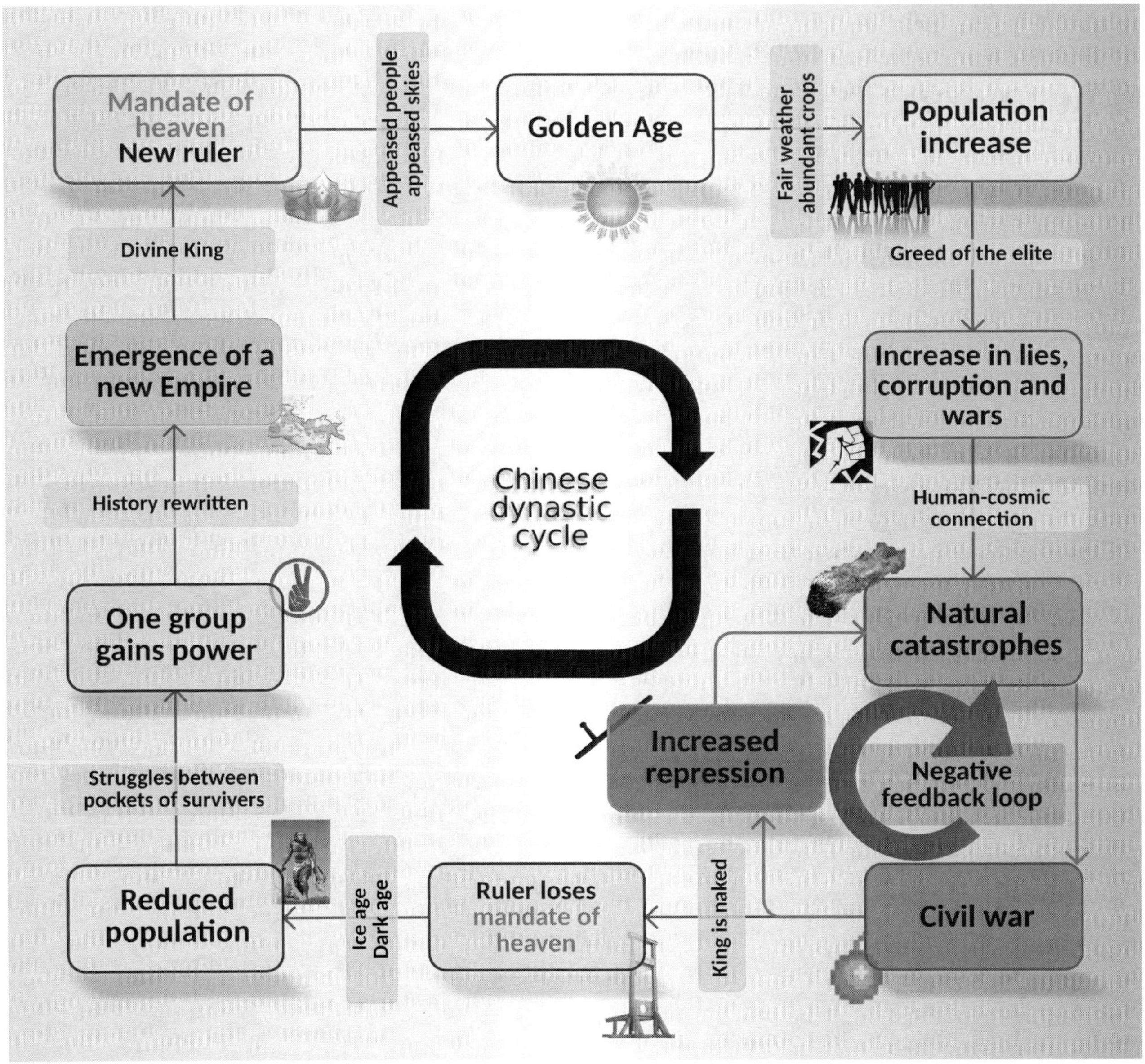

Figure 202: The Chinese dynastic cycle. (© Sott.net)

The frequency and intensity of cataclysms receded sometime after the major events that marked the 6th century and the beginning of the 7th century. Habitats slowly recovered while weather progressively improved. In 750, Justinian's plague finally abated after a wave of *18 epidemics spread over two centuries.*[742]

Around 900, a warm period began, the population grew and humanity redeveloped a rudimentary society.[743] Towns formed again, trade increased, technologies re-emerged and lost knowledge was partly rediscovered. Memories of the cataclysms were still fresh. In this sense, the negligible population had a fairly objective view of reality, which may partly explain the relatively quiet skies.

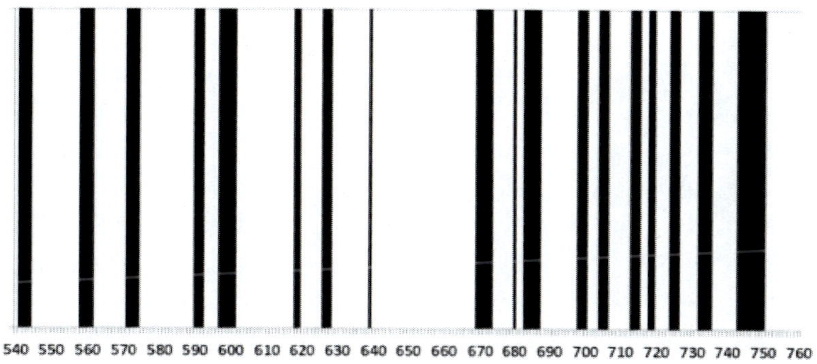

Figure 203: The 18 waves of plague epidemics in the Mediterranean and the Near East (540–750). (© Johannes Preiser-Kapeller, after Stathakopoulos, 2004)

However, a new elite soon began to dominate the surviving population (left side of the dynastic cycle diagram). In the 10th century, society reached a technological and organizational level that enabled a rewriting of history.

This rewriting was due to several factors. After cataclysmic events, historical records are typically destroyed, damaged or lost. Thus, the shocked and numerically reduced scholarly community has only sparse documents and personal memories to work with, often written in languages and contexts with which they are unfamiliar. All these difficulties help explain the numerous involuntary inaccuracies that are apparent in the subsequent rewriting or recopying of old texts with attempts to update them.[744]

Other, more *voluntary*, 'mistakes' served theological and political agendas of the day:

> Asteroid strikes, though important, are not the most serious short-term risk to mankind or civilization ... Mankind is subject to an increased risk of global insult through another kind of cosmic agency. This cosmic agency is a 'Shoemaker-Levy type' train of cometary debris resulting in sequences of terrestrial encounters with sub-km meteoroids. ... Subsequently perceived as pointless, *such transitions are commonly an embarrassment to national elites even to the extent that historical and astronomical evidence of the risk are abominated and suppressed.*[745]

The emerging elite, having no link whatsoever with the previous regime, needed to strengthen

742 Stathakopoulos, D., *Famine and Pestilence in the Late Roman and Early Byzantine Empire: A Systematic Survey of Subsistence Crises and Epidemics*

743 Mann, M., 'Medieval climatic optimum', *Encyclopedia of Global environment change*, pp. 514–516

744 For more details about the types of translation errors that can crop up, see: Carotta, F., *Jesus was Caesar*, pp. 350–380

745 The comments cited here can be found in a letter (dated June 4, 1996) from the office of Dr. P. A. Charles, then Head of Astrophysics of the University of Oxford, to Ms. Victoria Cox, then Chief, Physics and BMD Coordinator of the European Office of Aerospace Research and Development. The letter (written by V. Clube) summarizes the findings of USAF-funded research resulting in the following report: Charles, P. A., *The Hazard to Civilization From Fireballs and Comets*, Oxford University Dept. of Astrophysics, (4 June 1996)

its authority by establishing its historical legitimacy. They also had to progressively erase traces of the past cataclysm, knowledge of which would threaten their base of power, i.e. their alleged capacity to ensure a safe world for their people. After all, if you want to resurrect an empire, you don't want anyone to think it was evil to the core or that perhaps, God himself destroyed it!

During the 10th century, the new leaders, namely Constantine VII, Emperor of the Eastern Roman Empire, Pope Sylvester II and Otto III, King of the Franks,[746] began a concerted effort to establish a new history. They composed their history[747] such that the Carolingian and Byzantine leaders were direct descendants of the Roman emperors, the pope was a direct descendant of St. Peter, and Christianity was the one and only religion. *A new Mandate of Heaven had been revealed and sealed* (top left of the dynastic cycle diagram).

According to this revised history, the transition from the late Roman Empire to the Middle Ages was a smooth progression. But for this transition to occur, 'missing links' had to be invented, such as the lazy kings – who apparently didn't do much worth recording – the Merovingians, who served as the direct link between the contemporary Christian and Carolingian leaders and the long-dead Christian Roman emperors. A whole genealogy of kings and popes had to be created, because for three centuries there had been no king and no kingdom, no pope and no Christendom. In fact, all the rulers of this period either never existed or were simply small tribal leaders who were posthumously promoted to the status of kings or emperors by the freshly written official history.

The most striking example might be Charlemagne. According to official history, he was the first emperor of the Romans since the collapse of the Empire. Charlemagne was crowned in St. Peter's Basilica on December 25th, 800, exactly eight centuries after the birth of Jesus Christ. Thus, Charlemagne embodied the attributes of the dual powers of Christianity and the Roman Empire.

The revisionists downplayed the occurrence of any major cataclysm because they needed to defend Christianity against the pagan accusation that foreswearing the old gods had brought on the destruction.

So, they depicted an uninterrupted sequence of historical events and genealogies occurring in a stable and relatively undisturbed context. That's one of the reasons why they *fiddled with the time*. For present purposes, let us just quote Heribert Illig:

> In 1582 Pope Gregory ordered precisely ten days to be skipped in order to make the day count agree with the celestial situation. In passing it should be mentioned that the Pope introduced an improved intercalary rule which requires revision only every 2000 years. What is decisive for us is the following: the ten days that were skipped in October 1582 corrected the mistake that had accumulated in the Julian calendar over the previous 1,282 years. However, if you deduct these 1,282 years from 1582, you don't arrive in the year of Caesar's calendar reform, 45 BC.[748]

[746] The Franks originated in what is now Germany. Their survival rate seems to have been higher than other Western populations. Two factors may explain this fact: 1) sheer luck, with German territory avoiding the same level of destruction experienced by other territories, 2) nutritional and living habits may have made Germanic tribes more resistant to cosmic crises. According to Julius Caesar, they ate a lot of meat from a young age. They also bathed in cold water and wore almost no clothes. See: Caesar, J., *The Gallic Wars*, Book 4, chapter 1, and Book 6, chapters 21 and 22, respectively.

[747] De Sarre, F., *Mais où est donc passé le Moyen Âge?*, p. 53

[748] Illig, H., 'Anomalous Eras – Best Evidence: Best Theory', Toronto Conference, The Invented Middle Ages (28–30 June 2005)

Figure 204: The coronation of a Christ-like Charlemagne by Pope Leo III. (© Kaulbach)

The manipulation of historical timelines around this period will be the main topic of another volume of *The Secret History of the World*, so we won't develop it any further here. All we need to know for our subject is that timeline manipulation is one of the tools used by revisionists, and that our official historical timeline exhibits several chunks of time that were added or removed here and there.

Interestingly, the vast majority of 'antique manuscripts' date back to between the 9th and 11th centuries. Allegedly, this *systematic rewriting of history* occurred because the manuscripts were mostly written in Greek majuscules (large, capital letters), so they had to be rewritten using minuscule (and generally in Latin).[749] However, there were apparently less literary motivations involved than a simple 'mass system upgrade':

> *Clerics have only transmitted what they needed to.* During the so little known gap that separates the end of Antiquity and the Carolingian Renaissance, when Greek culture could have disappeared in the West, Vivarium and the Lateran have preserved what was considered essential to the clergy – and only that – namely the best exegetes of the Greek fathers and the works of Josephus because of his *Testimonium Flavianum.* Focillon surprises a monk, maiming manuscripts, cutting and destroying, profane parts. He says: 'We must give up the legend of monks spending their nights copying the ancient authors and saving them for posterity: the only writings they copied were the writings of the Fathers. During the tenth and the eleventh century, the ancient writers had no greater enemies than the monks.'[750]

The old saying that 'history is written by the victors' could easily be complemented by 'history is written by lucky and ambitious survivors'. In any case, having rewritten history, the ruling elites now had a solid power base and could exert their growing control over the masses, whose numbers were steadily growing thanks to fairer environmental conditions.

However, the greed-driven elites always want more: more power, more territory and more wealth. So, once again, as it was during the days of ancient Rome (leading cumulatively over hundreds of years to the ultimate evil, Justinian's reign), the lies, dominance and exploitation kept growing. A critical mass[751] was slowly building up (top right of the dynastic cycle diagram).

In 1209, Pope Innocent III, backed by the French monarchy, initiated the Cathar crusade.[752] In

[749] De Sarre, *op. cit.*, p. 62
[750] Gys-Devic, *From the war of the Titans to the battle of the manuscripts*, p. 4
[751] Population size x individual amount of lies and suffering. See Chapter 40: 'The truth factor'
[752] Bird, J., Peters, E. & Powell, J., *Crusade and Christendom*, p. 66

1229, the Inquisition was established.[753] Over the following years of barbarism, one million innocent French people were slaughtered[754] and the whole Occitan culture was swept away.[755] In 1252, Pope Innocent legalized torture.[756] The Inquisition engulfed Europe, leaving hundreds of thousands of victims,[757] while the 7th, 8th and 9th crusades were waged against the 'infidels'.

Figure 205: Scribal amanuenses writing in a gallery of the cloister of the Ecthernach Abbey (Luxemburg), 11th century manuscript.

Then, around 1300, the Medieval Warm Period ended, giving way to the Little Ice Age, and cataclysms started growing in frequency and intensity.[758] A few decades later, in 1347, the Black Death hit Europe.[759]

The Inquisition (1229–1821),[760] the Black Death (1347–1720)[761] and the Little Ice Age (1350–1800)[762] all synched up over almost the same five-century period.

Thus the loop was completed again. From the 6th century (Justinian's atrocities) to the 13th (Inquisition and Crusades) the dynastic cycle made a full revolution. If we ignore the three centuries of the Dark Age, it only took four centuries for Western civilization to redevelop, for a new ruling elite to gain power and rewrite history, to oppress the people and commit atrocities, and for 'cosmic wrath' to strike one more time; and that it did, with a vengeance.

The Black Death

The Black Death was one of the most deadly pandemics in human history, said to have killed possibly two thirds of the entire population of Europe, not to mention hundreds of millions all over

753 Tyerman, C., *God's War: A New History of the Crusades*, p. 602

754 Quénot, K., *Cathares. Le massacre oublié*, p. 11 (not translated into English)

755 In Beziers, the whole population, estimated to be 20,000, was slaughtered in an 'exercise of charity'. It was during this episode that the abbot of Citeaux, Arnaud Amaury, is reported to have said: "Kill them all. God will know his own." See: http://www.cathar.info/120502_arnaud.htm

756 Tracy, L., *Torture and Brutality in Medieval Literature: Negotiations of National Identity*, p. 22

757 The Spanish Inquisition alone tallied 340,592 victims (burnt, tortured, condemned) between 1480 and 1815. See: Pérez, J., *The Spanish Inquisition: A History*, p. 170

758 Baillie, M., *New Light on the Black Death*, pp. 136–160

759 *Ibid.*, pp. 145–155

760 The Inquisition was officially abolished six centuries after its creation, in 1821. See: Gallois, J., *History of the Spanish Inquisition*, p. 247

761 Kelly, J., *The Great Mortality: An Intimate History of the Black Death*, p. 135

762 Husher, J., *Facts and Myths Facing Today's World*, pp. 17–30

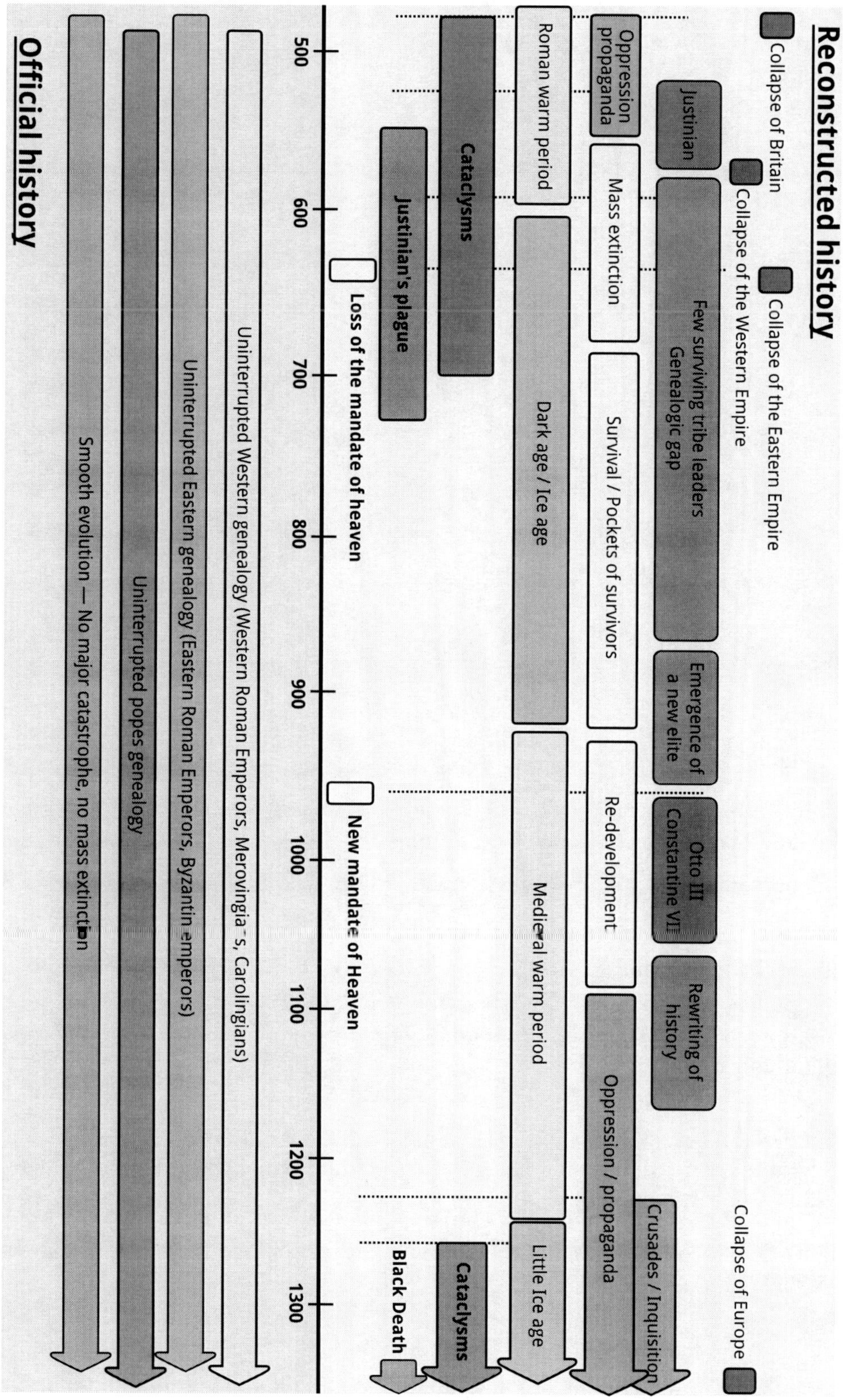

Figure 206: 500–1300 AD. Reconstructed history (top) vs. official history (bottom). (© Sott.net)

Figure 207: Peter Bruegel, 'The Triumph of Death', c. 1562. (Oil panel)

the planet.[763]

China, where the Black Death is said to have originated, lost around half of its entire population (going from around 123 million to around 65 million).[764] Recent research into European death tolls also suggest a figure of about 50% of the total population during a four-year period, though the figure fluctuated from place to place.[765] In Mediterranean Europe and Italy, the south of France and Spain, where the plague ran for about four years consecutively, it was probably closer to 70% to 75% of the total population.[766]

There is compelling evidence that the Black Death was *not an outbreak of bubonic plague.* This case is synthesized by Samuel K. Cohn, who discovered several irreconcilable inconsistencies in the bubonic plague hypothesis:

- Very different *transmission speeds* – the Black Death spread over 385 km in 91 days, compared to 12–15 km a year for the modern bubonic plague, which has the assistance of trains and cars

[763] The first outbreak alone (1347–1351) took some 75 million lives. See: Edwards, K., *And America Quaked: A Chilling Series of Visions of a Future American*, p. 100

[764] Baillie, *op. cit.*

[765] Daileader, P., *The Late Middle Ages*

[766] *Ibid.*

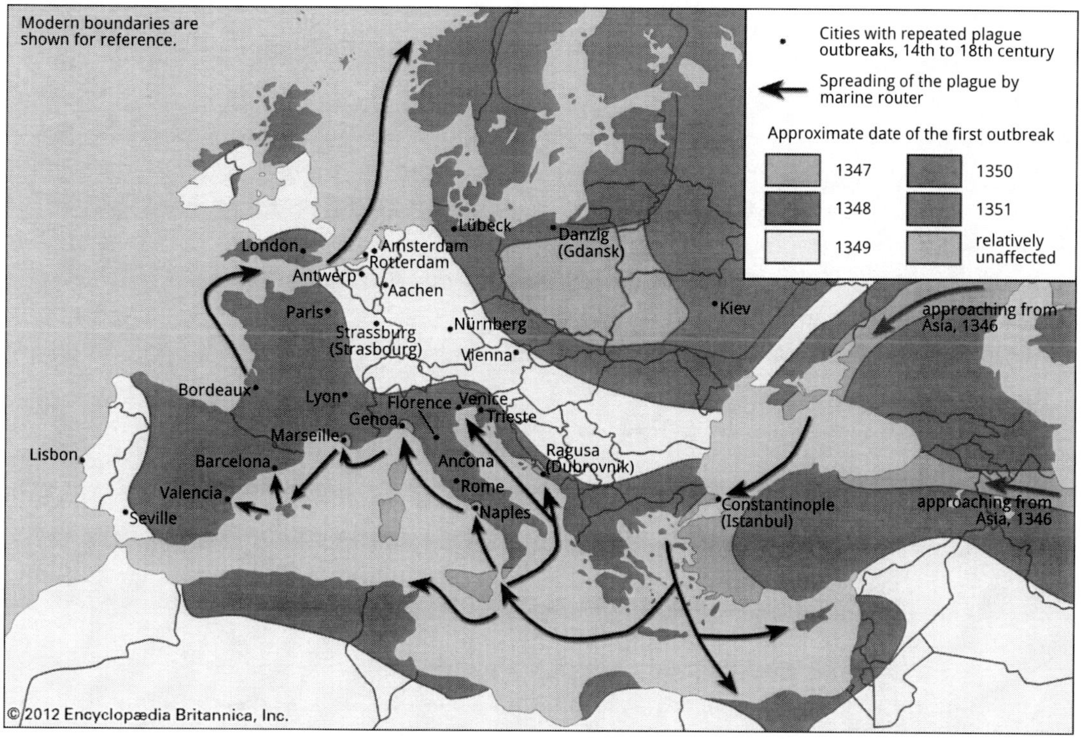

Figure 208: Reach of the Black Death in Europe from 1347 to 1351. (© Britannica)

- Difficulties with the attempt to explain the rapid spread of the Black Death by arguing that it was spread by the rare pneumonic form of the disease. In fact, this form killed less than 0.3% of the infected population in its worst outbreak in Manchuria in 1911
- Different *seasonality*. The modern plague can only be sustained at temperatures between 50 and 78 °F (10 and 26 °C) and requires high humidity, while the Black Death occurred even in Norway in the middle of the winter and in the Mediterranean in the middle of hot dry summers
- Very different *death rates*. In several places over 75% of the population appears to have died; in contrast the highest mortality for the modern Bubonic Plague was 3% in Mumbai in 1903

The last nail in the bubonic plague theory came from Gunnar Karlsson[767] who, in 2000, pointed out that the Black Death had killed between half and two-thirds of the population of Iceland, although there were *no rats* in Iceland at this time.

Researchers Duncan and Scott[768] carefully put all the available clues together, tracking the plague from its first appearance *out of nowhere* and chronicling its unprecedented catastrophic effects on European civilization – death on an unimaginable scale, but could very well happen again, at any time.

[767] Karlsson, G., *The History of Iceland*
[768] Scott, S. & Duncan, C., *Return of the Black Death*

They were able to not only surmise the amount of time from the appearance of symptoms to death, but also to establish the incubation period, latent period, period between symptoms and death, infectious period before the first symptoms, etc.

The authors were astonished when they were able to work out the duration of these statistics in more than 50 different plague outbreaks in England and to verify the length of the latent and infectious periods many times. The correspondence with the universal 40-day 'quarantine' period established as successful prophylaxis during the time of the plague supported their conclusions. From the data available in other countries, they argue convincingly that these statistics applied to the Black Death in all of Europe.

Duncan and Scott also analyzed and compiled the accounts relative to the plague symptoms:

- A victim usually displayed the symptoms for about five days before dying. But according to contemporary accounts, this period could have been between two to twelve days.

- The main diagnostic feature was the appearance of **hemorrhagic spots**, often red, but ranging in color from blue to purple and from orange to black. They often appeared on the chest, but were also seen on the throat, arms and legs, and were caused by bleeding under the skin from damaged capillaries.

- Various swellings were also characteristic of the disease: carbuncles, blains (burning ulcers) and the buboes, which were swollen lymph glands in the neck, armpits and groin. **If the buboes failed to rise and burst, there was little chance of survival, but if they broke the fever apparently declined.**

- Fever, continual vomiting, diarrhea and prolonged bleeding from the nose were additional characteristics. Also, blood-tinged urine, a burning thirst and, in some, madness and delirium.

- Autopsies revealed generalized necrosis of the internal organs. It was surely a gruesome way to die. **The victim was killed by the literal death and liquefaction of the organs.**

No known virus existing today is responsible for the Black Death, although the symptoms resemble those of Ebola, Marburg and the viral hemorrhagic fevers – diseases caused by filoviruses. They have a high mortality rate and tend to occur in explosive epidemics driven by person-to-person transmission. Outbreaks occur unpredictably and, as of yet, no animal reservoir is known.

Where did the disease go? Did the Black Death virus mutate, causing other fearsome diseases? Nobody knows. What we do know is that a more virulent form of smallpox came to the fore in the 1630s and, just as the Black Death disappeared from the stage of history, smallpox took its place as the most feared of human diseases. We can only speculate. Smallpox virus, as opposed to the causative agent of the Black Death, is very resistant to cold temperatures, making it a more viable virus. According to the data collected by Scott and Duncan which describe the disease process of the Black Death, **hemorrhagic smallpox is almost identical to the Black Death.**[769]

[769] Segura, G., 'New Light on the Black Death: The Viral and Cosmic Connection', *The Dot Connector Magazine* (2011) 1(13)

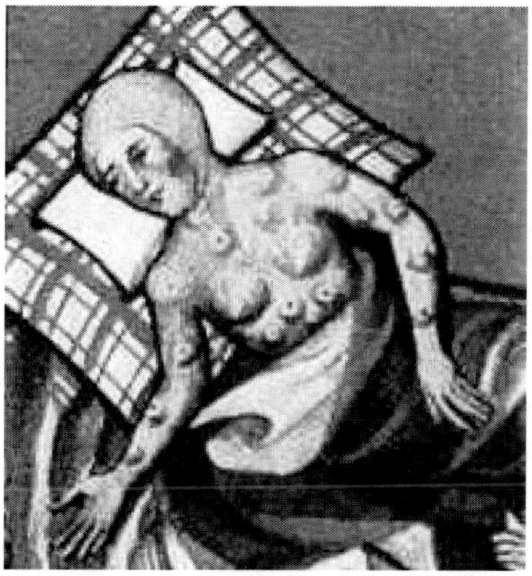

Figure 209: Detail of the illustration of the Black Death from the Toggenburg Bible (1411).

Where did the disease come from? As seen previously,[770] viruses can be carried by comets and this factor can come into play even without impact scenarios being involved. Coincidentally, Comet Negra was observed in 1347, around the time when the epidemics started. Several chroniclers reported this event:

> In France ... was seen the terrible Comet called Negra.[771]

Baillie presents a substantial body of evidence[772] showing that cometary events were indeed the initiators of the Black Death and many other cataclysms. Thus, the high mortality rates mentioned above were not solely due to the Black Death. This period of history was marked by various *interrelated catastrophes*, very similar to what was described in part II of this book.

Philip Ziegler collected many reports of these disasters from medieval chronicles:

> Droughts, floods, earthquakes, locusts, subterranean thunder, unheard of tempests, lightning, sheets of fire, hail stones of marvelous size, fire from heaven, stinking smoke, corrupted atmosphere, a vast rain of fire, masses of smoke.[773]

Amongst all those catastrophes, earthquakes were particularly prominent. In the 1340s there was a veritable rash of seismic activity, including the great earthquake on January 25th, 1348:

> In the thirty-first year of Emperor Lewis, around the feast of the Conversion of St. Paul there was an earthquake throughout Carinthia and Carniola which was so severe that everyone feared for their lives. There were repeated shocks, and on one night the earth shook 20 times. Sixteen cities were destroyed and their inhabitants killed.... Thirty-six mountain fortresses and their

[770] See Chapter 21: 'Increase in cometary activity' – Space-bound bio-hazard
[771] Short, T., *A General Chronological History of the Air, Weather, Seasons, Meteors, Etc.*
[772] Baillie, *op. cit.*
[773] Ziegler, P., *The Black Death*

inhabitants were destroyed and it was calculated that more than 40,000 men were swallowed up or overwhelmed[774].

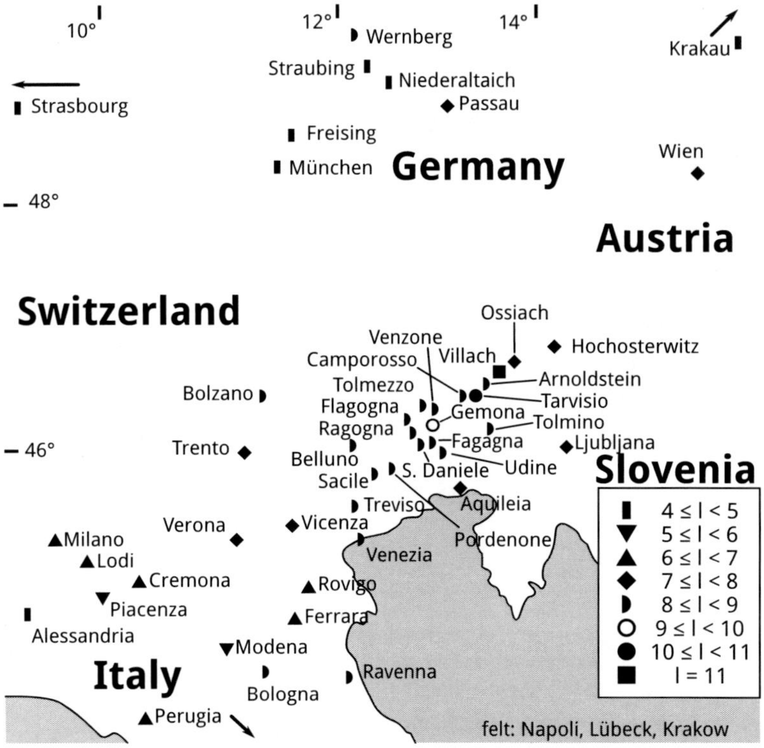

Figure 210: Effect distribution of the 1348 earthquake. The quake was felt in most of the European countries. (© Ambraseys, adapted by Sott.net)

Once again,[775] notice that earthquakes and cometary events can be closely related as Baillie[776] points out: *earthquakes could be caused by cometary explosions in the atmosphere or even impacts on the surface of the Earth.*

Baillie shows that the exact same signature is present at the time of the Black Death in both the tree rings and in the ice cores, *and* at other times of so-called 'plague and pandemics'.[777]

The ammonium signal in the ice-cores is directly connected to the earthquake that occurred on January 25th, 1348 – and Baillie discovered that there was a 14th-century writer who had written that the plague was a 'corruption of the atmosphere' having resulted from this earthquake. A German treatise states the same:

> Insofar as the mortality arose from natural causes its immediate cause was a corrupt and poisonous earthy exhalation, which infected the air in various parts of the world … I say it was the vapour and corrupted air which has been vented – or so to speak purged – in the earthquake that occurred on St. Paul's day [1348], along with the *corrupted air vented in other earthquakes and eruptions, which has infected the air above the earth and killed people in various parts of the world.*[778]

As Baillie notes, if this oft-cited earthquake was, in reality, the result of cometary impacts then the corrupted air could be from one or two causes: high-energy chemical transformations in the atmosphere or out-gassings from the Earth itself.

Jon Arrizabalaga compiled a selection of writings in an attempt to understand what educated people were saying about the Black Death while it was happening. Regarding the terms used by doctors and other medical people in 1348 to describe the plague, he writes:

[774] Horrox, R., *The Black Death*
[775] See Chapter 30: 'Geographic tilt and magnetic reversal'
[776] Baillie, *op. cit.*
[777] Including the 539 AD signal that closely preceded the outbreak of the Justinian's plague in 541.
[778] Medieval German treatise quoted in Horrox, *op. cit.*

> One ... Jacme d'Agramaont, discussed it in terms of an "epidemic or pestilence and mortalities of people" which threatened Lerida from "some parts and regions neighbouring to us" ... Agramont said nothing concerning the term epidemia, but he extensively developed what he meant by *pestilencia.* He gave this latter term a very peculiar etymology ... which came to be widely accepted throughout Europe during the Middle Ages. He split the term *pestilencia* up into three syllables, each having a particular meaning: pes = tempesta: 'storm, tempest'; te = 'temps, time', lencia = clardat: 'brightness, light'; hence, he concluded, *the pestilencia was 'the time of tempest caused by light from the stars.'*

In conclusion, the epidemics that ravaged Europe were not bubonic plague but very probably a virus of cometary origin. These epidemics were only one aspect of a series of cometary-induced cataclysms, including earthquakes and poisonous outgassing that erased the medieval European civilization almost entirely.

Other dynastic cycles and technological vulnerability

Justinian's plague occurred during the worst times of oppression exerted by the Roman Empire. Similarly, the Black Death appeared during the atrocities of the crusades and the Inquisition. So far we have focused on this single period of history comprised between the 6th century and the 14th century, and on one specific location, Europe.

This illustration of the dynastic cycle outlines a more general principle but should not be taken as a universal standard. Each empire exhibits its *own unique set of characteristics* affecting how exactly the dynastic cycle will manifest: population size and location, magnitude of historical rewriting, duration and extent of the lies and oppression, reaction of the people, duration of the build-up period, and the magnitude, duration and geographic scope of the cosmic events.

Economic historian David Hackett Fischer has shown how, during the last millennium, societal collapses have occurred *every three centuries* or so.[779] Although Fischer noted in passing that *climatic stresses punctuated these major socio-economic crises*, which affect every dimension of our society (demographics, trade, finance, economics, employment levels), his discovery of a pattern of 'price-revolutions' relied primarily on socio-economic indicators and not on environmental factors. Similarly, historian William R. Thompson noticed major collapses of empires every five centuries during the period 3500 BC–1300 AD, but he ascribed their causes to man-made ecological disruptions.[780]

When we take into account the archeological, historical, ice-core and dendrochronological records, *a strong correlation between imperial decadence and major climatic and cosmic disruptions (comets, climate change, volcanic eruptions and earthquakes) becomes apparent*, to the extent that *virtually every empire in question (Sumerian, Egyptian, Hittite, Assyrian, Greek, Roman, etc.) was apparently swept away by natural catastrophes in roughly the same way.*[781]

We tend to underestimate the influence of climate conditions on the rise and fall of civilizations. Their development is controlled by climate; climate is controlled by solar and cometary activity;

[779] Fisher, D., *The Great Wave*

[780] Thompson, W., *The Historical Evolution of World-Systems*, pp. 22–35

[781] For an extensive review of the link between the corruption of empires and the dramatic increase in natural catastrophes, see: Knight-Jadczyk, L., *The Secret History of the World* Series

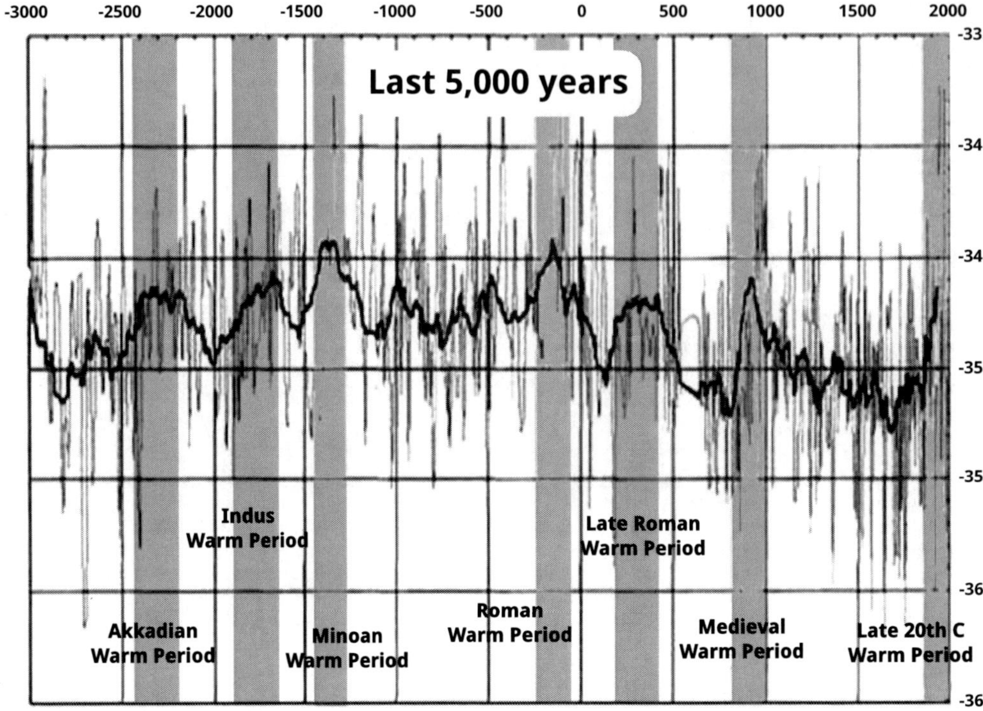

Figure 211: Warm periods over the last 5,000 years. (© Sott.net, adapted from Carter, 2007)

and solar and cometary activity can be closely coupled.

Figure 211 shows how the climate during recorded history has been characterized by long cold episodes (several centuries long), interspersed with short warm episodes lasting a few centuries (light grey columns). Most civilizations, including ours, bloomed during these ephemeral warm episodes. Most of them also collapsed during the ensuing cooling.

Our present civilization is characterized by a level of *technological dependency* that is probably unprecedented in history. Infrastructure features like power stations, dams or bridges, even roads and buildings, are designed within a specific, narrow range of environmental parameters (temperature, pressure, electromagnetic activity, seismic activity, etc.) Those ranges have been defined by examining historic records and building structures capable of facing extreme conditions (earthquakes, tsunamis, high winds, etc.) recorded during a period of time usually extending for only several decades to a few centuries. The historical records used to set those ranges are based on one of those quiet periods (the last one or two centuries), which is not representative of the chaotic periods our planet has experienced in the more distant past and what it is likely to experience in the future.

For example the Fukushima Daiichi nuclear plant was protected from the sea by a 5.7-meter-high dike.[782] This dike had been built to withstand the highest local tsunamis on record over the past 100

[782] Takeuchi, K. & Chavoshian, A., *Quick Report of Great Eastern Japan Earthquake and Tsunami*, p. 4

Figure 212: Aerial photo of the destroyed unit 3 and 4 of Fukushima nuclear plant. This photo was taken by a drone on March 24, 2011. (© Air Photo Service)

years.[783] Those tsunamis never exceeded 5.7 meters in height. But on Friday, March 11th, 2011, the Fukushima dike was submerged by a 14-meter tsunami triggered by a magnitude 9.0 earthquake.[784]

Similarly, the earthquake at Daiichi reached a maximum acceleration of 507 gal[785] at Unit 3, while the design basis for this building was that it would withstand a maximum 441 gal acceleration.[786]

In our modern civilization, most fundamental human needs like water supply, electricity, transportation, food production and long-distance communications are met thanks to the heavy use of technologies that were neither designed nor built to face unusual environmental events.

The argument that technological sophistication protects modern society from natural disasters is a grave misconception.[787] Without those technologies, our civilization would grind to a halt, teleporting us back to Stone Age conditions, assuming we actually manage to survive at all.

783 1896 and 1933 tsunamis.

784 'Fukushima faced 14-metre tsunami', *World Nuclear News* (23 March 2011). See: www.world-nuclear-news.org/RS_Fukushima_faced_14-metre_tsunami_2303113.html

785 The *gal*, also known as galileo, is a unit of acceleration. 1 gal is equal to 1 cm/s^2. See: 'Non-SI units associated with the CGS and the CGS-Gaussian system of units', *Bureau International des Poids et Mesures.* www.bipm.org/en/si/si_-brochure/chapter4/table9.html

786 *World Nuclear News, op. cit.*

787 Leroy, S., 'Natural hazards, landscapes, and civilizations', in: Shroder, J., *et al.*, *Treatise on geomorphology* Vol. 13

Chapter 36.

The rise and collapse of materialism

Since the Middle Ages the cyclical nature of our reality has steadily been whitewashed from human understanding. Along with the rewriting of history described in the previous chapter, religion also played a major part in the way people perceived their world and the role they played in it.

Until the 17th century, religion dominated and ruled the lives and minds of most, if not all, members of Western civilization. Considering the events the peoples of those times experienced, it's not hard to understand why. Destruction from the sky and mass death all around would have a tendency to induce extreme religiosity in an effort to find the right formula to appease the gods.

Figure 213: "God condemned the cities of Sodom and Gomorrah and turned them into heaps of ashes". 2 Peter 2:6 (New Living Translation). John Martin, 'The Destruction of Sodom and Gomorrah'. (Oil on canvas)

Religion permeated the medieval societies of Europe and all were ruled by a deep fear of damnation. All actions on Earth were aimed at gaining a seat in Heaven. All that mattered was to avoid earthly sin and/or to be forgiven, based on the assumption that the Bible was Truth, and that everything outside the Bible was heresy. And there was obviously a compellingly big stick being wielded by the environment.

In such a cosmogony, Christianity bred obscurantism[788] and people were kept in a state of ignorance, unable to understand their environment in general and its cyclical nature in particular.

But it had not always been like that. In the beginning of the Middle Age, paganism[789] was still a widespread form of religion. According to pagan beliefs, God is *immanent*, present in every creature, every constituent of the world. The supernatural and miracles were part of daily life.

[788] Obscurantism is that which 'prevents enlightenment, or hinders the progress of knowledge and wisdom.'

[789] Paganism encompasses several forms of religions like animism, shamanism, hermetism, catharism, bogomilism, etc.

But this worldview was in direct contradiction with the Church dogma and particularly the Bible, *according to which the monotheist God has the monopoly on miracles* and the supernatural.

So, to counteract paganism, the Catholic Church established a dualist dogma where human beings were separated from God. People lived in a material world solely ruled by natural laws while a remote, *transcendent* god intervened in the material world by performing miracles and supernatural feats. Thus, the (direct) *human-cosmic connection was removed and replaced by the arbitrary wrath of God*, interceding between human behavior and cosmic events. Notice, however, that at this point the possibility of dramatic cosmic events (cataclysms inflicted by God) still existed.

The human realm solely governed by natural laws as posited by the Church laid the foundations for the development of science during the 17th century. Science was tasked with figuring out those natural laws while the Church kept a tight grip on men's souls. Thus, after the 'medieval dark age' came the period of 'enlightenment',[790] or so we were led to think:

> You can, with some plausibility, represent the Renaissance as darker than the Middle Ages. Machiavelli, the Medicis, and the Borgias have long been regarded as sin incarnate in odious forms. Making all due allowances for exaggeration and perversion of truth, the Renaissance was not a golden age, and the dramas of horror are something more than the nightmares of a madman.[791]

Religion and science worked hand in hand. First, God was removed from the human realm by the Church that replaced the immanent god with a remote, transcendent god. Second, God was killed by philosophers and scientists who promulgated an atheistic, mechanistic paradigm.

Reason became the new religion. From the God who was preparing hell for the ones who dared ask what He was doing before He made Heaven and Earth,[792] to the uniformitarian, mechanistic science that anathematized anything or anyone who would dare question its diktats. Fundamentally, nothing had changed, and obscurantism (which ironically means the 'enemy of Enlightenment'), still prevailed:

> Just as the sun replaces the earth as the center of our cosmos in Copernicus' cosmological system, so *humanity itself replaces God at the center of humanity's consciousness in the Enlightenment.*[793]

With the reign of science, things became even worse, because any form of moral values (sense of good and bad, ethics) and transcendence (entities and principles that are higher than human beings) were erased. Man, in his totality, became considered as a spirit-less biochemical machine (figure 215).

---

[790] The enlightenment is a period stretching from the mid-17th century through the 18th century. It is associated with philosophers like Descartes, Voltaire, Rousseau, Kant, Diderot and Spinoza, and scientists like Newton, Leibnitz and Locke. It culminated with the French revolution, which marked the death of the Church and the aristocratic order while giving birth to a society ruled by capitalism, individualism and atheism. See: 'Enlightenment', *Stanford Encyclopedia of Philosophy* (20 August 2010). www.plato.stanford.edu/entries/enlightenment/

[791] Potter, M., 'History, III The Renaissance', *Lectures on the Harvard Classics*, The Harvard Classics, 1909–14

[792] Paraphrase of *The Confessions and Letters of Saint Augustine* (XII:14): "Behold, I answer to him who asks, 'What was God doing before He made heaven and earth?' I answer not, as a certain person is reported to have done facetiously (avoiding the pressure of the question), 'He was preparing hell,' saith he, 'for those who pry into mysteries.'"

[793] 'Enlightenment', *op. cit.*

From the 18th century onwards, quieter skies and the emergence of the uniformitarian/mechanistic paradigm would progressively and successfully remove the human-cosmic connection and its latest medieval avatar, the wrath of God, from memory.

One of the last written testimonies mentioning a link between human affairs and cosmic/cometary activity was made by a French alchemist called Sédir[794] at the end of the 19th century. At this time, the human-cosmic connection had been almost totally erased and only a few 'esoteric' circles retained shreds of this lost knowledge:

> The earth is not a homogeneous solid, and when the axis tilts too much, the pressure exerted by waters and rocks changes here and there, resulting in crust folds and cracks, this is what we call earthquakes. These pressure changes, these displacements of the globe relative to the sun and other planets, disrupt the underground magnetic currents; and as resistance to an electric current generates heat, volcanoes appear, like safety valves of the huge machine, and great eruptions terrorize populations ... Sunspots change, some disciples can distinguish *an unknown Sun on the horizon* ... But *such phenomena are not due to a hazard: They are willed by cosmic intelligences or else brought about as reactions to social or ethnic diseases.*[795]

Figure 214: Photograph of French alchemist Sédir.

Sédir established a clear parallel between human 'devolution' and cosmic events, including an 'unknown Sun'. But the modern age has made us orphans of this precious knowledge and blind to the major threat that looms over our heads.

By the 19th century, reason had fully triumphed: scientists firmly considered the universe as purely mechanistic. This was the era of the industrial revolution and the 'clockwork universe' model. According to this paradigm, the universe's only constituent was matter made of 'billiard ball' particles solely ruled by mechanical laws (also known as natural laws) like inertia, speed, mass, etc.

This rejection of any kind of 'action at a distance' naturally led to the conviction that human activity could have no influence whatsoever on cosmic activities. Even the human mind would soon be explained exclusively by electric signals and chemical reactions, or so the materialists hoped.

But by the end of the 19th century, some cracks began to appear in the edifice of materialism. In 1861, Maxwell[796] demonstrated the laws of electromagnetism. Matter was no longer alone. The scientific community had to acknowledge the existence of something more subtle: *energy.*

---

[794] Real name, Yvon Le Loup, 1871–1926.

[795] Sédir, excerpts from letters published in *Bulletins des Amitiés Spirituelles*, February 1919

[796] James Clerk Maxwell (1831–1879), Scottish mathematical physicist.

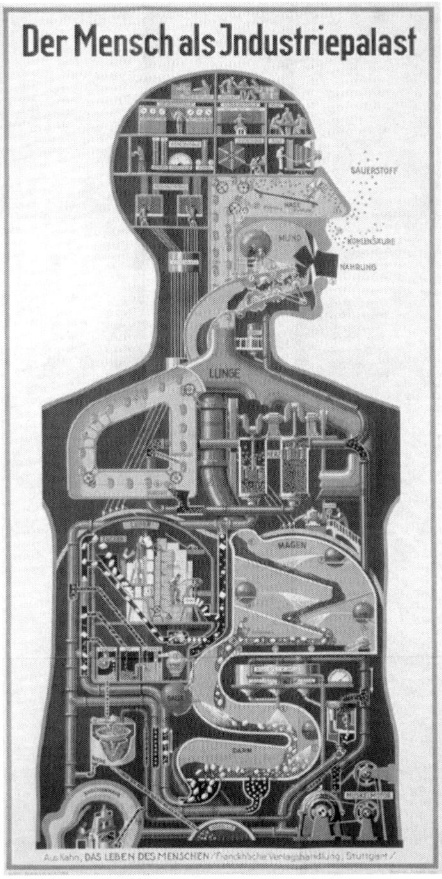

Figure 215: Fritz Kahn, 'Man as Industrial Palace'. (National Library of Medicine, Stuttgart)

In 1895, Röntgen[797] discovered radioactivity, a phenomenon where solid matter converts into radioactive waves. At this point, scientists began to realize that there was *no fundamental boundary separating matter and energy.* In his now famous $E=MC^2$ equation, Einstein showed how energy and matter were equivalent. Mass-energy had taken the place of mass alone. Hiroshima and Nagasaki would become the tragic proof of the principle of equivalence between mass and energy. With this principle came the 'field' character of matter, and scientists began wondering to what extent relativity theory implied a *de-materialization* of matter.

With energy introduced into the equation, the new model of the universe made of miniscule physical structures (particle matter) and waves (energy) worked much better than the purely mechanical one. However, many phenomena still remained unexplained, including what is condescendingly called 'the paranormal' or 'metaphysics'.[798]

The emergence of quantum theory at the beginning of the 20th century marked a second major setback to materialism and revolutionized our understanding of reality. According to quantum physics, our reality is not as tangible, nor as immutable, as previously thought.

One early discovery made by quantum physicists relates to properties of subatomic particles (electrons, protons, neutrons), the fundamental constituents of matter. For this purpose, a 'double-slit experiment' was conducted. As indicated by its name, this experiment consists of putting a double slit in the path of subatomic particles (figure 216).

In figure 216, we see photons behaving like waves. On the right of the slit openings, the light waves generate wave patterns similar to those created by a pebble thrown into water. Since there are two slits, there are two series of waves (as if you throw two pebbles into water).

Eventually those waves reach each other and begin to *interfere*. It means that in some places the waves amplify each other, but in others they cancel each other. The black curve on the right shows the brightness (quantity of photons) relative to the vertical location on the photon detector. For example, in the middle of the photon detector, brightness is at its maximum because the two series of waves fully add up.

However, those interference patterns only predict the *probability* that a photon will hit the detector in a specific location. The photon follows probabilistic rules. The place where it will hit the detector is not determinate; we can't know it in advance.

[797] Wilhelm Conrad Röntgen (1845–1923), German physicist.
[798] Talbot, M., *The Holographic Universe*, p. 2

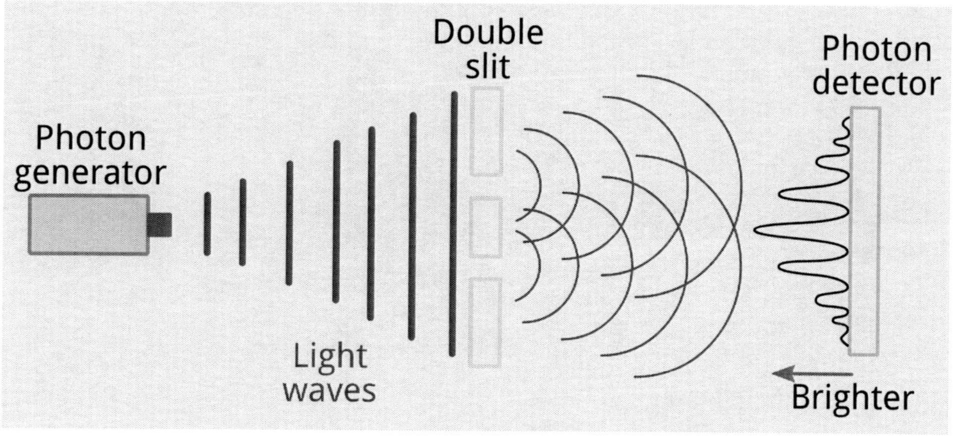

Figure 216: Double-slit experiment. (© Sott.net)

According to the standard interpretation in quantum physics,[799] the world is not fundamentally *determinate*. Heisenberg's[800] principle of indeterminacy means not only that we can't know the precise location and momentum of subatomic particles, but also that the particles simply do not *have* a precise location and momentum. In other words, quantum physics injected a dose of 'chance' into a world that was previously seen as totally determined.

The story doesn't end here. The results of the double-slit experiment become even stranger when an observer is introduced into this very same experiment, not at the level of the photon detector as before, but at the level of the slits. With the introduction of this observer, results change drastically; the photons stop behaving like waves and they start acting like particles.

As depicted in figure 217, each photon, now behaving like a particle (because it is observed at the level of the slit) has a 'choice' between two paths when encountering the double slit. It can go through slit A and then impact the photon detector at location A, or it can go through slit B and impact the photon detector at location B.

Austrian quantum physicists Brukner and Zeilinger summarized the phenomenon in these terms:

> The observer can decide whether or not to put detectors into the interfering path. That way, by deciding whether or not to determine the path through the two-slit experiment, he can decide which property can become reality. *If he chooses not to put the detectors there, then the interference pattern will become reality; if he does put the detectors there, then the beam path will become reality.* Yet, most importantly, the *observer has no influence on the specific element of the world which becomes reality.* Specifically, if he chooses to determine the path, he has no influence whatsoever which of the two paths, the left one or the right one, Nature will tell him ... where the particle is found. Likewise, if he chooses to observe the interference pattern he has no influence whatsoever where in the observation plane he will observe a specific particle. Both outcomes are completely random.[801]

---

799 'Copenhagen Interpretation', *Wikipedia*. See: en.wikipedia.org/wiki/Copenhagen_interpretation

800 Werner Heisenberg (1901–1976), German physicist, one of the creators of quantum mechanics.

801 Zeilinger, B., 'Young's experiment and the finiteness of information', *Philosophical Transactions of the Royal Society* (2002) 360: 1061–1069

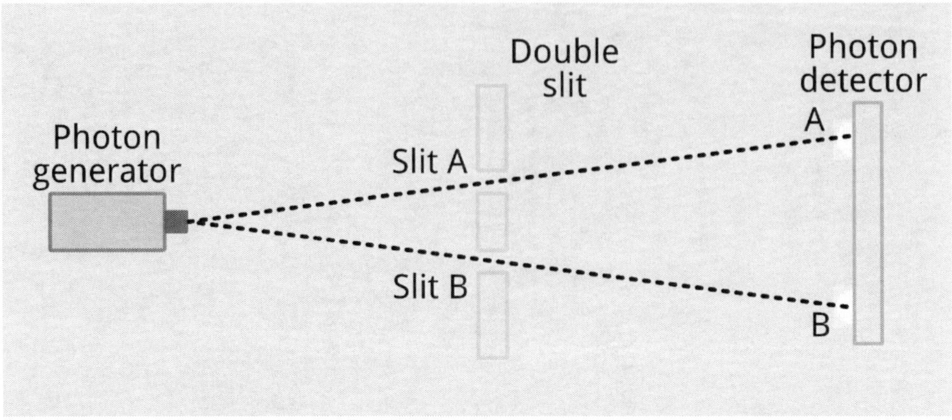

Figure 217: The double-slit experiment with photons behaving like particles. (© Sott.net)

This aspect of the quantum world is illustrated by the now famous Schrödinger's[802] cat experiment, which he described as follows:

> One can even set up quite ridiculous cases. A cat is penned up in a steel chamber, along with the following device (which must be secured against direct interference by the cat): in a Geiger counter, there is a tiny bit of radioactive substance, so small, that perhaps in the course of the hour one of the atoms decays, but also, with equal probability, perhaps none; if it happens, the counter tube discharges and through a relay releases a hammer that shatters a small flask of hydrocyanic acid. If one has left this entire system to itself for an hour, one would say that the cat still lives if meanwhile no atom has decayed. *The psi-function of the entire system would express this by having in it the living and dead cat (pardon the expression) mixed or smeared out in equal parts.*[803]

According to Schrödinger, the Copenhagen interpretation implies that the cat remains both alive and dead[804] (to the universe outside the box) until the box is opened:

> It is typical of these cases that an indeterminacy originally restricted to the atomic domain becomes transformed into macroscopic indeterminacy, which can then be resolved by direct observation.[805]

The key point here is that the 'indeterminacy' (the cat being both alive and dead to the universe outside the box) will be removed by the 'direct observation', i.e. *the exchange of information between the observer and the event.*

---

[802] Austrian physicist (1887–1961).

[803] Schrödinger, E., *The Present Situation in Quantum Mechanics*

[804] At that time, scientists believed that indeterminacy only applied at the particle scale. Large objects such as cats clearly couldn't exist in a superposition of two or more states. Microscopic objects obeyed quantum physics, while macroscopic things obeyed the classical physics rules. (See: Wogan, T., 'Coherent Schrödinger's cat still confounds', *IOP*, 23 November 2011) We will see in Chapter 39: 'Scaling it up from particles to people', that indeterminacy might also apply to large objects.

[805] Schrödinger, *op. cit.*

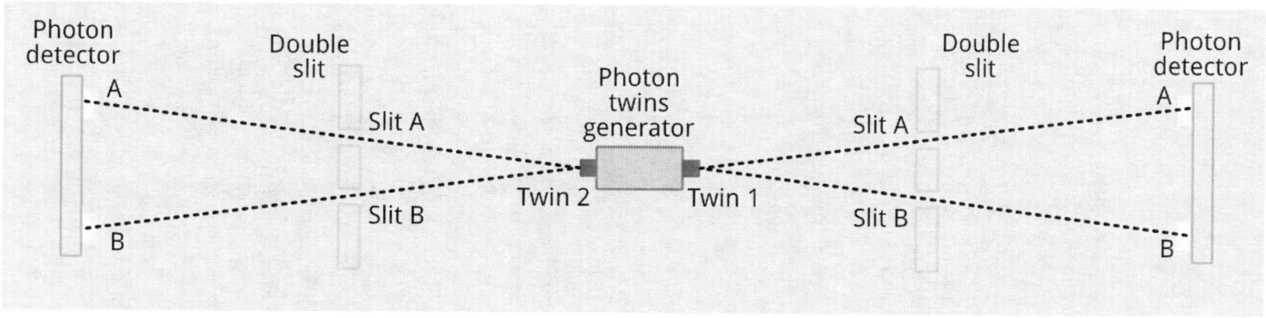

Figure 219: Two-double-slit experiments conducted simultaneously with two twin photons. (© Sott.net)

From this perspective it seems that *our reality emerges from an underlying dimension that comprises an infinite number of possible universes.* Only one of those possible universes materializes through what is called the 'wave collapse function', in the same way that one specific location or momentum of a particle emerges from the 'superposition' of all its possible states. For example, until a particle is observed, it can theoretically be in any position, some more probable than others. Once it's observed the particle 'freezes' in only one location.

Figure 218: Depiction of the 'Schrödinger's cat' thought experiment. (© thelifeofpsi.com)

Ironically, science, through quantum physics, had reintroduced the possibility of some interaction between mind (the observer) and matter (the observed event), and by extension the human-cosmic connection, two centuries after it had, through Newtonian physics, dismissed the very same phenomenon.

Schrödinger designed his cat thought experiment in the context of a discussion[806] about 'quantum entanglement', which is when a pair of particles (two particles coming from the same source or having interacted with each other) display the same properties (position, momentum, spin, polarization, etc.).

To understand the notion of entanglement better, scientists took the double-slit experiment one step further and set up a thought experiment where two double slits would be simultaneously fired with two twin photons.

As illustrated in figure 219, two paired photons (twin 1 and twin 2) are generated at exactly the

[806] The discussion was about a scientific article written by Einstein about quantum entanglement. See: Einstein A., Podolsky, B. & Rosen, N., 'Can Quantum-Mechanical Description of Physical Reality Be Considered Complete?', *Phys. Rev.* (1935) 47(777)

Figure 220: Aspect conducted his experiment in this lab at Orsay. The source of pairs of entangled photons consisted of several lasers, and a stream of atoms. Measurements were taken after they had travelled six meters in both directions from the source. (© GOA/LCFIO)

same time, and they reach their double slits at the same time. Because of *entanglement*, the two photons *should make the same choices* (slit A and slit A or slit B and slit B).

Discussion raged among scientists trying to make sense of entanglement. Some, like Einstein, argued that it was explained by what he called 'spooky action at a distance'. Somehow the first measurement made on twin 1 was *then* communicated to twin 2 which could mimic the behavior of the first photon.

For some other scientists, amongst whom was Niels Bohr,[807] the *communication between the two electrons was instantaneous* (non-local phenomenon). The debate between Bohr and Einstein lasted over 20 years until their deaths.[808]

The problem was, the two-double-slit experiment was nothing more than theoretical speculation. It was only in the 1980s that technology allowed for the generation of single photons and, thus, for an opportunity to conduct an experiment that would settle the entanglement case. In 1982, French physicist Alain Aspect[809] finally managed to design all the necessary equipment and he fired simultaneously one photon in one direction and its twin in the other direction.

The results were clear: both photons encountered their respective double slit simultaneously and chose the same corresponding A or B slit at the very same time.[810] Somehow, *the twins had managed*

[807] Danish physicist (1885–1962).

[808] Aspect, A., 'The Bohr–Einstein debate and quantum entanglement tested experimentally', *CNRS Communiqué* (1982)

[809] Ravaud, M., 'Alain Aspect: Shedding new light on light and atoms', *CNRS International Magazine*. See: www2.cnrs.fr/en/447.htm

[810] To make the experiment easier to understand, I've simplified its description, but the principle and the results remain

to instantaneously exchange information, even over large distances.[811]

Following the disclosure of Aspect's results, several scientists suggested that the experiments demonstrated superluminal (faster than light) action at a distance, therefore breaching Einstein's theoretical limits wherein light speed is the maximum possible speed.[812]

While quantum physics strongly suggests the central role played by information (observer-event interaction – entanglement), a connection between mind and matter (observer's role) and the existence of instantaneous interactions (two-double-slit experiment), it leaves many questions unanswered.

In the next three chapters we will focus on these questions:

- Can the connection between mind and particles be scaled up to macro events? How does the observer influence a 'random' event? (Chapter 37)
- What is the exact nature, and properties, of information? (Chapter 38)
- How can an information exchange transcending time and space occur? (Chapter 39)

the same. Alain Aspect didn't actually measure the slit chosen by the photons but their spin. He also used a set of mirrors to generate and direct the entangled photons. See: Schmüser, P., *The Strange Features of Quantum Mechanics in the Light of Modern Experiments*, pp. 24–37

811 In subsequent experiments, Aspect increased the distance travelled by the photons up to 10 km (about 6 miles). See: Davies, P. & Gregersen, N., *Information and the Nature of Reality*, p. 56

812 *Ibid.*

Chapter 37.

The role of the 'observer'

The term 'observer', as used by quantum physicists and in the title of this chapter, can be quite misleading. It suggests a purely passive state where the person in question only receives information. For example, visual observation is considered to be the passive reception of visual data, through photons emitted by an external source and hitting the retina.

However, recent research shows that vision is not a one-way process. While the eyes do receive photons, **they also emit photons**,[813] which are known to be prime *carriers of information.*[814] The eye is not an isolated organ in this respect. The brain[815] and most other organs[816] also emit photons. (We'll soon examine[817] the potential role played by these emitted photons.)

As shown in the previous chapter, the quantum principle of uncertainty is only supposed to apply at the particle scale. Although Schrödinger's experiment[818] was first a mere thought experiment about the influence of mind (the observer) on matter, it was later demonstrated in practical experiments at the level of particles and scaled up in later years.

Still far smaller than the human body,[819] by 1999 a molecule comprising 60 atoms of carbon[820] was observed exhibiting the same quantum behavior. Recent research shows that non-locality also occurs in macroscopic systems made of photons, nuclei and ions,[821] and in living systems such as photosynthesis[822] and DNA.[823]

---

[813] Wang, C., *et al.*, 'Spontaneous and visible light-induced ultraweak photon emission from rat eyes', *Brain Research* (19 January 2011) 1369: 1–9

[814] The photon is a fundamental carrier of information, comprising numerous information-carrying degrees of freedom, including frequency, phase, arrival time, polarization, orbital angular momentum, linear momentum, entanglement, etc. See: Kumar, P., 'information in a photon', DARPA, Defense Science Office. www.darpa.mil/Our_-Work/DSO/Programs/Information_in_a_Photon

[815] Kobayashi, M., *et al.*, 'In vivo imaging of spontaneous ultraweak photon emission from a rat's brain correlated with cerebral energy metabolism and oxidative stress', *Neuroscience Research* (July 1999) 34(2): 103–13

[816] Schwabl, H., *et al.*, 'Spontaneous ultraweak photon emission from biological systems and the endogenous light field', *Forsch Komplementarmed Klass Naturheilkd* (April 2005) 2: 84–9

[817] See Chapter 39: 'The field'

[818] See Chapter 36: 'The rise and fall of materialism'

[819] The human body is estimated to contain about $7x10^{27}$ atoms. Kross, B., 'How many atoms are in the human body?', *Jefferson Lab.* See: www.education.jlab.org/qa/mathatom_04.html

[820] Arndt, M., *et al.*, 'Wave-particle duality of C60', *Nature* (1999) 401: 680–682

[821] Vedral, V., Quantifying entanglement in macroscopic systems, *Nature* (19 June 2008) 453

[822] Sarovar *et al.*, 'Quantum entanglement in photosynthetic light-harvesting complexes', *Nature Physics* (2010) 6: 462–467

[823] Gutiérrez, R., *et al.*, 'Structural fluctuations and quantum transport through DNA molecular wires, a combined molecular dynamics and model Hamiltonian approach', *New J. Phys.* (2010) 12, 023 022

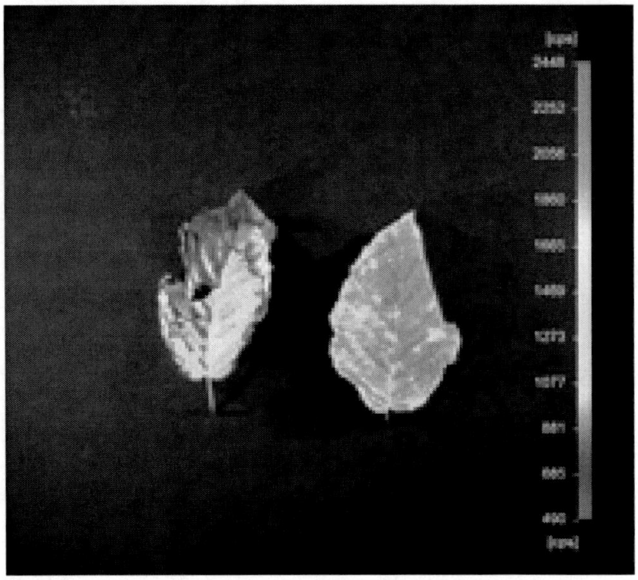

Figure 221: Ultraweak bio-photon emission measured in a leaf. (© Berthold Technology)

The scaling-up is probably not over. A survey conducted among 33 key physicists revealed that most of them believe that quantum physics do apply to large objects:

> More than two-thirds believed that there is no fundamental limit to quantum theory – that **it should be possible for objects, no matter how big, to be prepared in quantum superpositions like Schrödinger's cat**.[824]

Since the 1940s, hundreds of experiments[825] have been conducted showing that the human mind can indeed have an influence on matter at the macro scale. This phenomenon is called 'psychokinesis' ('PK') or 'telekinesis', which describes a person's capability of influencing a physical system without any physical interaction.

One of the first PK experiments involved dice-throwing, where individuals try to influence the result of a dice.[826] The first paper on the topic was published by L. E. Rhine & J. B. Rhine in 1943.[827]

Since then, hundreds of dice experiments have been conducted. In 1991 Princeton scientist Dean Radin directed an extensive meta-analysis of 179 dice experiments[828] and concluded that **the overall PK effect is statistically highly significant**.[829]

Along with the dice experiments, countless numbers of PK experiments have provided a body of evidence supporting the theory that humans have a non-material influence on various kinds of macro events, both non-living forms and living forms,[830] including other humans:

---

[824] Ball, P., 'Experts still split about what quantum theory means', *Nature* (11 February 2013)

[825] Braud, W., 'On the Use of Living Target Systems in Distant Mental Influence Research', in *Psi Research Methodology: A Re-Examination*

[826] Girden, E. & Helson, H., 'A review of psychokinesis (PK)', *Psychological Bulletin* (1962) 59(5): 353–388

[827] Rhine, L. E., & Rhine, J. B., 'The psychokinetic effect. The first experiment', *Journal of Parapsychology* (1943) 7: 20–43

[828] The meta-analysis covered experimental studies conducted between 1935 and 1987, which included almost 2.6 million die-casts. See: Radin, D. I. & Ferrari, D. C., 'Effects of consciousness on the fall of dice: A meta-analysis', *Journal of Scientific Exploration* 5(3): 61–84

[829] A controversy about the dice experiment had been raging for decades. Its detractors claimed a bias due to the fact that in some experiments the 'influencers' were free to choose the wished-for dice face and a majority chose the six-face which is lighter than the five other faces and therefore more likely to be casted. So Bösch reduced the database to the only 69 studies that had correctly controlled for die bias and the results were still 'statistically highly significant' See: Bösch *et al.,* 'Examining Psychokinesis: The interaction of human intentions with random number generators', *Psychology Bulletin* (July 2006) 132(4): 498

[830] The success of PK experiments using living targets might be due to the fact that living forms are better receptors of PK influence. See: Braud, W., 'On the Use of Living Target Systems in Distant Mental Influence Research', in *Psi Research Methodology: A Re-Examination*, p. 17

Figure 222: Joseph Bank Rhine and Louisa Rhine using a dice-casting machine. (© Foundation for Research on the Nature of Man)

- The behavior of dogs[831]
- Growth or viability of bacteria,[832] fungus,[833] yeast[834] and plants[835]
- Influence on the movements of protozoa, larvae, woodlice, ants, chicks, mice, rats, gerbils and cats[836]
- Sleeping and waking state in humans[837]
- The direction of eye movements in dreaming target persons[838]

---

[831] Bechterev, V. M., 'Experiments on the effects of the mental influence on the behavior of dogs', *Problems in the study and training of personality* (1920): 235–260

[832] Solfvin, G. F., 'Psi expectancy effects in psychic healing studies with malarial mice', *European Journal of Parapsychology* (1982)

[833] Nash, C. B., 'General and comparative study of the psychokinetic effect on a fungus culture', *J. Barry Journal of Parapsychology* (1968)

[834] Kriesa, C., *The Influence of Distance on Psychokinesis as Measured by Yeast Metabolism*, Sonoma State University, 1978

[835] Saklani, A., 'Psychokinetic effects on plant growth', *Research in Parapsychology 1989* (1990): 37–41

[836] Solfvin, J., 'Mental Healing', in *Advances in Parapsychological Research 4*, pp. 31-63. Benor, D. J., 'Survey of Spiritual Healing Research', *Complementary Medical Research (3 September 1990) 4: 9–33

[837] Kohsen, A., *Experiments in Mental Suggestion*

[838] Dean, D. E., 'Plethysmograph Recordings of ESP Responses', *International Journal of Neuropsychiatry* (2 September 1966)

- Human physiological activities, including electrodermal activity,[839] red blood cells,[840] blood pressure,[841] muscular tension, muscular tremor[842] and unconscious muscular movement.[843]

One of the most extensive and scientifically sound PK experiments was conducted by Dr. Robert Jahn, Dean of the School of Engineering and Applied Science at Princeton, and Brenda Dunne, a developmental psychologist at the University of Chicago.

Figure 223: Prof. Robert Jahn and Brenda Dunne, M.S.

In a 12-year period, Jahn and Dunne conducted nearly 2.5 million trials in which participants sitting in front of a carefully configured Random Event Generator (REG) would first attempt to 'will' the machine to produce more 1s than 0s, then the reverse, then try not to influence the machine in any way.[844] An 'REG' usually contains radioactive material that is supposed to decay randomly. Electronics within the machine transform the decay process into measurable signals (green/red light, for example). REG machines are calibrated to generate 50% red and 50% green responses. But, at the end of Jahn and Dunne's experiments, 52 per cent of all trials were in the intended direction and nearly two-thirds of the 91 participants had overall success in influencing the machines the way they had intended.

It should be emphasized that the research conducted by Jahn and Dunne is one of the very few parapsychological studies that has been recognized by a prominent scientific organization, the U.S. National Research Council, which concluded that the REG trials were rigorous and that its results could not be explained by chance.

Although the deviation (see figure 224) might seem limited (52% vs. 50%), statistically it is highly significant because the results were compiled from *millions of trials*, with dozens of correlating experiments, and conducted over several decades.[845] The odds of these results being produced by chance are one in a trillion.[846]

---

[839] Braud, W. & Schlitz, M., 'Psychokinetic influence on electrodermal activity', *Journal of Parapsychology* 47(2): 95–119

[840] Braud, W., 'Distant mental influence of rate of hemolysis of human red blood cells', *Journal of the American Society for Psychical Research* (1990) 84(1): 1–24

[841] Beutler *et al.,* 'Paranormal healing and hypertension', *British Medical Journal* 1998) 296: 1491–1494

[842] Braud, W., *Distant Mental Influence: Its Contributions to Science, Healing, and Human Interactions*

[843] Braud, W. & Jackson, J., 'The Use of Ideomotor Reactions as Psi Indicators', *Parapsychology Review* (1982) 13: 10–11

[844] McTaggart, *The Field,* p. 115

[845] "[PK experiments] compare favorably with what is typically observed in more conventional behavioral and biomedical studies.... The average effect size observed in these mental influence studies (0.32) is 10 times as great as that obtained in some representative medical study outcomes that have been heralded as medical breakthroughs (effect sizes of 0.04 and 0.03, obtained in two well-known studies of the effectiveness of propranolol and aspirin, respectively, in reducing heart attacks)." See: Braud, W., 'Wellness Implications Of Retroactive Influence: Exploring An Outrageous Hypothesis', *Alternative Therapies in Health and Medicine* (2000) 6(1): 37–48

[846] McTaggart, *op. cit.,* p. 117

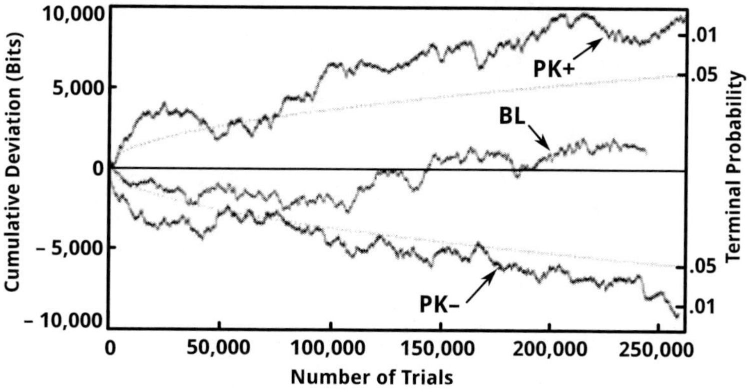

Figure 224: Cumulative deviations of mean shifts achieved. Top curve displays the shift for a '1' intent, bottom curve displays the shift for a '0' intent. (© Jahn & Dunne)

While the PK experiments quoted above show that individuals do have an influence on macro events, the example below suggests that they might even have a direct influence on events of a cosmic nature, the very core of the human-cosmic connection:

> Roy Sullivan (1912–1983) was a park ranger at Shenandoah National Park in Virginia for 36 years, beginning in 1940. On seven occasions between 1942 and 1977, six of them while in the park, Ranger Sullivan was struck by lightning and lived to tell about it.
>
> Sullivan earned his entry in Guinness World Records the hard way. Two of his ranger Stetsons, both with lightning-damaged crowns, are on display in Guinness World Exhibit Halls. ...
>
> In that first instance, back in 1942, Roy may not have been still in that fire lookout tower when the thunderstorm approached. He said on at least one occasion that he had bailed out of the tower (a lightning magnet) and was hot-footing it for a safer place when the bolt zapped him. He counted himself very lucky to still be alive, and indeed he was.
>
> In that second instance, the 1969 one, Roy was hit by lightning while driving his truck down a mountain road. Now stop right there. To say that's not supposed to happen is an understatement of stupendous proportions. Roy could probably have sat in that truck for a thousand years while lightning struck it a gazillion times and he'd never get hurt as long as he rolled up the windows, kept his arms inside, and didn't touch the metal exterior. That's how safe it is to be inside a vehicle during an electrical storm. (When a lightning bolt strikes a vehicle, the charge travels around the exterior, not through the interior, and just blows out the tires.) ...
>
> Here's what happened that time in 1969. Roy was tooling down a mountain road (presumably Skyline Drive) with his windows rolled down when lightning struck a tree next to the road and then went laterally through the cab of his truck and out the other side, singeing his eyebrows en route to zapping tree number two on the opposite side of the road. The odds of that happening to a motorist are almost indescribably small, but remember, we are talking about Roy Sullivan here.
>
> Strike number three (1970) occurred when Roy was walking across his yard to get his mail. Roy's property is not a good place to be in an electrical storm, folks. Roy's wife could have attested to that. She suffered nonfatal injuries when lightning struck her one day when she was hanging the wash in the back yard. (Roy was with her at the time.) ...

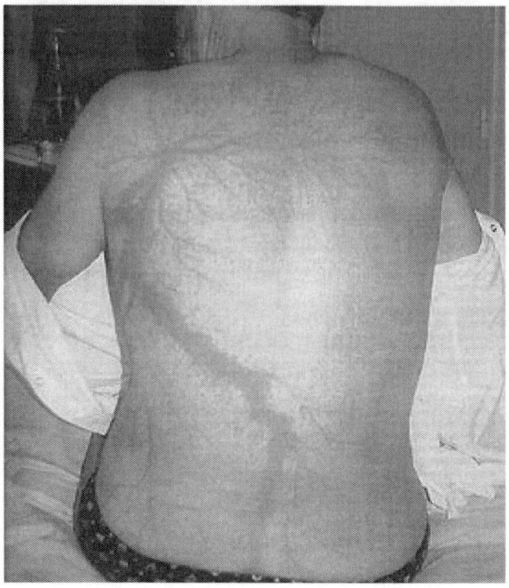

Figure 225: Burn mark left on Roy Sullivan's back after he was struck by a lightning bolt.

> In one of the instances, Roy was struck by the *only* bolt of lightning produced by the low cloud that passed overhead.[847]

Sullivan was avoided by people later in life because of their fear of being hit by lightning, and this saddened him. On September 28th, 1983, Roy Sullivan killed himself from a self-inflicted gunshot to the stomach. Six years had passed since he had last been hit by lightning.[848]

The odds of being struck by a lightning bolt once in 40 years is about 1/12,500.[849] So Roy Sullivan's chances of being struck seven times are 1 in $4.7^{\times}10^{28}$. [850] Of course, because of the nature of his work, Sullivan spent a lot of time outdoors and was therefore more exposed to lightning. However, after the first strike he became overly wary of lightning and tried to avoid it. Therefore the probability of him being struck should have been lower than average. Nonetheless, he was struck *six* more times.

Another striking element in Sullivan's story is that he *survived* all seven lightning strikes, while the death rate for people struck by lightning is 30%.[851] Obviously, the usual probabilities and death rates didn't apply to Sullivan.

While the Sullivan case puzzles modern scientists, it might not have done so earlier in history. Several books dealing with lightning were published at the beginning of the 18th century. According to those writings, "humans, by their presence, could attract lightning to a nearby object."[852]

---

847 Janiskee, B., 'Shenandoah National Park Ranger Roy Sullivan Set the World Record for Being Hit by Lightning' (1 August 2008)

848 Dunkel, T., 'Inside the life of the man known as the 'spark ranger'', *The Washington Post Magazine* (15 August 2013)

849 See: www.srh.noaa.gov/jetstream/lightning/lightning_faq.htm

850 Or 1 / $12{,}500^7$.

851 Copper et al., *Lightning injury*, p. 68

852 *Ibid.*, p. 69

Figure 226: Roy Sullivan wearing one of his two hats that were ripped through by a lightning bolt.

One is compelled to ask: what was the unique make-up of Roy Sullivan that led to these completely bizarre experiences? There is something deeply mysterious about a human being who can attract lightning *and survive* it. For Roy Sullivan, it was probably an accident of fate or genetics; but is it possible that such a state can be induced in persons and societies by external factors? We'll address this point soon.[853]

While most PK experiments show a clear connection between mind and matter, they generally don't explore what factors (beliefs, self-confidence, focus, willpower, etc.) modulate the psychokinetic phenomena.[854] Jahn and Dunne's studies provide a hint. Although most subjects had an influence on the 'random' event, in several cases the *effect was opposite to the intent.* So, is it really our conscious intent that influences 'random' events, or are some other parameters involved?

Researcher Helmut Schmidt selected a group of subjects who showed a tendency to influence 'random' events in a way opposite to their conscious intent.[855]

Then Schmidt tried to see if he could increase the 'negative' influence of his subjects. To do so, he subjected them to a negative environment: the experiment was conducted in a small dark closet, no encouragement was given, and instead the suggestion was made that they would fail. Schmidt found that in such a negative context the 'negative' influence of the subjects increased in a statistically relevant way.[856]

[853] See Chapter 42: 'Branching out Universe'

[854] Up to now, no PK meta-analysis has reported data on potential modulators like psychological variables. See: Bösch *et al.*, 'Examining Psychokinesis: The interaction of human intentions with random number generators', *Psychology Bulletin* (July 2006) 132(4): 498

[855] Schmidt, H., 'Mental Influence on Random Events', *New Scientist and Science Journal* (June 1971): 757–768

[856] In 32,768 trials testing subjects having an influence opposite to their intent, the deviation is so large that the odds against this, or a larger deviation being the result of a chance fluctuation, are more than 1,000 to one. See: Schmidt, *op. cit.*

Figure 227: Helmut Schmidt next to subjects using an RNG based on the atomic decay of strontium 90. (© Rhine Research Center)

Decreasing the subjects' self-confidence seems to have increased their 'negative' influence. Someone with low self-confidence holds the *belief* that he's not good enough, that he's condemned to failure. Thus, when asked to influence a 'random' event in a specific direction, he believes he will fail. And he is partly right. He fails to influence the 'random' event in the *consciously desired* direction. But he has *not really* failed because he *has* influenced the event, just in the *opposite* direction.

Conversely, research shows that subjects who *believe* in psi phenomena score higher in ESP (Extra Sensory Perception) tests[857] than non-believers, while people who *believe* in religion and an afterlife scored even higher.[858]

Rather than any given specific and conscious intent, what seems to really matter is the deeply held belief. The belief 'I will fail' doesn't cancel the influence; it reverses it. Belief in ESP strengthens ESP abilities. *Beliefs affect our influence on 'random' events.* The effectiveness of placebos[859] and hypnotherapy[860] provide further confirmation of the power of beliefs and suggestions.

This shows the limitation of the 'you-create-your-own-reality' concept spread in the New Age

[857] Schmeidler, G. R. & Murphy, G., 'The influence of belief and disbelief in ESP upon individual scoring level', *Journal of Experimental Psychology* (1946) 36(3): 271–276

[858] Haraldsson, E. et al., 'Are religiosity and belief in an afterlife better predictors of ESP performance than belief in psychic phenomena?', *Journal of Parapsychology* (1993) 57: 259–273

[859] Typically placebos follow the '55% rule', i.e. 55% of the subjects respond favorably to the placebo. This is quite a high figure comparable to the effectiveness to some prescription drugs. See: Maté, G, *When the Body Says No: Understanding the Stress-Disease Connection*, p. 139

[860] Talbott, M., *The Holographic Universe*, pp. 90–97, 104–110

movement.[861] We obviously don't *create* our own reality, although, as observers, we can have an *influence* on reality, albeit one that is usually tiny and sometimes diametrically opposed to our expectations (the outcome we consciously wish to bring about).

At this point, knowing more about interactions between the 'observer' and macro events and how the *information we believe* can affect these interactions, let's turn our attention to *information.*

[861] Knight-Jadczyk, L., *The Wave Series*, Vols. 1–8

Chapter 38.

Information theory and consciousness

Quantum physics repeatedly acknowledges the role played by 'information' but never really defines it.

Information theory was first popularized by Claude Shannon[862] in his paper 'A Mathematical Theory of Communication',[863] which focused on the efficiency of information transport and its application for informatics.[864] While this in itself is an interesting field of research, Shannon's take on information theory didn't address some important points:

> When the foundation for information theory was laid down by Shannon, he purposely left out of the account any reference to what the information means, and dwelt solely on the transmission aspects. His theory cannot, on its own, explain the semantics and communication of higher-order entities.[865]

In contrast, Tom Stonier[866] has dedicated most of his research and writing to the nature of information and has reached some groundbreaking conclusions. According to Stonier, information is a *fundamental constituent of the universe*:

> Matter and energy comprise the surface of the universe. The surface structure of the universe is readily perceivable to our senses. The internal structure is more subtle. It is organized in a manner not so obvious: *it consists not only of matter and energy but of information as well.*[867]

Naturally, one could wonder why this wasn't conceptualized earlier, to which Stonier answers:

> Matter is the ground we walk on ... Energy is what burns our finger ... Information is more subtle. It is true that it is also part of our daily experience. Every time we talk, read a newspaper, or watch television, we are busy absorbing or exchanging information. *But we have always associated information with activities inside our heads – not something that is real the way energy and matter are.*[868]

---

[862] Claude Elwood Shannon (1916-2001), American mathematician and electrical engineer.

[863] Two-part paper published in the July 1948 and October 1948 issues of the *Bell System Technical Journal.*

[864] Namely the creation of communication protocols that maximize the quantity of communicated information while minimizing used resources.

[865] Davies, P. & Gregersen, N., *Information and the Nature of Reality*, p. 4

[866] Biologist, philosopher, information theorist and pacifist (1927–1999). See: Goodman, G., 'Tom Stonier, a man of computers and peace', *Steinschneider.com* (28 June 1999). www.steinschneider.com/biography/tomstonier.htm

[867] Stonier, T., *Information and the Internal Structure of the Universe*, p. 1

[868] *Ibid.*

Tom Stonier, father of Information Theory (1927–1999). (© Valiant Technology)

Indeed, information is not exclusively inside our head: words written in a book are information, whether we read them or not. The DNA present in every living creature is information, fundamental information that controls the development of every cell. Without DNA, there can be no development. This development can't be accounted for solely by the matter (the nucleotides, proteins, water, etc.) that constitutes the cell; neither can it be explained exclusively by the energy produced or received by the cell (heat, electricity, etc.). Therefore, a third constituent must exist, beyond matter and energy, to account for the development and organization of the cell, namely information.

> ... information makes a causal difference to our world – something that is immediately obvious when we think of human agency. But even at the quantum level, information matters. A wave function is an encapsulation of *all that is known* about a quantum system. When an observation is made, and that encapsulated knowledge changes, so does the wave function, and hence the subsequent quantum evolution of the system. Moreover, informational structures also play an undeniable causal role in material constellations, as we see in, for example, the physical phenomenon of resonance, or in biological systems such as DNA sequences. What is a gene, after all, but *a set of coded instructions* for a molecular system to carry out a task?[869]

To explicate the relevance of information theory, Stonier addressed the famous question: 'If a tree falls in a forest and no one is around to hear it, does it make a sound?'[870]

For Stonier, to answer this question one must first define the term 'sound'. If 'sound' means 'vibrations in the air transduced by the human ear', then a tree falling on a desert island doesn't produce sound, but if 'sound' is defined solely as 'vibrations in the air', then a sound is produced by a falling tree whether or not an animal or a human hears it.[871]

Thus, whether an observer is present or not, any change or any event occurring in the universe modifies the state of the universe. More specifically, it alters its fundamental nature, i.e. *its information content.* This event modifies the universe from a universe with this specific tree standing in it to a universe where this specific tree has fallen.

This is a major difference to the position taken by most quantum physicists, for whom the presence of an observer is required for the probabilistic wave function to collapse and for reality to 'freeze' and materialize. For quantum physics, without an observer, the universe remains a sea of infinite potentials, a sea of abstract probabilities:

> ... Einstein quipped, 'Do you really believe that the moon is not there unless we are looking at it?' The adherents of quantum mechanics responded with a version of the old saw about a tree

[869] Davies & Gregersen, *op. cit.*, p. 7

[870] This question was first asked by philosopher George Berkeley in *A Treatise Concerning the Principles of Human Knowledge*, published in 1710.

[871] For Stonier, postulating that the tree doesn't produce a sound if there's no observer is egocentric and inadequate. It would imply that as soon as we shut off the radio, the room is no longer filled with radio waves. See: Stonier, T., *Information and the Internal Structure of the Universe*, p. 8

Figure 229: Whether or not an observer is present, any given event does modify the state of the universe and its information content. (© miriadna.com)

> falling in the forest: if no one is looking at the moon – if no one is 'measuring its location by seeing it'– then there is no way for us to know whether it's there, so there's no point asking the question.[872]

Stonier also brings a new interpretation to thermodynamics, a branch of natural science dealing with fundamental notions like heat, temperature, energy and work. One of the main concepts developed in thermodynamics is 'entropy', which is a measurement of the level of disorder in a system.

According to mainstream science and its second law of thermodynamics, any system tends towards a state of maximum entropy, i.e. maximum disorder.[873] Thus, any given entity is supposed to tend towards disorder and decay. That's why scientists predict the 'running down' of the universe and all its constituents, ultimately leading to entropic death and complete randomization.[874]

The problem is that this law only takes energy and matter into account. For example, a pot of water, when subjected to heat, will see its molecules begin to agitate, leading to an increased level of chaos, whereas cooling down this system would increase its level of organization and reduce its entropy.

However, as pointed out by Stonier, there is *another way to reduce the entropy of a system: by increasing its information content.* Despite the second law of thermodynamics, living forms become increasingly complex, and more organized over time. Instead of following the path of entropic death with an ever-increasing level of disorder, they follow the opposite path, exhibiting an ever-higher

[872] Greene, B., *The Fabric of the Cosmos*, p. 95
[873] Boltzmann, L., *The Second Law of Thermodynamics*, p. 20
[874] Thomson, W., 'On the age of the sun's heat', *Macmillan's Mag.* (1862) 5: 288–93; PL, 1: 394–68

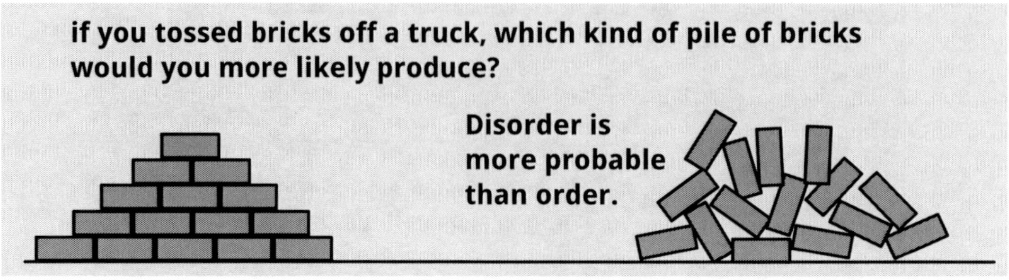

Figure 230: Order is highly improbable according to the laws of thermodynamics. Information can make the improbable happen. (© Hyperphysics)

level of organization and complexity.[875]

While mainstream science posits the existence of only one kind of entropy, solely driven by energy, for engineer Bryant M. Shiller, there are two kinds of entropy: energy-driven entropy and information-driven entropy. While non-living forms are exclusively controlled by energy-driven entropy and tend to lose energy and decay over time, living forms are subjected to both kinds of entropy.[876] While energy-driven entropy tends to increase with time (usual decay process), information-driven entropy can *decrease* over time if sufficient information or intelligence is present inside the system.[877] Hence the increase in complexity noticed among numerous living forms despite the energy-driven entropy that pushes them towards increased disorder.

The time arrow of entropy – the tendency of any system to lose energy over time and to eventually die – is counteracted by another arrow of time, one that's equally fundamental, where the universe is progressing through the growth of information, structure, organization and complexity to ever more elaborate states of matter and energy.

Figure 231 shows the curve of information (I) as a function of entropy (S). When information increases, entropy decreases (top left of the graph), yielding a highly organized state (illustrated here in the form of an improbably structured brick wall). When information decreases, entropy increases (bottom right of the graph), producing highly disorganized states (represented here by the more probable random heap of bricks).

Non-living forms follow the curve towards the right, the light grey arrow towards disorganization and decay. But living forms can travel this curve towards the left, the medium grey arrow towards higher levels of information, organization and complexity.

Not only is information a prime constituent of the universe and the core driver of living forms and creation, it also *intimately interacts with matter and energy* (figure 232).

Figure 232 shows three different forms of interactions:

1. A mixture of **energy and matter**, devoid of information, is represented, for example, by boiling water: energetic and disorganized water molecules.

[875] Stonier, *op. cit.*, pp. 41–57
[876] The boundary between living forms and non-living forms might not be as clear cut as it seems. This point is developed in Chapter 42: 'Branching out universe'.
[877] Shiller, B., *Origin of Life: The 5th Option*, pp. 305–315

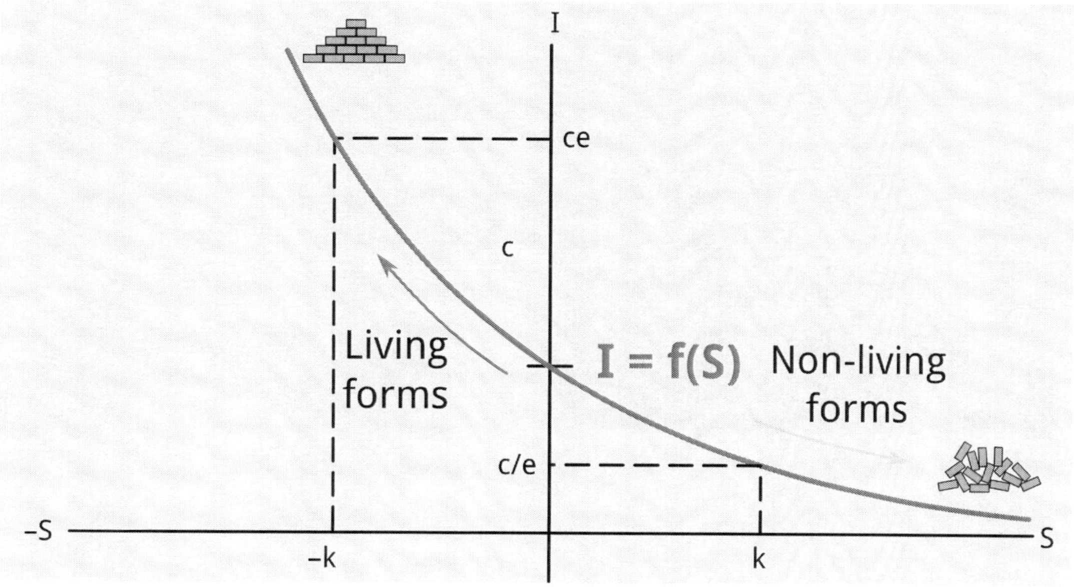

Figure 231: Information (I) as a function of entropy (S). (© Sott.net, adapted from Stonier)

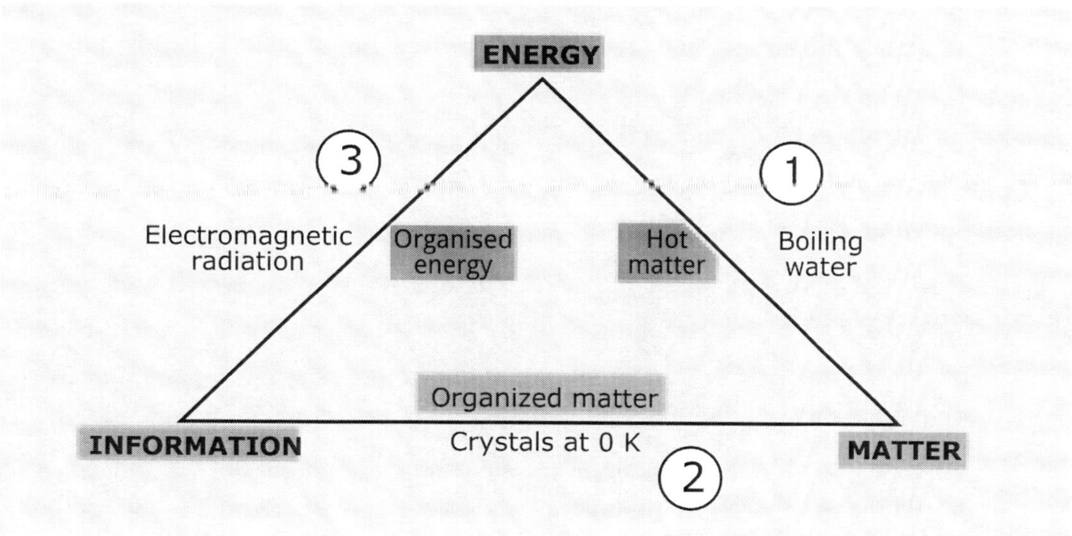

Figure 232: Information, energy and matter: Three interrelated prime constituents of the universe. (© Sott.net, adapted from Stonier)

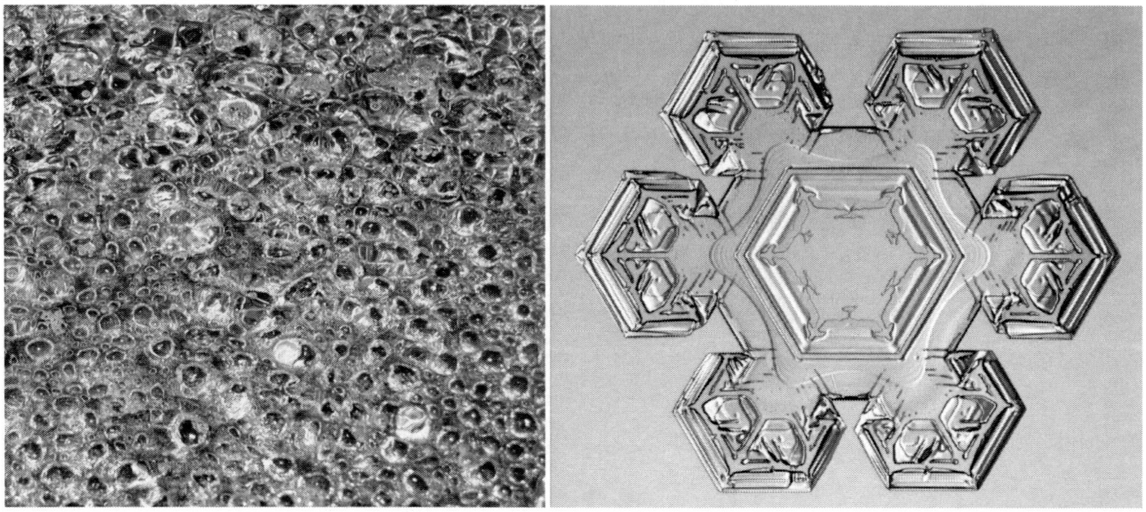

Figure 233: Left – Boiling water (high disorder, low information). Right – Water crystal (high level of order and information). (© Sott.net)

2. A mixture of **matter and information**, devoid of energy, is exemplified by a crystal at absolute zero, a highly organized and non-energetic ice crystal.

3. A mixture of **information and energy**, involving no matter, could be illustrated by light travelling in space: photons are energetic, massless particles moving in an orderly way and carrying information.[878]

From the above we can see that an energy input, like applying heating to a pot of water, causes water molecules to move more randomly. So, the level of disorder in the universe is increased and information is lost.

Conversely, if energy is removed from the system (cooling down the pot of water), randomness will be reduced. Eventually the water can freeze into highly structured and organized ice crystals. In the case of water, the cooling process – the removal of energy – is associated with an increase in information. Consequently:

> **Heat** is the product of **energy** interacting with **matter**. **Structure** represents the product of **information** interacting with **matter**.[879]

Physicists Seth Lloyd and Henry Stapp challenged some widely held beliefs about our reality and went one step further than Stonier:

> ... what happens if we do not assume that the mathematical relations of the so-called laws of nature are the most basic level of description, but rather if *information* is regarded as the

[878] The photon is a fundamental carrier of information, possessing numerous information carrying degrees of freedom including frequency, phase, arrival time, polarization, orbital angular momentum, linear momentum, entanglement, etc. See: Kumar, P., 'Information in a photon', *DARPA, Defense Science Office.* www.darpa.mil/Our_-Work/DSO/Programs/Information_in_a_Photon

[879] Stonier, *op. cit.*, p. 74

> **foundation** on which physical reality is constructed ... instead of taking mathematics to be primary, followed by physics and then information, **the picture should be inverted in our explanatory scheme, so that we find the conceptual hierarchy: information -> laws of physics -> matter**.[880]

If information is indeed the most fundamental constituent of our universe, higher even than energy and matter, then those *words* attributed to 'John the Evangelist' may take on a whole new meaning:

> In the beginning was the Word, and the Word was with God, and the Word was God Himself. He was present originally with God. All things were made and came into existence through Him; and without Him was not even one thing made that has come into being.[881]

[880] Davies & Gregersen, *op. cit.*, p. 3
[881] *John* 1:1–3

Chapter 39.

The Field

Now that we've sketched out the prime role played by information, how it interacts with matter and energy, and how it drives organization, let's try to better understand how information processes can transcend time and space, as exemplified by Aspect's two-double-slit experiments.[882]

Retrocognition describes knowledge of a past event which could not have been learned or inferred by normal means. It defies the commonly accepted past/present separation. Here is an example of retrocognition:

> In a series of experiments conducted in the 1960s, W. H. C. Tenhaeff, the director of the Parapsychological Institute of the State University of Utrecht, and Marius Valkhoff, dean of the faculty of arts at the University of Witwatersrand, Johannesburg, South Africa, found that the great Dutch psychic, Gerard Croiset, could *psychometrize even the smallest fragment of bone and accurately describe its past.*

In simple terms, by merely looking at a tiny bone sample, Croiset could determine the type of bone (tibia, humerus, etc.) and the species it belonged to.

Retrocognition famously led to the discovery of Glastonbury Abbey when *Captain John Bartlett made a sketch of it through automatic writing,*[883] and has been used since the 15th century.[884] Stephan A. Schwartz, former editorial staff member of *National Geographic* and member of the MIT Secretary of Defense Discussion Group on Innovation, Technology, and Society, dedicated a whole book to this topic.[885]

Perhaps more intriguing than accessing the past in this way is that future events can also be seen in advance. This is called *precognition*, where the usual separation between present and future is removed. By the 1930s, the previously mentioned Joseph and Louisa Rhine showed that volunteers could guess what cards would be drawn randomly from a deck with a success rate that was better than chance by odds of one million to one.[886]

Since the 1930s, precognition has been demonstrated in hundreds of studies. There is such a massive body of evidence that a meta-analysis[887] was conducted, covering 309 precognition experi-

[882] See Chapter 36: 'The rise and collapse of materialism'

[883] Goodman, J., *Psychic Archaeology: Time Machine to the Past*

[884] 'What is psychic archeology', *Ancient Digger* (27 July 2011). See: www.ancientdigger.com/2011/07/what-is-psychic-archaeology.html

[885] Schwartz, S., *The Secret Vault of Time: Psychic Archaeology and the Quest for Man's Beginnings*

[886] 'Who was J.B. Rhine?', *Parapsychological Association.* See: www.archived.parapsych.org/members/jb_rhine.html

[887] Honorton, C., & Ferrari, D., 'Meta-analysis of forced-choice precognition experiments 1935 – 1987', *Journal of Parapsychology* (1989) 53: 281–308

Figure 234: Glastonbury Abbey was excavated in 1909 by F. Bligh Bond after Captain John Bartlett made a sketch through automatic writing. (© 466ad.co.uk)

ments made between 1935 and 1987. The cumulative probability associated with the overall results was 1 in 10^{24} (when 1 in 20 is considered statistically significant).

Precognitive remote-viewing, where *space and linear time (present/future) are both transcended* also happens:

> In their remote-viewing experiments at Stanford Research Institute, Puthoff and Targ found that, in addition to being able to psychically describe remote locations that experimenters were visiting in the present, *test subjects could also describe locations experimenters would be visiting in the future, before the locations had even been decided upon.* In one instance, for example, an unusually talented subject named Hella Hammid, a photographer by vocation, was asked to describe the spot Puthoff would be visiting one-half hour hence. She concentrated and said she could see him entering 'a black iron triangle.' The triangle was 'bigger than a man' and although she did not know precisely what it was, she could hear a rhythmic squeaking sound occurring 'about once a second.'
>
> Ten minutes before she did this, Puthoff had set out on a half-hour drive in the Menlo Park and Palo Alto areas. At the end of the half hour, and well after Hammid had recorded her perception of the black iron triangle, Puthoff took out ten sealed envelopes containing ten different target locations. Using a random number generator, he chose one at random. Inside was the address of a small park about six miles from the laboratory. He drove to the park, and when he got there he found a children's swing – the black iron triangle – and walked into its midst. When he sat down in the swing it squeaked rhythmically as it swung back and forth.[888]

Even more impressive is retro-psychokinesis where not only does the influencer modify a 'random' event, but in addition, the event occurred and was *recorded* in the past:

> A random number generator is activated to produce a string of N binary numbers. These numbers are automatically recorded on magnetic tape, paper punch tape, or some other reliable

[888] Talbot, M., *The Holographic Universe*, p. 262

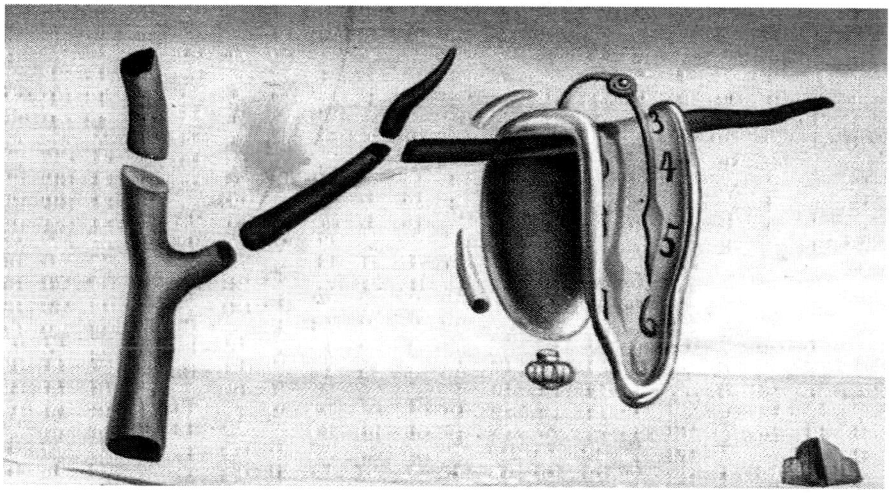

Figure 235: Salvador Dali, 'The Disintegration of the Persistence of Memory' (detail), 1954. (Oil on canvas, The Salvador Dali Museum). This painting was inspired by the discoveries of quantum physics and the non-linear nature of time.

> recording medium. Nobody is present during this generation and recording, and nobody looks at the data until at some later time the recorded sequence of 'heads' and 'tails' is played back to a subject in a PK test situation. During the slow playback each recorded head or tail makes a red or green lamp light up while the subject tries mentally to enforce an increased lighting rate of the red lamp.
>
> One might think that in this situation the subject could not succeed because the decision as to how many heads and tails will appear has already been made before the test session.[889]

It should be stressed here that these machines (random number generators) are programmed to simply play back the previously attained sequence of red and green lights. PK test subjects are somehow able to influence their outcome – recorded in the past – by 'willing' them to change as they are replayed in the present. The scoring rates of these tests on pre-recorded targets are significant, which suggests that ESP is non causal and that a past random process might be affected by PK efforts made in the present time.

Since the end of the 1970s, numerous retro-psychokinesis experiments have been conducted. Braud directed a meta-analysis which showed that the probability that the results were obtained by chance was 0.000,032%.[890]

To explain non-locality (phenomena transcending the space/time continuum), French physicist Bernard d'Espagnat devised the concept of 'quantum field theory'. For d'Espagnat, our reality is an emanation of a deeper, unreachable field. Individual subatomic particles being unique 'localizations' at a particular place and time, d'Espagnat argues that it is no longer accurate to understand objects as objects; they should really be understood as properties of a field:

---

[889] Schmidt, H., 'PK effect on pre-recorded targets', *J. Am. Soc. Psychical Research* (1976) 70(3): 267–291

[890] The meta-analysis covered 233 experimental retro-PK sessions conducted between 1979 and 1998. See: Braud, W., 'Wellness implications of retroactive influence: exploring an outrageous hypothesis', *Alternative Therapies in Health and Medicine* (2000) 6(1): 37–48

> According to d'Espagnat's controversial book,[891] the only possible answer is that the quantum state vector expresses properties of some *deeper underlying reality.* Since we know its manifestations to us – we know what it is like when measured – and since quantum physics forbids us to speak about what it is 'really like' when not measured, d'Espagnat speaks of it as a 'veiled reality.'[892] His is a sort of realism-at-a-distance: we cannot say that reality is 'just this way or that,' since our observations and what we observe are intertwined; and yet we *can* say that the-world-as-observed is a manifestation of the real; reality really takes this or that form in our observations. Unfortunately for traditional theories of matter, however, *this 'veiled reality' can be neither mental nor material, insofar as it precedes the mind–matter distinction altogether.*[893]

A *field* of this proposed type would be a space that has the same value for each point in space and time. Thus a field is non-local: properties are established and changed instantaneously for all the points covered by the field. In this sense, fields contradict Einstein's postulate according to which light speed is the maximum possible speed. If a change occurring in a field affects all its points at the same time, some sort of superluminal (faster-than-light) process must be occurring.

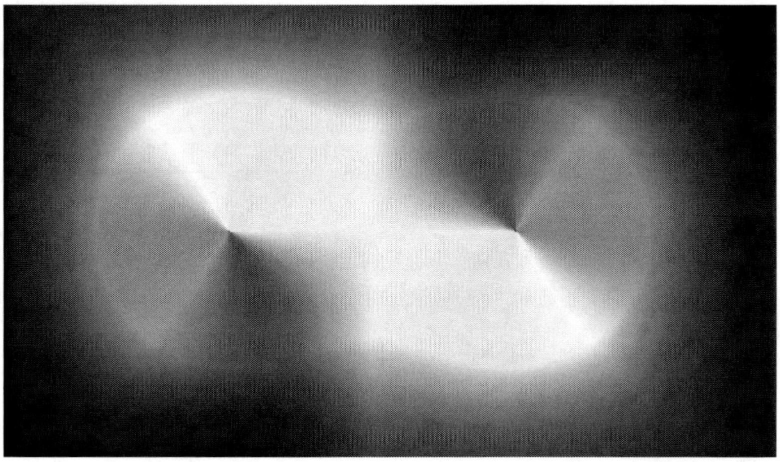

Figure 236: The magnitude (brightness) and direction (hue) of an electric field generated by two oppositely charged particles. (© Vegard)

In recent decades, a number of scientists from various research domains have independently identified an unknown 'dimension' in relation to matter, energy and information. This hidden 'dimension' exhibits the properties of a field, a deeper reality which transcends time and space and is *permeated with information.*

British biologist Rupert Sheldrake's *morphogenetic field*[894] holds the *information* relative to the differentiated development of individual cells. Sheldrake's *morphic field*[895] holds all the *information* relative to the collective knowledge held by a specific species/element. Michael Talbot's *holographic field*[896] possesses a fantastic capacity for *information* storage. Harold Puthoff's *zero-point field*[897] holds *information* states interacting with

[891] D'Espagnat, B., *In Search of Reality*

[892] D'Espagnat, B., *Veiled Reality: An Analysis of Present-Day Quantum Mechanical Concepts*

[893] Davies & Gregersen, *op. cit.*, p. 56

[894] Sheldrake, R., *Morphic Resonance: The Nature of Formative Causation*, pp. 65–78

[895] *Ibid.*, pp. 163–188

[896] In holographic imagery, by changing the angle at which the two lasers strike a piece of photographic film, it is possible to record many different images on the same surface. Any image thus recorded can be retrieved simply by illuminating the film with a laser beam possessing the same angle as the original two beams. By employing this method researchers have calculated that a one-inch-square of film can store the same amount of information contained in fifty Bibles! See: Collier, J., *et al., Optical Holography*

[897] "Matter is a selective frequency filter (resonator) that is orchestrated by the zero-point field and generates information states in the zero-point field ... The carrier of primordial consciousness is the zero-point field. Our individual

matter and consciousness. Finally, d'Espagnat's *quantum field*[898] contains the quantum wave function which holds all the measurable *information* about a particle within it.

Data gathered in this chapter and in the two previous ones strongly suggest the existence of a deeper, veiled reality, that could be termed the 'Cosmic Information Field', exhibiting the following characteristics:

- **Immanent** (permeating the whole space)
- **Connected** to all non-living and living forms, including humans
- Made of **information**
- **Transcending time** (instantaneous effects + connecting past, present and future)
- **Transcending space**
- Representing a **deeper level of reality** from which our reality emerges.

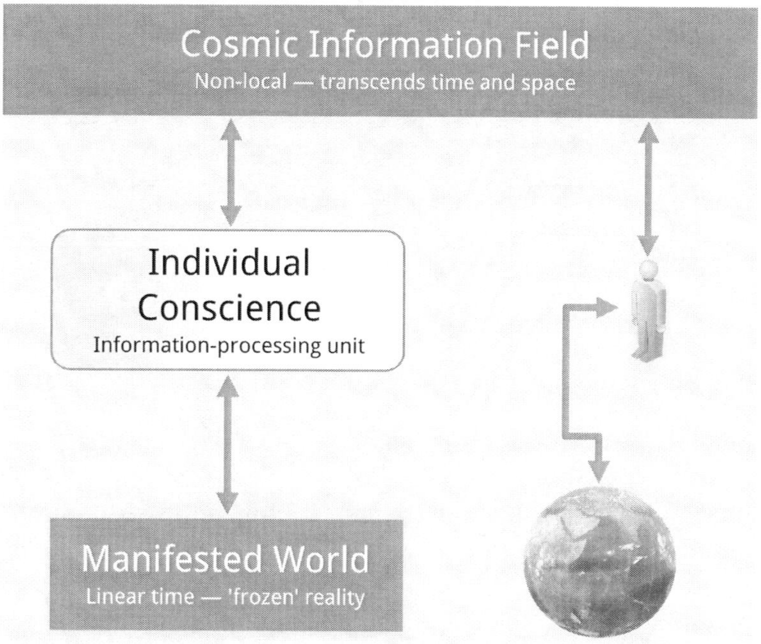

Figure 237: Interrelations between the 'Cosmic Information Field', 'individual consciousness' and the 'manifested world'. (© Sott.net)

Notice that the Cosmic Information Field seems to be not only involved in the non-local mind-matter interaction described above (psychokinesis, remote viewing, ESP, etc.) but also in our very own memory and thinking processes.

For the longest time, memories (i.e. memorized information) were thought to be stored in 'memory cells' located in the brain. But this perspective changed in 1929, when neuropsychologist Karl Lashley[899] conducted an experiment that was supposed to confirm the hypothesized location of memory cells. Lashley designed his rat memory test as follows: he built a jumping stand for his

consciousness is the result of a filtering process which causes the realization of information states in the zero-point field. Such information states have external as well as internal aspects. The external aspects are physical and manifest themselves as dynamic coherence patterns (NCC), while the internal aspects are phenomenal, i.e., a conscious moment is a zero-point information state experienced from inside." See: Keppler, J., 'A View Behind the Scenes of Matter Provides Valuable Clues for the Development of a Theory of Consciousness', *Talk at the TSC 2012 Conference*, Tucson (April 2012)

[898] Lee, C., 'The insanely weird quantum wave function might be 'real' after all', *Ars Technica* (22 November 2011)

[899] Beach, F. A., *Biography of Karl Spencer Lashley 1890-1958*, National Academy of Science

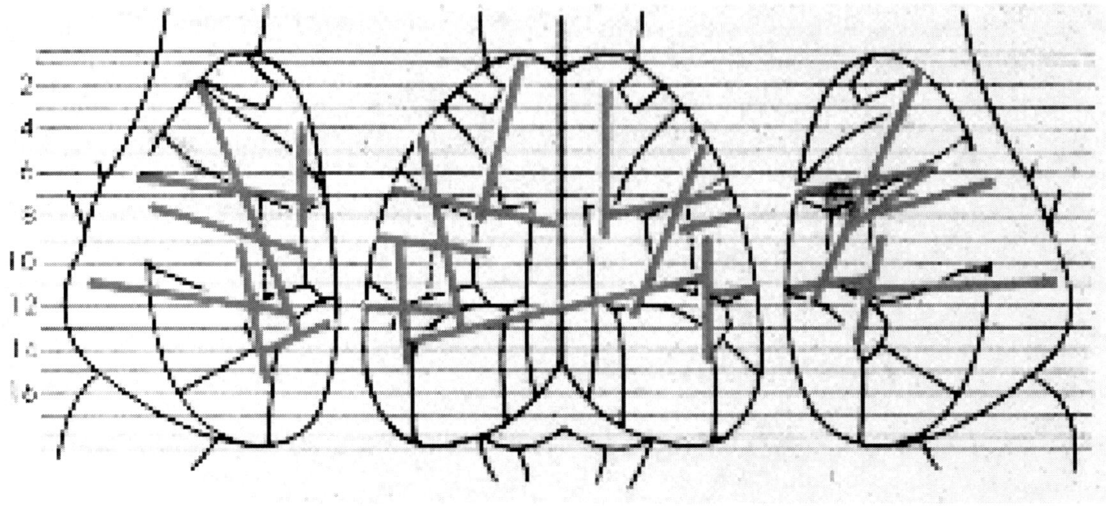

Figure 238: Diagram showing some of the multiple incisions Lashley made in rats' brains. (© Lashley)

rats, from which the rats learned to jump through miniature doors, and behind which food rewards were placed. If the rats followed the right routine and went to the right door, they got food (positive reinforcement). If they followed the wrong routine, they fell into water (negative reinforcement).

Once the rats had learned the routine, Lashley systematically tried to surgically remove that memory. Lashley's objective was to deactivate certain portions of the rats' brains to discover which part held the specific memories. To accomplish this task he chose as his surgical instrument his wife's curling iron (!) and simply burned off the part he wished to remove.[900]

No matter how much brain tissue Lashley removed from these poor rats, they still remembered which doors to take and which to avoid. Determined to find the location of memory cells, Lashley burned ever larger brain areas. *In the end, he had burned every single brain region, but despite major motor impairment the rats still remembered their way to the right door.*

Notice that those discoveries don't just apply to rats. Apparently humans don't need a brain either to think and remember. There are documented cases of hydrocephalus where people have virtually no brain, but still function normally:

> There's a young student at this university who has an **IQ of 126**, has gained a first-class honors degree in mathematics, and is socially completely normal. And yet the boy has virtually no brain ... When we did a brain scan on him, we saw that instead of the normal 4.5-centimeter thickness of brain tissue between the ventricles and the cortical surface, there was just a thin layer of mantle measuring a millimeter or so. *His cranium is filled mainly with cerebrospinal fluid.*[901]

After these discoveries, several scientists, including systems theorist Ervin Laszlo, would go on to argue that the brain is simply the retrieval and read-out mechanism of the ultimate storage medium,[902] what we coined the 'Cosmic Information Field'.

[900] McTaggart, L., *The Field*, p. 78

[901] Lewin, R., 'Is your brain really necessary?', *Science* (12 December 1980) 210: 1232(3)

[902] Lazlo, E., *The Interconnected Universe: Conceptual Foundations of Transdisciplinary Unified Theory*, p. 100

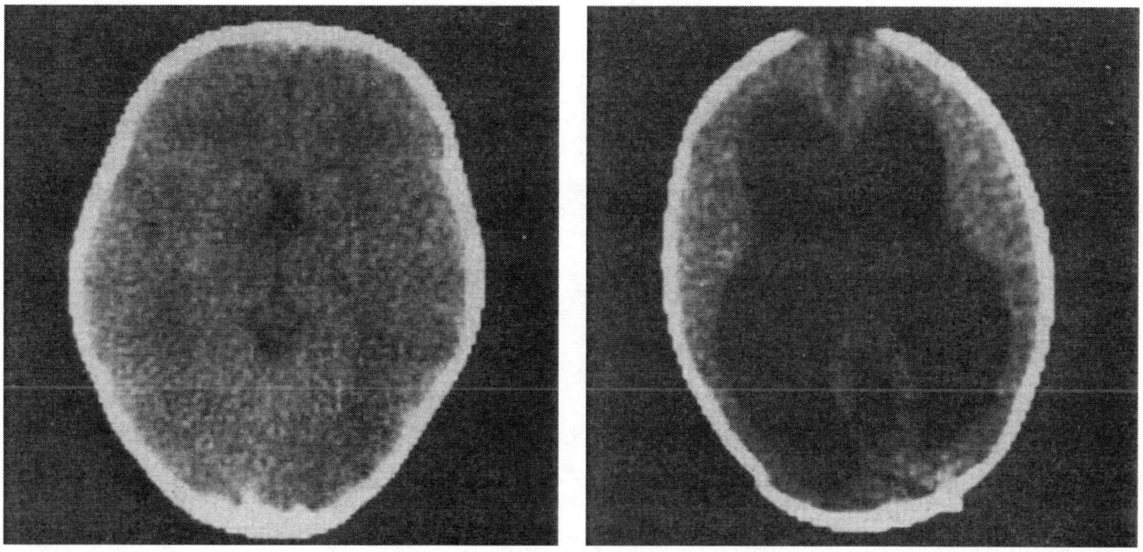

Figure 239: A horizontal scan across the brain shows the ventricles as narrow slits in a normal individual (left) and large cavities in a hydrocephalic patient (right). (© Lorber)

It would explain why one tiny association often triggers a riot of sights, sounds and smells. It would also explain why, with long-term memory in particular, recall is instantaneous and doesn't require any scanning mechanism to sift through years and years of memory.[903]

Thus, the brain may not be an information-storage device like a hard drive, but rather an interface connecting what an individual learns, experiences or perceives with a 'delocalized' database, a part of the Cosmic Information Field. Therefore, removing this or that part of the brain doesn't hinder memory since it is not stored 'within' brain cells *per se*. Thus the brain would act more like an emitter/receiver device, or an antenna. Even if an antenna is shortened (via removal of most of the brain, for example), it can still emit and receive information (thinking and memorizing processes).

Laszlo's findings led to theories[904] proposing that memories – stored information – did not occur locally in some specialized brain cells, but that brain waves were interacting through *interference*[905] processes with external waves related to the Cosmic Information Field:

> Our brains mathematically construct objective reality by interpreting frequencies that are ultimately projections from another dimension, a deeper order of existence that is beyond both time and space.[906]

As I mentioned earlier, the brain produces photons.[907] It is also a highly electric organ, where

903 McTaggart, *op. cit.*, p. 95

904 Pribram, K., *Brain and Perception*, p. 27

905 Wave interference patterns can hold much more than the 2.8×10^{20} bits of information held in a human brain. The entire U.S. Library of Congress, which includes almost every book ever published in English, could be stored in a volume as small as a sugar cube if wave interference patterns could be used for data storage. See: McTaggart, *op. cit.*, p. 86

906 Talbot, *op. cit.*, p. 54

907 See Kobayashi research in Chapter 37: 'The role of the 'observer''.

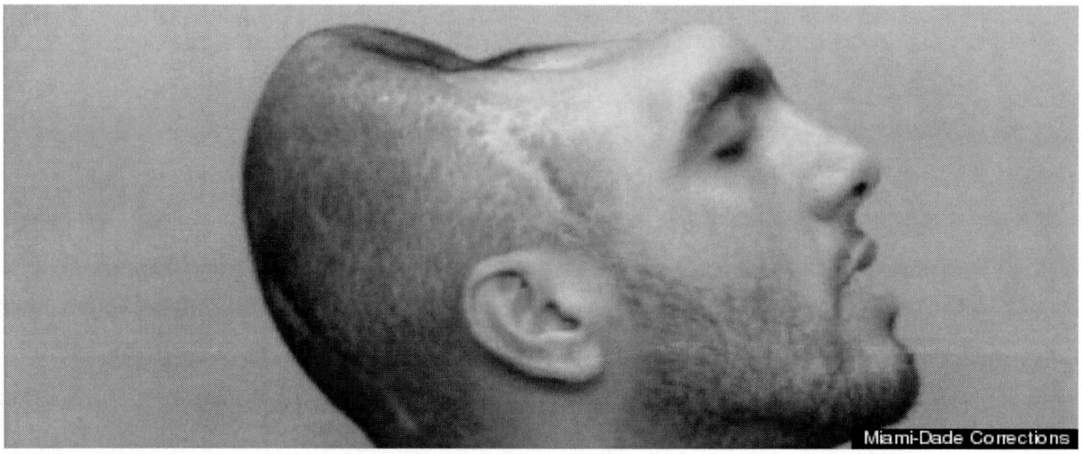

Figure 240: After a car accident, Carlos Sosa had more than half of his brain removed. However his intellectual capacity remained intact. (© Miami Dade Corrections)

neurons and axons communicate through constant electrical signaling made of various kinds of electromagnetic waves. (See figure 241.)

Are photons (proven fundamental carriers of information)[908] and/or electromagnetic waves (the core of wireless information exchange)[909] the media through which our minds interact with the Cosmic Information Field?

Light (photon movement) and electricity (electron movement) are intimately related. When light shines upon a piece of metal, it emits electrons. This is called 'photoelectric effect'.[910] Conversely, when electrons flow through a piece of metal, it emits photons (that's what happens in an incandescent bulb). In addition, electricity, the very source of electromagnetic waves, is intimately related to order, the hallmark of information. Matter is a mix of electrons and protons, but when an electric field is applied, *charge separation* occurs: positive charges group on one side while negative charges group on the other side, i.e. order increases.

The same applies at the molecular level when water molecules align along an electric field (figure 242).

In the absence of an electric field (left image), the H2O molecules are *randomly* oriented. When an electric field is applied (right image), the molecules align themselves with the field, such that the positive end of the molecular dipole points toward the negative and *vice versa*. The electric field increases the level of order/organization amongst the water molecules.

As we know,[911] plasma is matter that has become ionized (charge separation) due to electric

[908] A photon is a fundamental carrier of information because it possesses numerous information carrying degrees of freedom including frequency, phase, arrival time, polarization, orbital angular momentum, linear momentum, entanglement, etc. See: Kumar, P., 'information in a photon', DARPA, Defense Science Office. www.darpa.mil/Our_-Work/DSO/Programs/Information_in_a_Photon

[909] AM radio, FM radio, cellular phones, GPS systems, satellite broadcastings, walkie-talkies, etc., are all wireless communication systems based on electromagnetic waves carrying information. See: Toupin, L. P., 'Riding the electromagnetic wave', *Houghton Mifflin Science*. www.eduplace.com/science/hmsc/6/f/cricket/cktcontent_6f18.shtml

[910] See: Elert, G., 'Photoelectric effect', *The Physics Hypertextbook*. physics.info/photoelectric/

[911] See Chapter 3: 'What is plasma?'

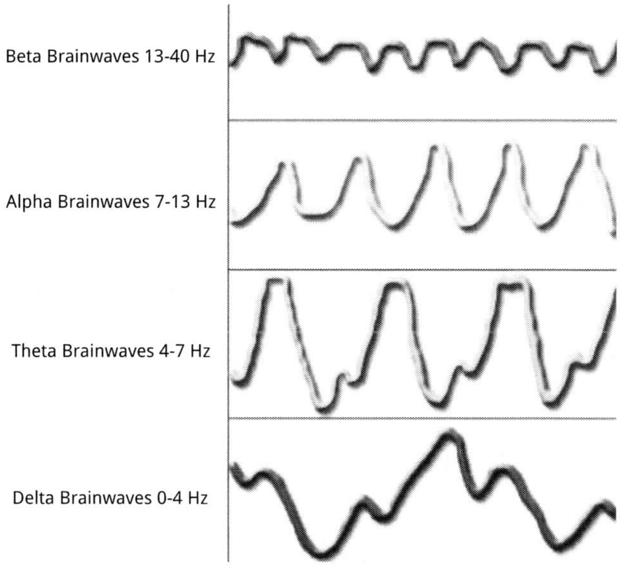

Figure 241: Beta, Alpha, Theta and Delta are four of the usual frequency ranges covered by the electric activity of the brain. (© Sott.net)

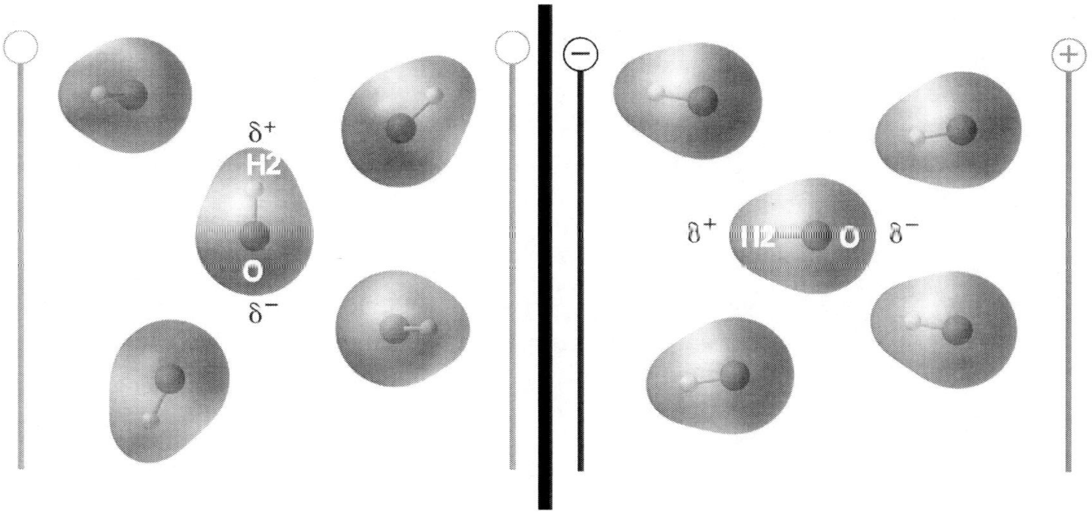

Figure 242: Influence of an electric field on the alignment of water molecules. (© Sott.net, adapted from Averill and Eldredge)

stress and it exhibits a very high level of organization[912]:

> [In plasma] electrons are not scattered because, through the action of the quantum potential, the whole system is undergoing a *coordinated movement* more like a ballet dance than a crowd of unorganized people.[913]

Could it be that the order displayed by electric phenomena (charge separation, plasma) is a tangible manifestation at the particle scale of more subtle phenomena occurring at the level of the Cosmic Information Field?

In any case, at this point, we can reasonably hypothesize – even assume – the existence of a Cosmic Information Field transcending time and space. We've also seen that human beings, *on an individual basis*, can have an influence on matter and macro events. The next logical questions are: What happens on a collective level? What happens when several observers/influencers are involved?

[912] According to quantum physicist David Bohm: "Like some amoeboid creature, the plasma constantly regenerated itself and enclosed all impurities in a wall in the same way that a biological organism might encase a foreign substance in a cyst. So struck was Bohm by these organic qualities that he later remarked he'd frequently had the impression the electron sea was 'alive'." Talbot, *op. cit.*, p. 38

[913] *Ibid.*, p. 41

Chapter 40.

Collective resonance

What interests us now is how a group of individuals might 'connect' with each other and interact with reality.

Let's start by focusing on how two minds can connect. That's actually the essence of telepathy, also known as ESP. Telepathy is defined as:

> ... the 'transmission of *information*[914] from one person to another without using any of our known sensory channels or physical interaction.[915]

Telepathy has been the topic of six meta-analyses[916] and a recent 'meta-meta-analysis'.[917] That is what was required for the scientific community to finally recognize that telepathy was real and scientifically proven.[918]

While telepathy experiments show how two minds can interact with each other, in their previously mentioned experiment,[919] Jahn and Dunne went one step further and studied how a pair of subjects influenced 'random' events:

> Jahn and Dunne had wondered if the tiny effect they were observing with individuals would get any larger if two or more people tried to influence the machine in tandem. The PEAR lab ran a series of studies using pairs of people, in which each pair was to act in concert when attempting to influence the machines.
>
> Of 256,500 trials, produced by fifteen pairs in forty-two experimental series, many pairs also produced a 'signature' result, which didn't necessarily resemble the effect of either individual alone.
>
> Being of the *same sex tended to have a very slight negative effect.* These types of couples had a worse outcome than they achieved individually; with eight pairs of operators the results were the very opposite of what was intended.
>
> Couples of the opposite sex, all of whom knew each other, had a powerful complementary effect, producing more than three and a half times the effect of individuals.
>
> However, *'bonded' pairs, those couples in a relationship, had the most profound effect,* which was nearly *six times as strong* as that of single operators.

---

914 During telepathy experiments, the information usually communicated is names, numbers, letters, symbols, etc.

915 'Telepathy', *Wikipedia.* See: www.en.wikipedia.org/wiki/Telepathy

916 Those six meta-analyses cover 108 scientific papers published between 1974 and 2008.

917 Tressoldi, P. *et al.*, 'Extrasensory Perception and Quantum Models of Cognition' *NeuroQuantology* (2010) 4: S81–87

918 Tressoldi, *op. cit.*, p. 84

919 See Chapter 37: 'The role of the 'observer''

Figure 243: Telepathy is typically tested via the 'Ganzfeld technique' where the receiver is put in a state of mild sensory deprivation. (© Jeremy Walker)

> If these effects depended upon some sort of resonance between the two participating consciousnesses, it would make sense that stronger effects would occur among those people sharing identities, such as siblings, twins or couples in a relationship. Being close may create *coherence.* As two waves in phase amplified a signal, it may be that a bonded couple has an especially powerful resonance, which would enhance their joint effect on the machine.

Jahn and Dunne showed that *closeness* increases the influence exerted by a pair of individuals. It's obviously not about genetics, since couples show the greatest influence. However, what is usually shared by couples, relatives and siblings are *topics of interest*, discussions, cultural background, past experiences and *emotional bonding.* Close people usually *share a similar information set* (intellectual and emotional). So, one can hypothesize that it is shared information that makes people resonate on a similar frequency.

By the way, such a conclusion might give a new meaning to the following quote:

> If two make peace with each other in a single house, they will say to the mountain, 'Move from here!' and it will move.[920]

Jahn and Dunne have evidenced a form of *resonance* between two people who are close to each other. In a similar vein, while explaining psychokinetic phenomena, physicist David Bohm suggested that 'resonance' was indeed the main driver:

> On this basis, psychokinesis could arise if the mental processes of one or more people were focused on *meanings that were in harmony* with those guiding the basic processes of the material systems in which the psychokinesis was to be brought about. It is important to note that this kind of psychokinesis would not be due to a causal process, that is, a cause-and-effect relationship

[920] Gospel of Thomas, verse 48, translated by Stephen Patterson and Marvin Meyer.

Figure 244: REG Testing at the Princeton Engineering Anomalies Research Program. (© Princeton University)

> involving any of the known forces in physics. Instead, it would be the result of a kind of nonlocal 'resonance of meanings', or a kind of nonlocal interaction.[921]

Resonance permeates the whole universe, it occurs with all types of vibrations (mechanical, acoustic, electromagnetic, etc.), from the atomic scale[922] to the celestial scale.[923]

Resonance occurs between two waves when their signals are in sync (see figure 245). There might, after all, be a reason why we use the expression 'we are on the same wavelength' to express that we think/feel the same as another.

Figure 245 shows that when the two primary waves are in sync, the resultant wave reaches its maximum amplitude (case 5 – bottom of the image). However, this analogy only reveals part of the process. In this case, two signals that are in sync fully add up, nothing more. We'll see in the next chapter that *synergies* can occur, where the interaction between two people or more can produce an effect much greater than the sum of their individual effects.

While Jahn and Dunne's experiments, as well as most telepathy experiments, focused on pairs of subjects, Robert D. Nelson[924] decided to test the effect of whole crowds on 'random' events. For that purpose he brought a portable REG (Random Event Generator) to all kinds of venues (stand-up comedy performances, concerts, theatrical events).[925] He then monitored the deviations displayed by the machines.

In 1995 and 1996, Nelson used his REG during numerous shows in 10 different cities. A very consistent pattern emerged: each time the show triggered peak attention from the audience, the

921 Bohm, D., 'A new theory of the relationship of mind and matter', *Journal of the American Society for Psychical Research* (April 1986) 80(2): 128

922 Hornak, J. P., *The basics of nuclear magnetic resonance*

923 Orbital resonance as exemplified by Jupiter's moons Ganymedes, Io and Europa. See: Pogge, R., *The moons of Jupiter*, Ohio State University (17 November 2007)

924 Nelson was a colleague of Jahn and Dunne. He was the Coordinator of Research at the Princeton Engineering Anomalies Research (PEAR) laboratory at Princeton University from 1980 to 2000.

925 Nelson, R. D., 'Field REG anomalies in group situations', *Journal of Scientific Exploration* (1996) 10(1): 111–41

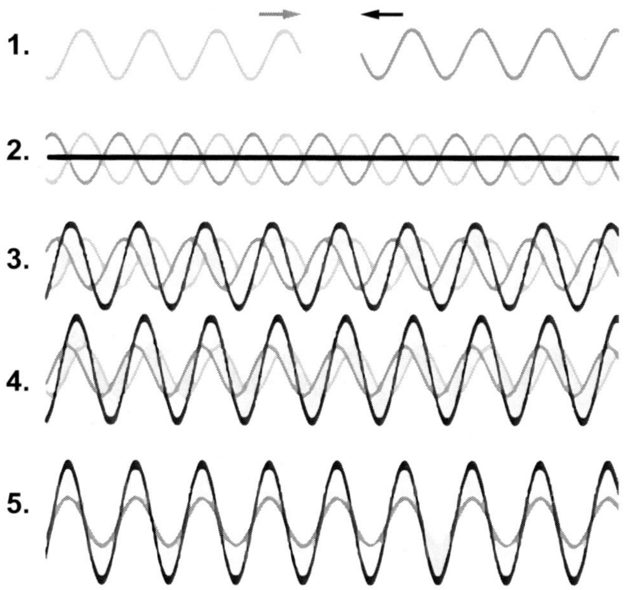

Figure 245: Resultant wave (black) as a result of the interaction between two primary waves (light grey and medium grey). (© A-level Physics tutor)

REG would deviate from its random movements and display a deviation much larger than the one induced by a single individual.[926] The nature or the location of the show didn't matter. What was important was that everyone was thinking the same thought.[927]

Dean Radin, one of Nelson's colleagues at Princeton, went one step further and monitored the influence of the world population, or most of it. For that purpose, he set 27 REGs all around the world during the Y2K New Year transition. New Year is a worldwide event when people typically share similar thoughts and focus. Radin had predicted a relevant deviation during a ten-minute period around midnight. Data confirmed his prediction. (See figure 246.)

Sheldrake theorized and tested this strong but invisible connection within whole species, including humans. He has been researching morphogenesis,[928] which raises challenging questions: how do cells, which have exactly the same genes, develop to form an arm on one side of the body and a leg on the other? How do animals like the salamander regenerate perfectly shaped limbs after amputation? Clearly, there must be some form of infusion of information in order to coordinate the specific development of cells, which will eventually lead to a fully integrated, harmonious, operational system made of billions of cells.[929]

To explain this elusive phenomenon, Sheldrake developed the concept of *formative causation*, which posits that the form of living things – from the smallest bacterium to the greatest mammals – is shaped by morphic fields. He coined the term 'morphic resonance' to describe what occurs

[926] Radin, D., *The Conscious Universe*, pp. 157–174
[927] McTaggart, L., *The Field*, pp. 204–205
[928] Morphogenesis studies cell differentiation: how, for example, a human egg will transform into an adult body.
[929] The human body is made up of about 100 trillion cells (10^{14} cells).

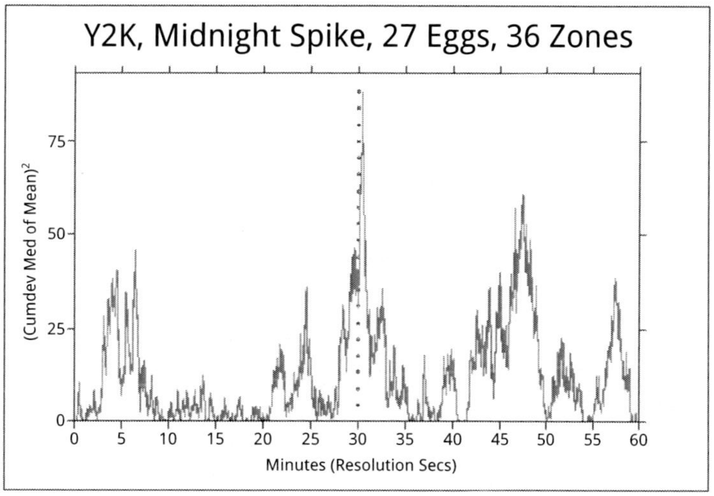

Figure 246: One hour of data surrounding midnight, Y2K. Cumulative deviation across 27 REGs, plotted with all time zones superimposed. (© Nelson)

between these fields, i.e. '*the influence of like upon like through space and time*'.

In terms of space, morphic resonance is non-local, meaning its effects are not attenuated or delayed by distance.[930] It is just as effective over 10,000 miles as it is over an inch. Although morphic resonance is attenuated by time, it is also cumulative. It means that the first generation influences the second, the first and second generations influence the third, and so on.

Sheldrake argues that morphic fields not only affect form and development, but also behavior and perception, learning and memory.[931] Thus the greatest interest of his research applies to psychology:

> According to the hypothesis of morphic resonance, human beings draw upon a *collective memory*: something learned by people in one place becomes easier for others to learn all over the world.[932]

Nelson, Sheldrake and Radin, to name a few, came to the same conclusion: every member of the human population is connected to each other via what we are calling here the Cosmic Information Field, through which each of us provides his or her own information and taps into the information gathered by the others.

[930] Sheldrake, R., *Morphic resonance*, pp. 84–99
[931] *Ibid.*, pp. 144–188
[932] *Ibid.*, p. xxv

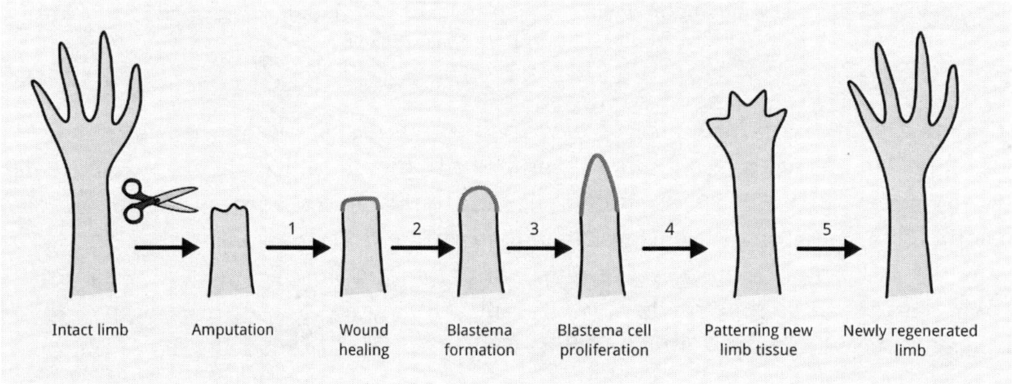

Figure 247: Steps in salamander limb regeneration. (© Whited et al., Journal of Biology, 2009)

Chapter 41.

The truth factor

At this point we know that the whole human population interacts with the Cosmic Information Field. We also know that individuals can add up their influence and that the information we *believe in* plays a major role. So, the next logical question is: what is the difference, on an individual level and on a collective level, between believing in something that is true as opposed to believing in something that is false?

To deal with the properties of true information vs. false information, let's start with the effects of true and false information on an individual level.

Truth factor and lie factor on an individual level

This is not the place to engage in philosophical or semantic discussion as to whether or not information, by definition, can even be 'false information'. The dictionary gives two definitions of information:

1. Facts provided or learned about something or someone;
2. What is conveyed or represented by a particular arrangement or sequences of things.[933]

By the first definition, information would necessarily have to be true. By the second definition, obviously, anything 'conveyed' could be called 'information'.

We can point to Plato's Cave (figure 248) as an example of how unsatisfactory the second definition is: the shadows may convey or represent things, but they can be easily misunderstood or taken to be the whole of the matter, when they are actually only a representation – or worse, pure fabrication. The representations can be further obscured by making up stories about the moving shadows that have no foundation in any direct knowledge. An unscrupulous individual might even use the shadows to terrify the gullible and prevent them from leaving the cave to find out for themselves.

Facts directly learned about something or someone have a different character, a depth, an objectivity and relevance that goes beyond a 'representation.' This difference is similar to that between the action of neurochemicals produced in the body vs. drugs that imitate neurochemicals imperfectly. Witness the different actions of adenosine and caffeine: when caffeine binds to the adenosine receptors, it blocks the natural neurochemical which would otherwise lodge there and deliver important information into the cell. The caffeine is information-less, so to say. In a similar vein, Stonier's

[933] *Oxford Dictionary Online.* See: www.oxforddictionaries.com/definition/english/information

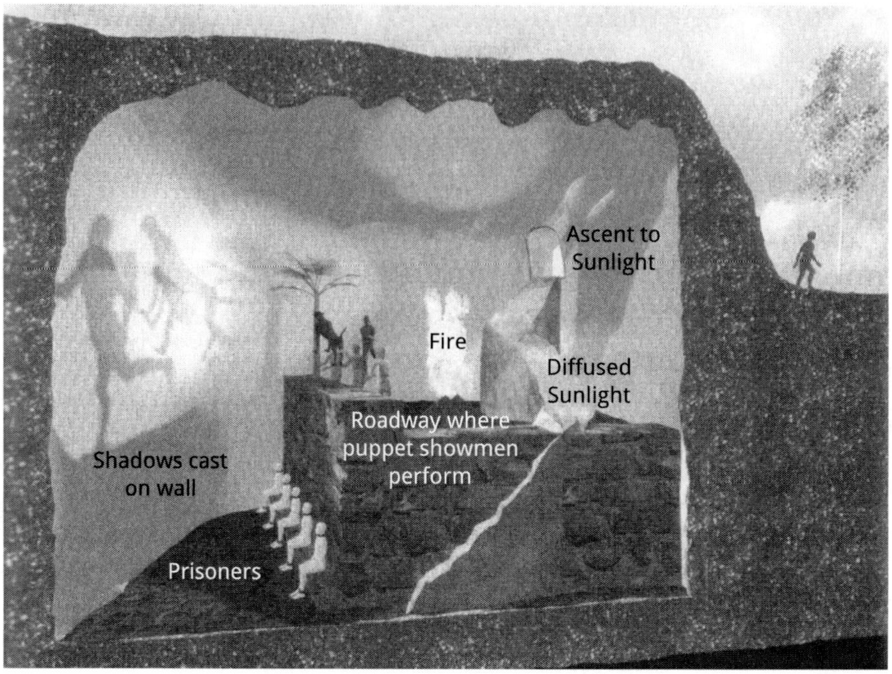

Figure 248: Plato's Allegory of the Cave: Puppet showmen project all kinds of shadows on a wall and make the prisoners believe that this is reality. (© Gneiting)

information brings order, therefore it is intrinsically true. The rest is gibberish, noise, non-sense, blockages, only adding chaos to any given system, i.e. non-information.

Nevertheless, in the following discussion, in the interest of clarity and the true-false duality, we will adhere to the convention that information can either be true or false, with the above proviso in mind.

Now let's return to truth and lies on an individual level. When several true pieces of information ($1 + 1 = 2$, $2 + 2 = 4$, $4 + 4 = 8$, etc.) are held by an individual, they present a high level of coherence; they are *consistent with each other*. They allow the *emergence of a higher-level order*. In this case, the *law* of mathematical additions (see figure 249) is information of a higher order that can be *inferred* from the collection of true information ($1 + 1 = 2$, $2 + 2 = 4$, etc.) of a lower order.

A set of false pieces of information ($1 + 1 = 3$, $2 + 2 = 5$, $4 + 4 = 9$, etc.) doesn't exhibit internal consistency; the individual pieces are incompatible, they don't 'resonate' with each other, they partly cancel each other and *hinder access* to any higher level of order (the law of mathematical addition).

To illustrate this concept further, let's take stereo conversion as an example. During this process, multiple 2D images are compiled together in order to reconstruct a 3D image (figure 250). Thus, the combination of 2D projections will reveal an extra third dimension (depth) that was not visible in any of the 2D pictures. Notice that the 3D representation of the object will only emerge if each 2D image is a faithful (true) projection of the 3D object.

That's exactly what happens with our vision. If we look with one eye only we get a 2D picture,

+	0	1	2	3	4	5	6	7	8	9
0	0	1	2	3	4	5	6	7	8	9
1	1	2	3	4	5	6	7	8	9	10
2	2	3	4	5	6	7	8	9	10	11
3	3	4	5	6	7	8	9	10	11	12
4	4	5	6	7	8	9	10	11	12	13
5	5	6	7	8	9	10	11	12	13	14
6	6	7	8	9	10	11	12	13	14	15
7	7	8	9	10	11	12	13	14	15	16
8	8	9	10	11	12	13	14	15	16	17
9	9	10	11	12	13	14	15	16	17	18

Figure 249: Addition table from zero to nine. (© TVDSB)

but if we look with two eyes and get two converging, consistent 2D representations, they combine and reveal an extra 'depth dimension'. But for that to happen you must have two eyes that *work together simultaneously as a coordinated team.*[934]

Truth and lie factors on a collective level

To illustrate the influence of true information and false information on a collective level, let's start with a simple example (see figure 251).

Case 1: Two individuals hold the same true information ($1 + 1 = 2$). This information is depicted by the two upwards white arrows. The arrows are collinear; they fully add up to each other (light grey upwards arrow). In addition, their final result (dark grey arrow) is even longer than the sum of the two arrows because of *synergy*. We will develop this point very soon.

Case 2: Two individuals hold two *different* false pieces of information ($1 + 1 = 3$ and $1 + 1 = 4$). These two pieces of information are depicted by the two divergent white arrows pointing down. This represents the fact that, since those two propositions are not collinear, they partly cancel each other and don't fully add up (light grey downward arrow). The final result (dark grey arrow) is equal to the sum of its two parts because there is no synergistic effect.

Case 3: Two individuals share the same false information ($1 + 1 = 3$). These two pieces of information are depicted by the two collinear white downwards arrows. They fully add up *to each other* because they are collinear (light grey arrow). The final result (dark grey arrow) is exactly equal to the sum of its two parts because there is no synergistic effect.

[934] 'Avatar 3D Movie Falls Flat for People Who Can Not See with Both Eyes at the Same Time', *Optometrics Network*. See: www.vision3d.com/whycant.html

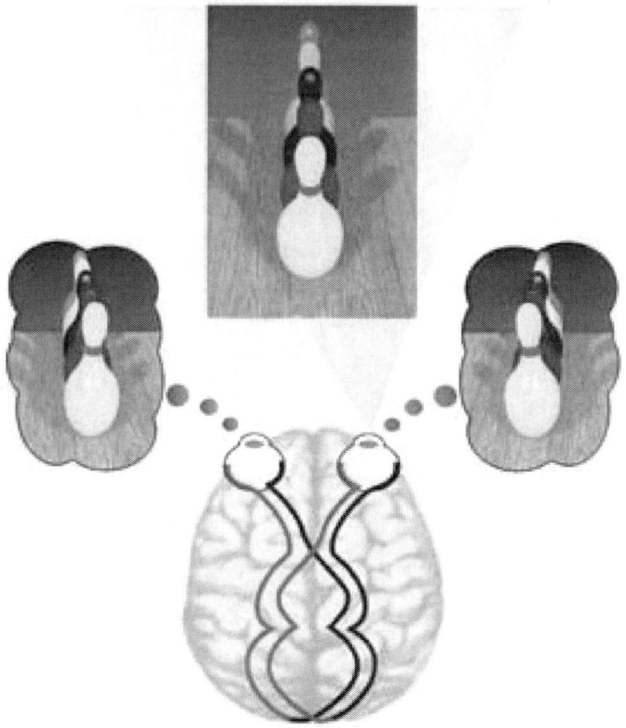

Figure 250: 2-eye view and stereovision. (© Visio 3D)

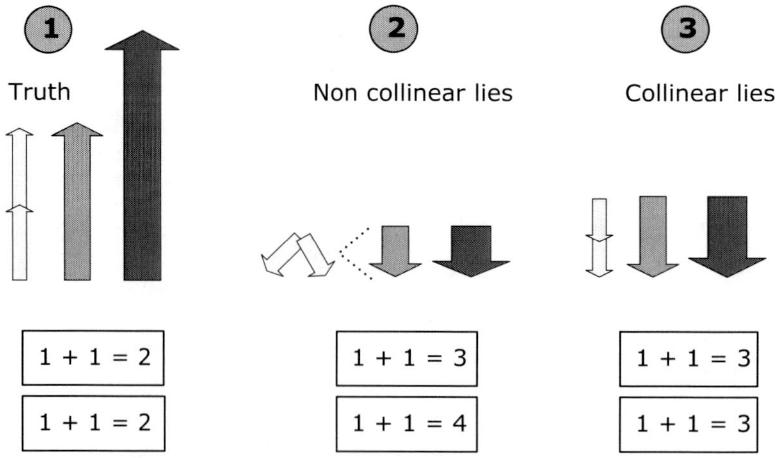

Figure 251: Interactions between true or false information believed by two individuals. (© Sott.net)

Truth factor on a collective level

True information interactions (Case 1 in figure 251) exhibit two fundamental differences when compared with false information:

1. As previously shown, the individual signals (white arrows) are longer when they are true.
2. The total effect is greater than the sum of the individual effects (synergy).

Synergy in a human group is relatively rare, but there are documented instances where the group achieves collectively something which could not have been achieved by any member working alone or by a combination of individual efforts.[935] Those few instances occurred most of the time when the group members were *highly motivated*, which leads to a shift where:

> The group abandons incorrect answers when better ideas are proposed by someone in the group.[936]

The rarity of real-life examples of synergy occurring in human groups might be due to the lack of *co-linearity*: shared motivation, shared objectives, shared vision of the world. Also, *individual interest must come second to group interest* for synergy to occur.

Unlike humans, ants are not subjected to lies (propaganda, disinformation, etc.). In this sense they exhibit a higher level of co-linearity.

An ant by itself is almost nothing, but with the other members of its colony it can achieve results that far exceed the sum of the individual contribution. Synergy is fully at work and leads to what some call a super-organism.[937]

Contemplating the image depicted in figure 252, the documentary commentator said in awe:

> Everything looks like it has been designed by an architect: a single mind. This colossal and complex city was created by the collective will of the ant colony: the super-organism.[938]

However, co-linearity, sharing similar true information, might not be all that is required to reach full synergy. The John and Dunne research mentioned above[939] showed that while bonded couples of different sex had a synergetic influence (six times more than a single individual), *couples of the same sex tended to have a very slight negative effect* (they had a worse outcome than what they achieved individually).

This suggests that synergic individuals not only need to be collinear (shared information, vision and objective) but also need to be *complementary*. Perhaps in the case of bonded couples of different sex, one brings a more emotional take on things to the equation, while the other brings a more intellectual approach. Both takes can be true, but alone they form only part of the overall perspective, hence their complementarity.

---

[935] Collins, B. E. & Guetzkow, H., *A Social Psychology of Group Processes for Decision-making*

[936] Forsyth, D., *Group Dynamics*, p. 303

[937] Holbrook, T. & Clark, R., 'Secrets of a Superorganism', School of Life Science, Arizona State University. See: askabiologist.asu.edu/explore/secrets-superorganism

[938] Forchi, L., *City of Ants*, video documentary. See: www.youtube.com/watch?v=7ZiKfbs1f8M

[939] See Chapter 40: 'Collective resonance'

Figure 252: After injecting cement in an abandoned ant colony, then excavating the surrounding dirt, researchers reveal a marvel of engineering: interconnected chambers, fungus gardens, highways, garbage pits, and ventilation tunnels. (© Luis Forchi)

Lie factor on a collective level

The lie factor on a collective level will reach its maximum when *a maximum number of people believe a maximum number of the same lies* (Case 3 in figure 251). For example, during the Middle Ages the world population was relatively small and scattered.[940] Every group held a distinct set of beliefs and cultural references. Because of the variety of beliefs held by people in the past, the 'signal' sent by humanity was confused: on a cosmic scale, the level of incoherence was so high that all those signals mostly cancelled each other out. In addition to being incoherent, the signal sent in the past was relatively weak because the population was scarce. However, the weak and incoherent signal emitted in the past was still sufficient to be correlated with several major cataclysmic episodes.

The cataclysms that occurred during the Middle Ages[941] were relatively minor compared to the catastrophes that wiped out the Bronze Age civilization[942] and the Roman Empire.[943]

Compared to the Middle Ages, the Roman Empire exhibited a larger population and more structured means of logistics and cross-civilization communication. In this setting, there is greater potential for co-linearity, which is not necessarily a bad thing, but by the time the elites' greed kicks in on the dynastic cycle[944] and lies, corruption and wars resulted, this co-linearity produced collinear lies (Case 3), greater lie resonance and, therefore, a greater cosmic reaction.

Plato made a very clear reference to 'Atlantis'.[945] Was it a 'great' civilization comprising billions of people that was largely defined by lies and suffering? Was it annihilated by catastrophes of the greatest magnitude? Is the story of Atlantis today remembered, if only in myth, by peoples the world over, the most well-known version being the Noachian Flood story?

According to Firestone *et al.*,[946] there *was* an Atlantis civilization and it *was* wiped out by a series of major cometary bombardments c. 12,500 BC, as evidenced by the Carolina Bays and its 500,000 elliptical craters concentrated along the Atlantic seaboard.

We can't speculate much further than that because the core problem with such hypothetical 'total' global catastrophes during human history is that the magnitude of the destruction would mean that most, if not all, traces of such a civilization would have been erased.

Nowadays, the situation is much worse than during the Middle Ages: the world's population exceeds 7 billion, a figure unprecedented in documented history. Mass media reach almost every human being through the Internet, newspapers, books, radio and TV, spreading propaganda and disinformation about any and all topics (economics, nutrition, weather, politics, science, finance, etc.) 24 hours a day and 7 days a week.

One consequence of the development of mass media is that *the same lies are spread and repeated throughout the entire world population*, which could lead to a more coherent lie signal, represented by a shift from Case 2 to Case 3 in the 'truth factor diagram' found at the beginning of this chapter.

---

940 In 1340, just before the Black Death, the world population was estimated to be 450 million. That's about 15 times less than the population today. See: www.census.gov/population/international

941 See Chapter 34: 'Historic evidence of a human-cosmic connection'

942 See Chapter 30: 'Geographic tilt and magnetic reversal'

943 See Chapter 34: 'Historic evidence of a human-cosmic connection'

944 See Chapter 35: 'Dynastic cycle'

945 Plato, *Timaeus*, 24e–25a, R. G. Bury translation

946 Firestone, R., West, A. & Warwick-Smith, S., *The Cycle of Cosmic Catastrophes: How a Stone-Age Comet Changed the Course of World Culture*

Figure 253: Detail (left) and geographic scope (right) of the 'Carolina Bays' cometary bombardments. (© LIDAR / Terrametrics)

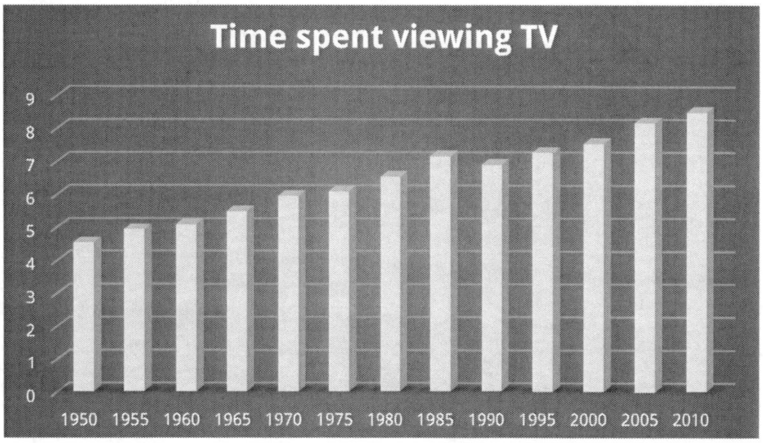

Figure 254: Average daily time spent viewing TV in the USA between 1950 and 2010. (© Sott.net, adapted from The Nielsen Company)

However, this point is generally mitigated because, in order to obfuscate the truth, the mainstream (and sometimes 'alternative') media tend to disseminate several contradictory lies, i.e. they 'muddy the waters'. Take the September 11th, 2001, attacks in NY and Washington, for example. Several false scenarios about the events of that day have been spread by the mainstream and 'alternative' media (Osama Bin Laden and his 19 hijackers, the no-planes theory, the use of mini-nukes, etc.) in the hope that this or that person will cling better to this or that lie because it fits better with his or her belief system.

As a result, while an important percentage of the global population believes a false narrative about how and why the 9/11 attacks happened, which tends to increase the lie factor, because of the *multiplicity of the lies* that are being spread, the coherence of the lie signal is reduced.

Despite this potentially mitigating factor, never before in written history has humanity emitted such a strong and coherent *lie signal* overall because of its exceptional size and the universality and homogeneity of its wrong beliefs.

This signal might indeed *increase* the probability of the occurrence of cataclysmic events, but humanity is only one factor in the cataclysm equation: the last five past mass extinctions occurred well before the appearance of human beings.

A close study of the interrelatedness of all life on Earth suggests that, even in those times, something was amiss, something that needed correction or re-balancing. The Cosmic Information

Figure 255: The Vatican Library was established in 1475. At that time knowledge was scarce and not unified. Books were rare and most people couldn't read. (© Idlespeculations)

Field found it necessary to blast through blockages or terminate failed evolutionary experiments that were going nowhere and could not contain or express more ordered or higher-level information. In this sense, it could be suggested that human beings with self-consciousness and awareness were desirable outcomes from regular pruning of the evolutionary tree. However, it should also not be forgotten that human beings themselves are part of this tree and that those who do not become adequate information-processing units might also be pruned.

Of course the '1 + 1 = 2' example mentioned above is simplistic. Often the truth is not so obvious. In many instances, the absolute truth can't even be reached by human minds. We can only try to strive towards it. However, it's not because we don't know the truth that we should accept a lie. By not believing a lie, even if one doesn't know the exact truth, an individual removes his or her personal contribution to the lie factor (at the collective level) and at the same time reduces the level of fragmentation in his or her own mind (individual level).

As a side note, believing a lie today is not exactly like it was, say, five centuries ago. In the past, information was scarce, books were rare and illiteracy high. Things have changed drastically. With the development of the media in general, and the Internet in particular, we've shifted from an information-deficient world to a world overloaded with information. Today, numerous explanations are available for most events, including many false ones, *but also the true one.* The truth generally being available, believing a lie today amounts to – consciously or not – *choosing the lie over the truth*, something that could be considered an 'aggravating circumstance'. Our ancestors didn't have this choice.

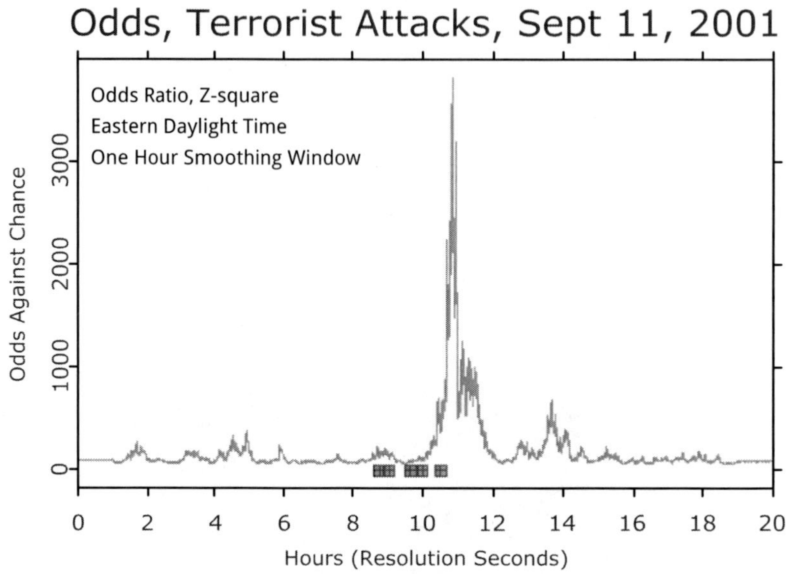

Figure 256: Cumulative deviation over the 20 hours of data surrounding the WTC event. The black squares show the time of hits and collapse of the towers. (© Nelson)

In addition to beliefs, another element affecting the lie factor on a collective level is emotions. They do play an important role in mind/matter interactions. Telepathy experiments show that not only are we connected on an intellectual level but also on a physiological and emotional level:

> A light flashed in a test subject's eyes will register in the EEG readings of a test subject isolated in another room, and even the blood volume of a test subject's finger changes – as measured by a plethysinograph, a sensitive indicator of autonomic nervous system functioning – when a 'sender' in another room encounters the name of someone they know while reading a list composed mainly of names unknown to them.[947]

After testing the effects of whole crowds on 'random' events at popular venues, the previously mentioned Princeton scientist Robert Nelson[948] tested events broadcast worldwide. For that purpose he set up a network of 40 researchers equipped with REGs (Random Event Generators) spread all over the world. Their task was to monitor the deviations displayed by the machines during widely broadcast events.[949]

While the REGs displayed statistically significant deviation spikes for the O. J. Simpson trial, the Concorde crash, the impeachment of Clinton, elections, volcanic eruptions, earthquakes, et cetera.[950] the deviations recorded at the time of the death of Princess Diana[951] and the collapse

[947] Talbot, M., *The Holographic Universe*, p. 143

[948] Nelson was mentioned in Chapter 40: 'Collective resonance'. He was a member of the PEAR team in Princeton University along with Jahn, Dunne and Radin.

[949] Radin, D., *The Conscious Universe*, pp. 157–174

[950] McTaggart, L., *The Field*, p. 200

[951] Nelson, R., *et al.*, 'Global resonance of consciousness: Princess Diana and Mother Teresa' *Electronic Journal of parapsychology* (1998)

Figure 257: Flower tribute to Diana in front of Kensington Palace (September 1997). (© Corbis)

of the WTC were unparalleled.[952] What those two events have in common, and what makes them stand out, is the collective tragedy, compassion and *suffering that was felt* by a large percentage of the global population.

Princess Diana was a beloved icon whose kindness and compassion touched most of us. Literally *billions mourned her death* and joined each other in a communion of sympathy, sadness and thoughts.

From these two examples (the 9/11 attacks and the death of Diana) we can deduce that *collective suffering is a strong modulator of 'random' events.* From the previously mentioned Helmut Schmidt trials[953] and lie factor on a collective level,[954] we see that collective false beliefs are *also* a strong modulator of random events. Interestingly, collective suffering and collective false beliefs are the two main markers of imperialism.

Lies and suffering go pretty much hand in hand. Empires involve both. Lies and their corollaries – propaganda, censorship, scapegoating, smokescreens and newspeak – are the trademarks of imperial domination. The reality 'created' by empires is so grim, so detrimental to people, with its massive suffering, exploitation, inequality and destruction, that only lies (if believed widely enough) can make people accept such an unacceptable state of affairs.

[952] Nelson, R., 'Terrorist disaster, September 11, 2001', *Global Consciousness Project.* See: noosphere.princeton.edu
[953] See Chapter 37: 'The role of the observer'
[954] See Chapter 40: 'Collective resonance'

Chapter 42.

Branching universe

In this final chapter we'll continue our quest and try to better understand the fate and reality of a group that resonates on the truth frequency in comparison to another group, the vast majority of the population that resonates on the lie frequency. Of course this is not black and white; no group holds the whole truth, while other groups believe only lies. But, sadly, we can safely state that today the amount of lies believed by most of the population has reached unprecedented levels.

Now, what do we suppose would happen when the majority of humanity shares the same beliefs based on false information – lies? Would humanity's influence on 'random events', including cosmic ones, still be minute, as suggested by the studies quoted previously?[955] And what would be the influence of the truth-resonant group?

At the end of the 19th century, the previously mentioned alchemist called Sédir[956] made another statement mentioning a link between human affairs and cosmic/cometary activity:

> A few minutes later, Stella came in bringing some café-au-lait, and we commented on the disaster which was ruining Paris and had ravaged the suburbs for the past two weeks.
>
> 'These modifications can only result from the precession of Equinoxes or from a subterranean eruption, or still, from the birth of a new magnetic center such as the proximity of a comet could bring about. But such phenomena are not due to a hazard: they are willed by cosmic intelligences or else brought about as reactions to social or ethnic diseases, you might say. Hence, wisdom is a-priori to let them alone.' ...
>
> 'What about us; can't we do anything against such cataclysms?'
>
> '*It is a little late. There should have been a few courageous men fifty or one hundred years ago.* Unless an innocent being, hidden somewhere, is willing to sacrifice himself, there is nothing else but to submit to it.'[957]

Sédir's postulation that 'a few courageous men' could help prevent a cataclysmic future strongly resonates with Gurdjieff's words, as recorded by Ouspensky[958]:

> Two hundred conscious people, if they exist and if they find it necessary and legitimate, could change the whole of life on the earth. But either there are not enough of them, or they do not want to, or perhaps the time has not come, or perhaps other people are sleeping too soundly.[959]

---

955 See Chapter 37: 'The role of the observer'

956 Yvon Le Loup (1871–1926). See Chapter 36: 'The rise and collapse of materialism'

957 Sédir, *Initiations*, chapter XXXIII

958 Pyotr Demainovitch Ouspensky (1878–1947). Esotericist, mathematician and writer, he was one of the principal exponents of G. I. Gurdjieff's teachings. His book titled *In Search of the Miraculous* is an authorized account of his early years with Gurdjieff. See: Chryssides, G. D., *Historical Dictionary of New Religious Movements*, p. 265

959 Ouspensky, P., *In Search of the Miraculous*, p. 310

Figure 259: Hanabusa Itcho (1652–1724), 'Blind monks examining an elephant'. (© Wikimedia Commons)

George Gurdjieff (1866–1949) was Pyotr Ouspensky's teacher. (© bigjoebuck)

Those two similar quotes make us wonder how this group of conscious individuals could change the world. What would happen if a group of individuals seeking truth were to resonate enough on the same frequency?

In the previous chapter we saw the example of stereo conversion. During this process the combination of 2D images creates a 3D picture and reveals one extra third dimension (depth) that was not visible in any of the 2D pictures.

In a similar way, but at a collective level, the story of the blind men and the elephant illustrates this point: a *group* of blind men touch an elephant to learn what it is like. Each one feels a different part, but only one part, such as the trunk or the leg. They *share* their observations, *listen* to each other and *collaborate* to assemble a full picture of the elephant (see figure 259).[960]

[960] In some versions, the blind men end up arguing, reach violent conflicts, stay in total disagreement, don't listen to each other, etc. Depending on the philosophical or religious agenda involved, this tale was given various interpretations: The manifold nature of truth (see: Hughes, M., *The voice of Prophets*, Volume 2 of 12, pp. 590–591).
Men are by nature blind, quarrelsome and ignorant (see: Wang, Randy. *The Blind Men and the Elephant*).
Individual perceptions are limited (see: Arberry, A. J., *The Elephant in the dark, on the reconciliation of contrarieties*).

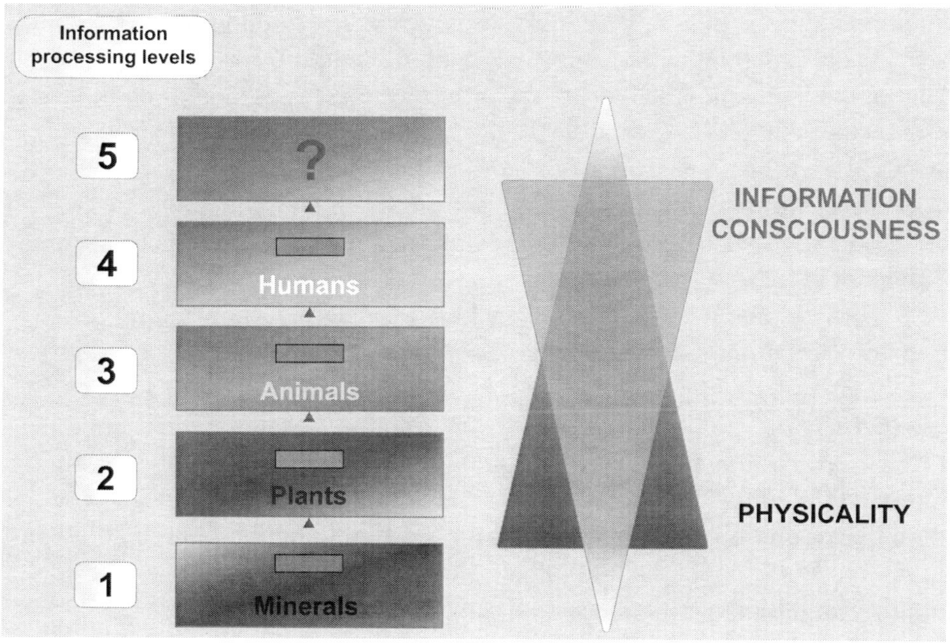

Figure 260: Five information-processing levels. (© Sott.net)

Thus, by sharing and comparing limited but *objective and complementary information*, the blind men gained access to a concept (a full 3D elephant) far beyond what they could have accessed individually (an isolated trunk, leg or tusk...).

Note that for a full assessment of the elephant to be made, several conditions are required:

- Each individual assessment must be objective, true
- Assessments must be shared
- A *critical mass* of *complementary* assessments must be conducted in order to fully 'map' the elephant.

In the three previously mentioned examples (stereo vision, the elephant and the law of mathematical addition) the same fundamental principle seems to be at work: the sharing of objective complementary information enables the observer(s) to reveal a hidden dimension, to have access to laws or concepts of a higher order. Does it mean that such a group somehow attains a higher level of reality/knowledge/conscience?

Figure 260 represents this hypothesized process. Five levels of information-processing capabilities are assigned to various entities (minerals, plants, animals, etc.). The higher the information-processing level, the more efficient the entity is at collecting, comprehending and processing information.

At the lowest level (minerals), physicality is predominant and consciousness almost nonexistent. At the second level (plants), consciousness increases slightly. Plants exhibit some primitive communication and sensory abilities,[961] while physicality remains predominant. Animals on the third level have an increased level of consciousness, emotional life and information-processing capabilities. Humans, at the fourth level, have self-awareness and face the fundamental choice between good and evil (serving themselves or serving others).

Notice that small arrows connect one box embedded within each level to the level above. This box represents a hypothesized group of entities from their existing information-processing level (plants, animals) holding a critical mass of true information that enables it to 'graduate' to the next information-processing level.

The group of 'truth-resonant' individuals mentioned previously could be represented by the small rounded box within the human level (information-processing level four).

The closest example of this 'graduation process' that can be found in nature might be what is called *superconductivity.*

At normal temperatures, matter, like metal, is not superconductive, just conductive, i.e. current flow is characterized by *chaotic* electron movement, low information level and low efficiency: a lot of energy is lost because of the collisions occurring between the disordered electrons. This is also known as *electric resistance* (transformation of electric energy into heat).

In addition, at normal temperatures, electrons are considered to be fermions,[962] which means they are ruled by the Pauli exclusion principle.[963] In simple terms, they are not free and not organized either: they behave individually, they can't bond and they can't be homogeneous, i.e. their state (mass, orbit, spin) must differ.

But when the temperature drops, i.e. when information increases, a threshold is eventually reached, superconductivity occurs almost instantaneously as if a critical mass of information was reached and the electrons abruptly switched to superconductivity.[964] Superconductivity can be reached at relatively high temperature,[965] which tends to suggest that, ultimately, superconductivity is not so much a temperature-driven phenomenon as an information-driven organization.

When superconductivity happens, *everything* changes. Electrons start to behave in a totally different way. Even their identities change; they're not fermions anymore but bosons.[966] Electrons free themselves from the Pauli exclusion principle, they start to form pairs, they stop moving chaotically and follow orderly patterns, and the level of organization and efficiency dramatically increases to the point that the collective electron movement becomes perfectly smooth, collisions stop and energy loss disappears. The once heterogeneous electrons now share similar mass, spin and locations.

---

[961] Sheldrake, R., *Morphic Resonance*

[962] Fermions are a class of particles including quarks, baryons, leptons, neutrons and protons. Fermions obey the Pauli exclusion principle. See: 'Fermions and Bosons, The Particle Adventure', Berkeley Laboratory. particleadventure.org/fermibos.html

[963] The Pauli exclusion principle is the quantum mechanical principle stating that two identical fermions (particles with half-integer spin) cannot occupy the same quantum state simultaneously. It has been shown that the Pauli exclusion principle is responsible for the fact that ordinary bulk matter is stable and occupies volume.

[964] 'Superconductivity', *alternativephysics.* See: http://www.alternativephysics.org/book/Superconductivity.htm

[965] Mercury barium calcium copper oxide (HgBa2Ca2Cu3O8) reaches superconductivity at around 133 K. See: Schilling, A., *et al.*, 'Superconductivity in the Hg-Ba-Ca-Cu-O system', *Nature* (1993)

[966] Bosons are a class of particles including photons and mesons. Unlike fermions, they don't obey the Pauli exclusion principle. See: 'Fermions and bosons', *op. cit.*

Figure 261: Sample of a 'high-temperature superconductor' named bismuth strontium calcium copper oxide. (© Creative Commons)

In fact, superconductive electrons defy most fundamental laws of physics, they exhibit no energy loss whatsoever (in contradiction with the laws of thermodynamics), they generate never-ending electric currents[967] and provide permanent magnetic fields,[968] despite the absence of any power source. What we are describing here is nothing less than *perpetual motion.*

Somehow, through an information increase (drop in temperature), electrons have reached a 'higher level of reality', allowing more freedom, more organization and more efficiency, while defying the limitations (Pauli exclusion principle, resistance) that previously restricted them.

Could a group of truth-resonant individuals undergo a similar process? In such a case, how could an 'augmented' reality modify/mitigate the way cataclysms can be experienced? Would the truth-resonant group only change their collective reality? Would they change the whole world, with seven or so billion people living in it, and resonating on a totally different frequency? Would they not change their world *per se* but the way they perceive it? Would they change the existing world or 'move' to another one?

Two *non-mutually exclusive* scenarios seem possible. The truth-resonating group could:

1. Perceive the world in a much different way (perception change).

2. Have an influence on the world they live in and modify it to some extent (reality change).

Perception change

Let's take the example of two individuals, where individual A believes all the lies, all the official stories, while individual B only believes *half* of those lies. For the latter, it could be reasonably

[967] Gallop, J. C., 'SQUIDS, the Josephson Effects and Superconducting Electronics', *CRC Press* (1990): 3–20
[968] Dutoit, B., *Superconductivity I, high field superconducting solenoids*, pp. 17–19

Figure 262: Levitating magnet above a superconductor in which the persistent electric current doesn't require a power source. (© Mai-Linh Doan)

said that he perceives reality in a way that is drastically different from individual A.

Their respective perception of fundamental topics like politics, war, governing elites, history, terrorism, finance, big corporations, etc., would be so different that, indeed, they would be living in two different *perceived* worlds.

Nonetheless, they would still live in the same world and they would both suffer from the same abuse, pollution and violence perpetrated by the elites, although individual A might not be fully aware of – or willing to acknowledge – his suffering and *its causes.*

They might however experience cataclysm in a different way because their perception and the information they hold differs drastically.

An analogy might help clarify this point. According to the 'five information-processing levels' chart (figure 260), butterflies are creatures of the third information-processing level, while humans are creatures of the fourth information-processing level.

While butterflies and humans possess different knowledge and perceptions, they live in the same world and experience the same natural events. A rain episode, for example, would be experienced in a very different way.

For butterflies, rain can be cataclysmic, or even deadly. Human beings, thanks to their knowledge, have weather forecasts, sturdy houses, waterproof clothes, they know how to swim, etc. This cumulative theoretical and applied knowledge makes such an event far less dramatic.

From this perspective, a *higher-level of awareness can indeed mitigate cataclysms* and their effects.

Reality change

Knowing that groups have an influence on reality according to the information they believe, what would happen when the perceptions held by two groups become virtually opposite? Would reality split?

Figure 263: The 'Schrödinger's cat' paradox according to the 'many worlds' interpretation. (© Wikimedia Commons)

To answer those questions, let's first go back to the Schrödinger experiment. While quantum physicists were arguing endlessly about the state of this poor cat, physicist Hugh Everett hypothesized that at the moment when the cat's box was open, i.e. the moment when information was exchanged between the observer and the observed event, the universe split in two:[969] one universe where the cat is alive and one universe where the cat is dead.[970]

Let's now apply Everett's hypothesis to macro events, specifically a human life. During our lifetime we encounter some crucial moments when we face important decisions. These are 'nodal points', as depicted by ellipses on the floor in figure 264. We make decisions based on the information (whether true or false) that we believe.

The diverging personal timeline shows the path(s) followed by an individual. The black icon represents the actual timeline of the individual; the grey icon represent 'non-actualized' timelines.

A timeline can be schematized as a series of decisions. The journey starts with a boy (bottom left of the drawing). He faces a first nodal point: suicide vs. no suicide. If the boy chooses not to commit suicide, he will eventually face a second nodal point: marrying or not marrying. If he marries he will face the choice between having children or not, etc.

The series of information-based decisions will lead to very different futures: happy husband, murderer ...

Of course, the choices given in this scenario are deliberately simplistic. In real life, every moment is a potential universe-splitting moment. At every moment we make decisions (a non-decision being a decision to not do anything), with each decision being based on our level of awareness, the amount of information we have access to and/or hold within.

Of course, in real life one single decision can lead to numerous different results. In addition, some results are immediately perceivable while others exhibit delayed effects.

[969] Everett, H., 'Theory of the Universal Wavefunction', *Thesis*, Princeton University, (1956, 1973), pp. 1–140

[970] For more than a decade Everett's thesis was ignored. It was theoretical physicist Bryce DeWitt who popularized and coined the term 'many worlds', which is also known as the Everett-Graham-DeWitt interpretation.

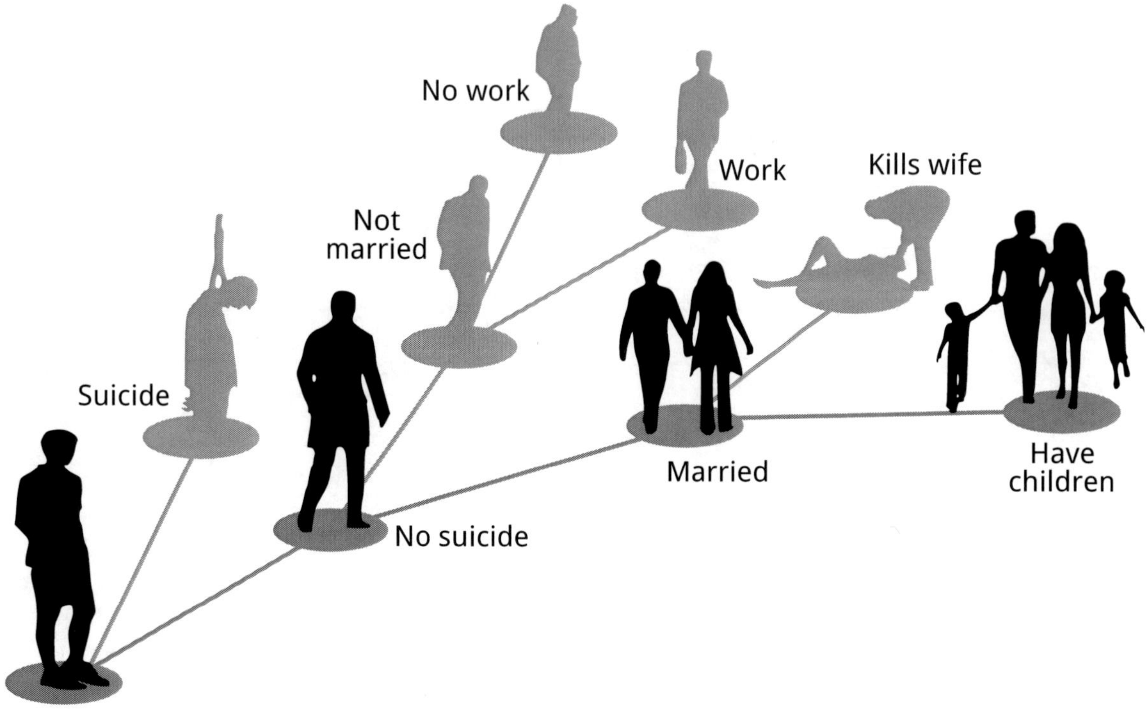

Figure 264: Illustration of diverging personal timelines. (© Sott.net)

The example above is also deliberately dramatic in order to show that in some cases the decisions will lead to a dead end: to timelines that are not viable.

This being said, the important point in Everett's multiverse theory is that the choices made by the individual gave birth to several universes: the *information-based decisions changed not only the way the individual perceives reality but also the very reality in which he lives.*

We have applied Everett's theory to an individual scenario; let's now apply it to a collective level, even to the whole human population.

Today we have, on one side, a group that believes lies and makes decisions based on these lies; on the other side, a minority that doesn't believe lies and acts accordingly. Truth-resonant people and lie-resonant people have a drastically different perception of the world, but they still live in the same physical world.

Could it be that the growing difference in perception and information held, and possibly catalyzed by increasing cosmic events and their electromagnetic effects, will reach a threshold and lead to a reality split? Could the truth-resonant group access a reality that matches the information they hold and the decisions they made, while the lie-resonant group remains in the collapsing world they believed in and chose to embrace?

Consider Roy Sullivan, the poor guy who was struck by lightning seven times. We left that mystery dangling, but it does point to a possibility that the internal state of a human being can attract very real cosmic events. And that was just one guy. Imagine the magnitude of attraction that might be generated by billions of human beings.

Figure 265: Maple tree. Only one secondary branch is alive while the main branch is dead. (© Sott.net, adapted from GardenWeb)

From an anthropocentric perspective, a partial or even total extinction is absolutely dramatic, but from a cosmic perspective, is it the same? Sédir wrote the following about this point:

> Comets are like missionaries from the stars community, they are bringers of hope, carriers of strength. *They arise from the depths of space, accomplish their mission of regeneration or medication for a world where life weakens*, then they return back into the space from which they came.
>
> Comets, because they were noticed before disasters, wars, epidemics, have kept a negative meaning. Don't all energetic remedies lead to the expulsion of organic waste?
>
> If *comets are a cure for the world, their role is to get rid of what is no longer useful to the world, and it is always a blessing.* Moreover, be certain that God never punishes, we attract negative reactions the same way indigestion occurs in the stomach of the child who was too greedy.[971]

While true information brings order, false information/lies bring chaos. Today, seven billion people resonate on the lie frequency, generating an unprecedented level of chaos. Extinctions decrease the lie factor and the subsequent chaos. From this perspective, comets can indeed be seen as a means of simply reducing chaos, as tools of a necessary cosmic rebalancing process.

[971] Sédir, Letters sent to the *Bulletin des Amitiés Spirituelles*, 1919

Conclusion

Our modern civilization suffers from amnesia. The psychopathic elites have stolen our heritage: the wisdom of our ancestors who knew about the cyclical nature of cosmic events and the role played by corruption in those catastrophes. Revision after revision, lie after lie, year after year, a mechanistic, linear, uniformitarian model of the universe has been imposed upon us.

This new paradigm has erased any trace of spirituality and responsibility. God is dead, cosmic events are wholly independent phenomena and humans are nothing more than biochemical machines. We were freed from any conscience, any ethical and cosmic responsibility, *but to what end?*

A period of relative cosmic quietude helped with the implementation of this new materialist cosmogony. But this period of relative calm seems to have reached its end. Since the turn of the 21st century, the signs are everywhere that this is the case, and they keep growing.

As we have seen in respect of the Roman Empire, collapses can take some time to fully develop. Nowadays, in our world of instant global communication and extremely rapid transit abilities, processes on the planet occur much more quickly. If that is the case, perhaps the feedback loop with the cosmos is such that the reaction will be correspondingly fast and globally intense.

The Roman Empire is strikingly similar to our present civilization. The Romans were certainly rational and scientific in many respects. They had factories that produced tableware which has been found at the farthest reaches of the Empire, even in peasant homes. They manufactured roof tiles that covered the heads of even the poorest workers and their livestock. A cache of letters was found in northern Britain where soldiers wrote home to have socks sent to them which were, apparently, mass produced. There were roads, sanitation systems, *haute cuisine*; in short, everything that we take for granted as essential to civilization.

One important difference between our modern civilization and the Roman Empire is that we have harnessed sources of power that enable our civilization to aspire to control nature and become 'god'. Yet, 'science without conscience is the soul's perdition.' This conscienceless science may have made us bigger and 'badder', but, as the saying goes, 'the bigger they come, the harder they fall.'

If that is so, we have indeed a big fall coming soon with over 7 billion individuals swimming in a sea of lies.

But is the situation utterly hopeless? Have we passed the point of no return? Is there any solution? One way to mitigate such events is to first rediscover the true history of humanity, punctuated as it is by dramatic cometary encounters, and the main cause of these cataclysms that have repeatedly befallen humanity. That main cause is the 'psychopath-ization' of the human population. Unless we do so: *'Those who do not learn from history are doomed to repeat it.'* Once we have learned from history and we know about cyclical catastrophes and the human-cosmic connection, the next step, for any purported rational race of beings, should be to avoid repeating history.

That is where every one of us carries a great responsibility. Obviously, the psychopathic elites won't change. Because of their very nature, they can only pursue power, greed and wealth without

Figure 266: John Martin, 'The Great Day of His Wrath', 1853. (Oil painting on canvas)

seeing the disastrous consequences of their deeds. Only people of conscience can change and begin the laborious work of becoming able to see reality as it is, acknowledge the lies and free themselves from the illusion that pervades the reality the elites have created.

The key point seems to be to *see the world as it is.* An objective assessment of reality is the best way to send a truth signal to the *Cosmic Information Field.* This assessment includes not only the world that surrounds us, but also what's going on within us: sure we are told many lies, but we also tell many lies to ourselves.

If the human-cosmic connection is valid, *all data should be inspected carefully before we consider it as true or false.* This applies to every piece of information, including the content of this book. It certainly offers a more objective assessment of reality than the mechanistic uniformitarian paradigm hammered into us by the elites and their minions. However, it does not mean everything is true. So, check the sources, read other material, think, compare, discern, share information with collinear people, and finally build your own assessment of reality, *as objectively as possible.*

In this quest, several psychological obstacles stand between us and the Truth. We are hardwired for 'sheepdom' by way of the *Stockholm syndrome*, a term coined in the early '70s to describe the puzzling reactions of four bank employees to their captors.

On August 23, 1973, three women and one man were taken hostage in one of the largest banks in Stockholm. They were held for six days by two ex-convicts who threatened their lives but also showed them kindness. To the world's surprise, all of the hostages strongly resisted the government's efforts to rescue them and were quite *eager to defend their captors.*[972] Indeed, several months after

[972] Nissen, V., *Masters of Manipulation*, p. 29

the hostages were saved by the police, they still had warm feelings for the men who threatened their lives. Two of the women eventually got engaged to the captors.

Psychologist Dee Graham has theorized that the *Stockholm syndrome occurs on a societal level.*[973] As victims of societal Stockholm syndrome, we are encouraged to develop psychological characteristics *pleasing to the system.* These include: dependency, lack of initiative, inability to act, decide, think; strategies for staying alive, denial, attentiveness to the system's demands, wants and expressions of approval of the system itself. We are taught to develop fondness for the system accompanied by fear of interference on the part of anyone who challenges the system's perspective.

Figure 267: For five days, the hostages were held in the vault of the bank in Stockholm, Sweden. (© AFP)

Most of all, we are conditioned to be overwhelmingly grateful to the system for giving us life. We *focus on the system's kindnesses, not on its acts of brutality.* Denial of terror and anger, and the perception of the system as omnipotent, keep us psychologically attached to the system. High anxiety functions to keep us from seeing available options. *Victims have to concentrate on survival, requiring avoidance of direct, honest reaction to destructive treatment.*

We are also naturally affected by the *normalcy bias*, which causes our brains to assume that things will be predictable and normal all the time. Instead of springing to action when something unexpected happens, our brain kind of 'shrugs' and figures that what is going on cannot be so bad, because truly bad events are so out of the ordinary, and they only happen to 'other people':

> Normalcy bias manifested itself in dramatic fashion during a plane collision in 1977 that killed *583 people* – the worst aircraft disaster in history. Two 747 jumbo jets collided with each other over Tenerife airport. After the collision, one jet fell to the ground and exploded, killing all 248 passengers on board.
>
> The other jet crash-landed, but didn't explode. The collision sheared away the top of the jet and flames began to take over the aircraft. Passengers who survived the initial collision *could have escaped unharmed, but they had to act fast.* Paul Heck, a passenger on the burning plane sprang to action. He unbuckled his seatbelt, grabbed his wife's hand, and hightailed it to the nearest exit. They, along with 68 other passengers, survived, while 328 died.
>
> In an interview after the disaster, Mr. Heck noted how most people just sat in their seats acting like everything was fine even after colliding with another plane and seeing the cabin fill with smoke. Researchers believe that passengers had a little over a minute to escape before being consumed by the flames, and are convinced that if more people had taken immediate

[973] Dee Graham has expanded on her theories in *Loving to Survive: Sexual Terror, Men's Violence, and Women's Lives*

Figure 268: Photo of the Tenerife crash shot a few minutes after the disaster.

> action instead of remaining in their seats *pretending like things were okay*, the survival rate would have been much, much higher.[974]

We are the passengers of that 747 jumbo jet, and the plane is in flames, the explosion is imminent. Are we going to normalize the event and sit down quietly believing the reassuring messages from the captain? Or, are we going to open our eyes, assess the situation objectively and take appropriate measures?

Are we going to wake up to and throw off our collective Stockholm syndrome? Will we stop denying our anger? Will we overcome our fears? Will we cease seeking approval from the system? Will we stop believing the illusion of our 'great democracies'?

Will we reclaim our freedom? Will we start thinking, deciding and acting by ourselves? Will we open our eyes to the gigantic lies we are fed daily?

At this point in the dynastic cycle, our only hope is the Truth. To seek, find and embrace the Truth. The Truth about our history, the Truth about our 'elite', the Truth about ourselves. *And the Truth, only the Truth, shall set us free.*

[974] McKay, B. & K., 'How to Survive a Plane Crash: 10 Tips That Could Save Your Life', *Art of Manliness* (30 July 2013)

Bibliography

- Abell, George O., *Exploration of the Universe*, (3rd edition), Harcourt Brace College Publishers, 1975
- Abreu, J. A., *et al.*, 'For how long will the current grand maximum of solar activity persist?', *Geophys. Res. Lett.*, 2008
- Adams, M., 'MSFC Solar Group', *NASA*.
- Albarelli, H.P., *A Secret Order,* Trine Day, 2013
- Alfred, J., 'Plasma life forms', *Unexplained mysteries,* 7 August 2008.
- Alfvén, Hannes *et al.* (1978). 'Interstellar clouds and the formation of stars'. *Astrophysics and Space Science* 55 (2): 487–509.
- Allan, D. & Delair J. *Cataclysm!*, Bear & Co, 1997
- Meichsner, J., *Nonthermal Plasma Chemistry and Physics, CRC Press 2012*
- Anderson, D., 'Giant quakes increase: Seismic storm Lull Ends', *Los Angeles Times,* 29 August 1976
- Arberry, A.J., *71 – The Elephant in the dark, on the reconciliation of contrarieties,* 2004
- Archer, C. L. & Caldeira, K., 'Historical trends in the jet streams', *Geophysical Research Letters*, vol. 35
- Arndt, M. *et al.*, 'Wave-particle duality of C60', *Nature,* 401, 680-682 (1999)
- Arrian, *Campaigns of Alexander*, Penguin Classics, 1976
- Aspect, A., 'The Bohr–Einstein debate and quantum entanglement tested experimentally', *CNRS Communiqué,* 1982
- Atkinson, N., 'Space Shuttle-Sized Asteroid 2013 XY8 to Fly Past Earth on Dec. 11', *Universe Today*, 10 December 2013.
- Bagley, M. 'Krakatoa Volcano: Facts About 1883 Eruption', *Livescience*, 26 March 2013.
- Baillie, M., *Exodus to Arthur* and *The Celtic Gods.* Batsford; 1999
  _________, *New Light on the Black Death: The Cosmic Connection,* The History Press, 2006
  _________, *World Archaeology,* 1991, 33:233-43
- Bainbridge, K.T., *Report LA-6300-H*, Los Alamos Scientific Laboratory. May 1976
- Bakich, M., 'Comet ISON will light up the sky'. *Astronomy Magazine*, September 25, 2012.
- Ball, P., 'Experts still split about what quantum theory means', *Nature*, 11 February 2013
- Barbosa, R. & Overstreet, O., 'What Is Soil Electrical Conductivity?', *LSU AgCenter Pub.* 318
- Barnes-Svarney, Patricia & Svarney, Thomas E, *Skies of Fury: Weather Weirdness Around the World,* 1999
- Barriopedro, D., Garcıa-Herrera, R., & Huth, R., 'Solar modulation of Northern Hemisphere winter blocking', *J. Geophys. Res.*
- Barry, P., 'Saturn's rotation puts astronomers in a spin', *New Scientist*, 3 May 2006.
- Battams, K., 'ISON and Encke in STEREO: Nov19-23, 2013', *NASA Comet ISON Observing Campaign*
  _________, 'Very quick update', *NASA Comet ISON Observing Campaign*, November 27, 2013.
- Bechterev, V. M., 'Experiments on the effects of the mental influence on the behavior of dogs', *Problems in the study and training of personality,* 1920, pp. 235-260

- Ben-Menahem, A. *Historical Encyclopedia of Natural and Mathematical Sciences*, vol. 1
- Bennet, A. J., 'Variability in surface atmospheric electric field measurements', 2008, *J. Phys.*: Conf. Ser. 142 012046
- Bennett, A. J. & Harrison, R. G., 'Variability in surface atmospheric electric field measurements'. *Journal of Physics: Conference Series,* 142, 2008
- Benor, Daniel J., 'Survey of Spiritual Healing Research', *Complementary Medical Research 4*, 3 September, 1990, pp. 9-33
- Berger, A., 'Obliquity and Precession for the Last 5,000,000 Years', *Astronomy and Astrophysics*, pp. 127–135
- Berlin, J., 'Sinkhole Science: A primer', 5 March 2013, *National Geographic.*
- Beutler *et al.*, 'Paranormal healing and hypertension'. *British Medical Journal*, 1998, 296. 1491-1494
- Bevan A., De Laeter J., *Meteorites: A Journey Through Space and Time*, 2002
- Bird, J., Peters, E. & Powell, J. *Crusade and Christendom,* University of Pennsylvania Press, 2014
- Blöss, Christian, Niemitz, Hans Ulrich, 'The Self-Deception of C-14 and Dendrochronology', *Zeitensprünge 8*, 1996, Berlin
- Bohm, D., 'A new theory of the relationship of mind and matter', *Journal of the American Society for Psychical Research* 80, n°2, April 1986, p.128
- Boltzmann, L., *The Second Law of Thermodynamics*, Princeton Press, 1886
- Borenstein, Seth, 'Weather wimps?', *Salisbury Post.* January 10, 2014
- Bösch *et al.*, 'Examining Psychokinesis: The interaction of human intentions with random number generators', *Psychology Bulletin,* July 2006, 132(4), p.498
- Boyett, J., 'Are the 'Strange Sounds' Videos a Hoax?', *Huffington Post*, 8 February 2012.
- Bradley, N., 16 September 2012. 'Incoming! Meteor or Comet Fragment Explodes Above Southwestern US, Prompting US Army 'Missiles' Cover-up', *Signs of the Times.*
- Brand, J.C., 'Comets', *Scient Am,* 1981
- Braud, W., *Distant Mental Influence: Its Contributions to Science, Healing, and Human Interactions*, Hampton Roads Publishing, 1 December 2003
  _________, 'Distant mental influence of rate of hemolysis of human red blood cells', *Journal of the American Society for Psychical Research*, 1990, 84 (1): 1-24
  _________, 'On the Use of Living Target Systems in Distant Mental Influence Research', in *Psi Research Methodology: A Re-Examination*, New York, NY, 1991
  _________, 'Wellness implications of retroactive influence: exploring an outrageous hypothesis', *Alternative Therapies in Health and Medicine.* 2000; 6(1): pp. 37-48
- Braud W. & Jackson J., 'The Use of Ideomotor Reactions as Psi Indicators', *Parapsychology Review* 13 (1982), pp. 10-11
- Braud, W., Schlitz M., 'Psychokinetic influence on electrodermal activity', *Journal of Parapsychology.* 47(2): 95-119
- Brennan, M., 'Billionaire Bunkers: Beyond the Panic Room, Home Security Goes Sci-Fi', *Forbes Magazine*, 16 December 2013
- Briffa, K. *et al.* 'A 1,400-year tree-ring record of summer temperatures in Fennoscandia', *Nature* 346, 434-439 (02 August 1990)
- Brignell J., 'How we know they know they are lying', January 2009, *Numberwatch,*
- Bruce A., *2012: Science or Superstition*, Disinformation Books, 2009
- Byrd, D., 'Which spiral arm of the Milky Way contains our Sun?', *Earthsky*
- Caesar, J., *The Gallic War,* Harvard University Press, 1919

- Campanelli L. *et al.*, 'Ellipsoidal universe can solve the cosmic microwave background quadrupole problem'. *Phys Rev Lett.* 2006 Sep 29;97(13):131 302.
- Campbell, A. 'Killer Sinkholes: Unexpected Holes Swallow Entire Families, Cars And Homes', *Huffington Post*, 1 March 2013
- Carotta, F., *Jesus was Caesar,* Aspekt, 2005
- Caulfield, P. 'Bankrupt Harrisburg, Pa., can't fix the 41 sinkholes plaguing its streets', *New York Daily News*, 1 February 2013
- Cebik, L. B., 'Some (Old) Notes on Home-Brew Parallel Transmission Lines', *QSL.*
- Chandrasekar, *Basic of atmospheric science*, EEE, 2010
- Chang, K. 'First Mission to Explore Those Wisps in the Night Sky', *NY Times,* 24 April 2007
- Charles P A, *The Hazard to Civilization From Fireballs and Comets*, Oxford University - Dept Of Astrophysics, 4th June 1996
- Chavalier, G., *The Earth's electrical surface potential,* 2007
- Christian *et al.*, 'Global frequency and distribution of lightning', *Journal Of Geophysical Research*, vol. 108
- Christou *et al.*, 'Comparative studies of meteoroid-planet interaction in the inner solar system', *Planetary and Space Science*, 2007, vol. 55(14), pp. 2049-2062
- Chryssides, George D., *Historical Dictionary of New Religious Movements*, p.265
- Clube, V. & Napier, W., *Cosmic Winter*, Blackwell Pub, 1990
  _________, 'The Microstructure Of Terrestrial Catastrophism', *Monthly Notices Of The Royal Astronomical Society*, 211: 4, pp. 953-968
- Clube, V., Bailey, M.E. & Napier, W.M., *The Origin of Comets*, Pergamon Press, 1990
- Clube, V., *The Hazard to Civilization from Fireballs and Comets,* 1996
- Cochran, M., 'What causes sinkholes to form', *Accuweather*
- Coe, R. *et al.* 'New evidence for extraordinarily rapid change of the geomagnetic field during a reversal', *Nature* 374, pp. 687 - 692 (20 April 1994)
- Collier J. *et al.*, *Optical Holography*, New York, Academic Press, 1971
- Collins B. E., Guetzkow H, *A Social Psychology of Group Processes for Decision-making*, 1964
- Committee on Scientific Accomplishments of Earth Observations from Space, *Earth Observations from Space: The First 50 Years of Scientific Achievements* (2008)
- Courtillot, V. & Le Mouel, J. 'Time variations of the earth's magnetic field - From daily to secular', *Annual review of earth and planetary sciences.* vol. 16, 1988, p. 389-398
- Cruttenden, W., *Lost Star of Myth and Time,* St Lynn's Press, 2005
- Curie, R., 'Detection of the 11-yr sunspot signal in Earth rotation' *Science* 23, January 1981: vol. 211, no. 4480, pp. 386-389
- Dachille, 'Axis Changes in the Earth from Large Meteorite Collisions', *Nature*, 198, 176, April 1963
- Daileader, Philip, *The Late Middle Ages*, The Teaching Company, 2007
- D'Arigo, S. *et al.* '1,738 years of Mongolian temperature variability inferred from a tree-ring width chronology of Siberian pine', *Geophys. Res. Lett.* 2001, 28:543-46
- Davies, P. & Gregersen, N., *Information and the Nature of Reality*, Cambridge University Press, 2010
- Dean, D.E., 'Plethysmograph Recordings of ESP Responses', *International Journal of Neuropsychiatry*, 2 September 1966
- De Graham, *Loving to Survive: Sexual Terror, Men's Violence, and Women's Lives*, NYU Press, 1995
- De Grazia, A. & Milton, E.R., *Solaria Binaria,* Metron Publications, 1984

- Delingpole, J., 'Global Cooling and the New World Order', *The Telegraph*, 26 September 2010
- Demidov, B.A. *et al.*, 'Anomalous Resistance And Microwave Radiation From A Plasma In A Strong Electric Field', *Soviet Physics*, vol. 21, no. 2, August 1965
- Denman, K.L. et al (2007) 'Couplings between changes in Climate System and the Biogeochemistry', *IPCC*, Cambridge, UK
- De Sarre, F., *Mais où est donc passé le Moyen Âge*, Hades Editions, 2013
- D'Espagnat, B. (1995) *Veiled Reality: An Analysis of Present-Day Quantum Mechanical Concepts.* Reading, MA, Addison-Wesley
  ________, *In Search of Reality,* Springer Verlag, 1983
- Dunkel T., 'Inside the life of the man known as the 'spark ranger'', *The Washington Post Magazine,* 15 August 2013.
- Dutoit B., *Superconductivity I, high field superconducting solenoids*, 2007 Ecole Polytechnique Federal De Lausanne, pp.17-19
- Eden, D., 'Chemtrails, what's going on?', *Viewzone.* July 2013
- Edwards, K. *And America Quaked: A Chilling Series of Visions of a Future American*
- Einstein A., Podolsky B., and Rosen N., 'Can Quantum-Mechanical Description of Physical Reality Be Considered Complete?', *Phys. Rev.* 47, 777 (1935)
- Elert G., 'Photoelectric effect', *The Physics Hypertextbook*
- Elizabeth, E. & Janay, N. *Where is the Family in medieval and early modern Scotland?*
- Emric, A., 'Bosnians panic as sinkhole swallows village pond', *Associated Press*, 26 November 2013
- EPA, *Guidance Manual Alternative Disinfectants and Oxidants*, April 1999, chapter IV
- Erman, G.A. 'Kritische Beitrage zur atmospharischen Electromie', *Annln Phys*, vol. 15, 1803
- European Space Agency, 'Could Venus be shifting gears?', 10 February 2012.
- Evans, J. A., *Justinian, De Imperatoribus Romanis*, University of British Columbia
- Evans, L., 'The growth and fragmentation of ice crystals in an electric field', *J. Atmos. Sci.*, 30, 1657–1664.
- Everett, Hugh, 'Theory of the Universal Wavefunction', *Thesis*, Princeton University, (1956, 1973), pp. 1–140
- Ezer, T. *et al.*, 'Gulf Stream's induced sea level rise and variability along the U.S. mid-Atlantic coast', *Journal of Geophysical Research:* oceans, vol. 118
- Fahey *et al.*, 'Emission Measurements of the Concorde Supersonic Aircraft in the Lower Stratosphere', *Science,* 6 October 1995, vol. 270, no. 5233, pp. 70-74
- Fearn, D. R., *The Geodynamo*, August 19, 2004
- Felix, R. W., *Magnetic reversals and evolutionary leaps*, p.77, Sugarhouse Publishing, 2009.
- Ferrin, I., 'Comet ISON is in Outburst', *The Comets Page of Ignacio Ferrin.*
- Firestone, B. & Topping, W., 'Carbon and radiocarbon dating: A primer', *Mammoth Trumpet*, March 2001
- Firestone R., West A. & Warwick-Smith, S., *The Cycle of Cosmic Catastrophes: How a Stone-Age Comet Changed the Course of World Culture*, 2006, Bear & Company
- Fischetti, M., 'Was Typhoon Haiyan a Record Storm?', *Scientific American*, 12 November 2013
- Fisher, D. *The Great Wave*, Oxford University Press, 1999
- Flannery, T., Beale, R. & Hueston, G. *The critical decade: International action on climate change,* Climate Commission, 2012
- Fleming, J. R., 'The-Climate-Engineers', *The Wilson Quarterly*
- Flottorp, Gordon, 'Effect of Different Types of Electrodes in Electrophonic Hearing', *J. Acoust. Soc. Am.* vol. 24, no. 4, pp. 447-448

- Fomenko, A., Nosovskiy, G., Tamdhu, F. & Zinoviev, P. *History: Fiction or Science?* 2008
- Forsyth D., *Group Dynamics*, 5 Edition, 2010, Belmont
- Frank A. Beach, *Biography of Karl Spencer Lashley 1890-1958*, National Academy of Science
- Freidenreich, S. *et al.* 'Solar radiation absorption by CO2, overlap with H2O, and a parameterization for general circulation models', *Journal of Geophysical Research: Atmospheres*, vol. 98, pp. 7255–7264
- Freund, T.F., 2003. 'Rocks That Crackle and Sparkle and Glow: Strange Pre-Earthquake Phenomena', *Journal of Scientific Exploration*, vol. 17, no. 1
- Gallet, Y., 'Possible impact of the Earth's magnetic field on the history of ancient civilizations', *Earth and Planetary Science Letters,* vol. 246, nos. 1–2, 15 June 2006, pp. 17–26
- Gallois, J., *History of the Spanish Inquisition*, Brissot-Thivars,1826
- Gallop John C., 'SQUIDS, the Josephson Effects and Superconducting Electronics', *CRC Press*, 1990, pp. 3, 20.
- Gary, B. L., 'Comet ISON Observations by Three Non-Professional Observers', *Bruce Gary personal website.*
- Gasperini L. *et al.*, 'Possible impact crater for the 1908 Tunguska Event', *Terra Nova,* vol. 19, no. 4, pages 245–251, August 2007
- Geertz, B., 'Report: China building electromagnetic pulse weapons for use against U.S. carriers', *Washington Times*, 21 July 2011
  ________, 'Report: Possible EMP Strike on Iran?' *Washington Free Beacon*, 29 August 2012.
- Geoman, 'What is the best estimate of the densities of the various layers of the Earth?' *University of Oregon.*
- Ghys, E., 'The butterfly effect', *12 International Congress on Mathematical Education*, 8-15 July 2012, COEX, Seoul, Korea
- Girden, E.; Helson, H., 'A review of psychokinesis (PK)', *Psychological Bulletin*, 1962, vol. 59(5), pp. 353-388
- Glasstone, S. & Dolan, P., *The Effects of Nuclear Weapons*, United States Department of Defense and the Energy Research and Development Administration
- Glionna, J., 'Colorado firefighters hampered by winds, heat – and meteors', *L.A. Times*, 21 June 2012
- Goeschen C. & al., 'Damage of aromatic amino acids by the atmospheric free radical oxidant NO in the presence of NO, NO, O and O', *Org Biomol Chem.*, 7 May 2011; 9(9):3380-5
- Goodman, G., 'Tom Stonier, a man of computers and peace', *Steinschneider.com*, 28 June 1999.
- Goodman, J., *Psychic Archaeology: Time Machine to the Past*, Berkley Publishing, 1977
- Greene, B. *The Fabric of the Cosmos*, Vintage, 2005
- Grego, P., 'New comet might blaze brighter than the full Moon', *Astronomy Now*, September 25, 2012.
- Gribbin, J. *et al.*, 'Discontinuous Change in Earth's Spin Rate following Great Solar Storm of August 1972', *Nature*, vol. 243, May 4, 1973
- Gribbin, J., *Beyond the Jupiter effect,* 1983
- Gubbins, D. *et al.*, 'Can the Earth's dynamo run on heat alone?', *Geophys. J. Int.,* 2003 155: 609-622.
- Gutiérrez, R. *et al.*, 'Structural fluctuations and quantum transport through DNA molecular wires, a combined molecular dynamics and model Hamiltonian approach'. *New J. Phys.*, 2010, 12, 023 022.
- Gys-Devic, *From the war of the Titans to the battle of the manuscripts*
- Haigh, J D., 'Climate variability and the influence of the sun', *Science*, 2001, 294: 2109–2111
- Hall, R., 'Land Grabbing in Africa and the New Politics of Food', *Future Agricultures*, June 2011.
- Hammel, H. & Lockwood, G. 'Suggestive correlations between the brightness of Neptune, solar variability, and Earth's temperature', *Geophysical Research Letters*, vol. 34, 8 April 2007.
- Handwerk, B., 'Sea Levels Rose Two Feet This Summer in U.S. East', *National Geographic*, 10 September 2009

- Haraldsson, E. *et al.*, 'Are religiosity and belief in an afterlife better predictors of ESP performance than belief in psychic phenomena?' *Journal of Parapsychology*, 1993, 57, 259-273
- Hardouin, J., *The Prolegomena*, Books on Demand, 2010
- Harl, W. H., 'Early Medieval and Byzantine Civilization: Constantine to Crusades', *Tulane University*
- Harrison, R. G.; 'Atmospheric Electricity And Cloud Microphysics', University of Reading, 2001
- Harris, T., 'How stun gun work', *How Stuff Works.*
- Hathaway, D. H., 'The Solar Cycle', *Solar Physics.*
- Hattori, Katsumi, 'ULF Geomagnetic Changes Associated with Large Earthquakes', *TAO*, vol. 15, No. 3, 329-360, September 2004
- Haywood, J., 'Global Dimming', *MetOffice UK*
- Hegerl, G. C. *et al.* (2007), 'Chapter 9, Understanding and Attributing Climate Change' in *Climate Change 2007: The Physical Science Basis*, *IPCC*, Cambridge, UK
- Henri, B., 'Sinkhole Shock', 30 April 2013, *WSFA*,
- Hill, E., 'On the possibility of changes in the Earth's axis', *Geol Mag*, vol. V, pp. 262-266
- Holbrook T., Clark R., 'Secrets of a Superorganism', School of Life Science, *Arizona State University*
- Hollenbach, D. F. *et al.*, 'Deep-Earth reactor: Nuclear fission, helium, and the geomagnetic field', *PNAS,* vol. 98 no. 20
- Holt, L., *Daniel's Key: Exposing a Dark America at the Epicenter of Apocalypse,* IUniverse, 2005
- Homer, *The Iliad,* Penguin Classic, 1998
- Honorton, C., & Ferrari, D., 'Meta-analysis of forced-choice precognition experiments 1935–1987', *Journal of Parapsychology*, 1989, vol. 53, 281-308
- Hornak J.P., *The basics of nuclear magnetic resonance*, Center for Imaging Science, Rochester Institute of Technology, New York, 1997
- Horrox, R., *The Black Death*, Manchester University Press, 1994
- Hough, A., 'Record 110,000 lightning bolts strike during UK 'superstorms'', *The Telegraph*, 29 June 2012
- Howard, J., 'Asteroid 2012 DA14 Will Miss Earth By Just 15 Minutes', *Huffington Post*, 12 February 2013
- Howell, E., 'Sun's 2013 Solar Storm Peak Expected to Hit Century Low'. *Livescience*
- Hoyle, F. & Wickramasinghe, C., 'SARS – a clue to its origins?', *The Lancet*, vol. 361, no. 9371, p 1832, 24 May 2003
- Hughes, Marilynn, (2005) *The voice of Prophets*, vol. 2 of 12
- Hulot, G. *et al.* 'The Magnetic Field of Planet Earth', *Space Sci Rev*
- Husher, J. D., *Facts and Myths Facing Today's World,* IUniverse, 2008
- Idso C. D., Carter R. M., Singer S. F., and Soon W., 'Scientific Critique of IPCC's 2013 'Summary for Policymakers'', *Watts Up With That?,* 19 October 2013
- Illig, Heribert, (2005) 'Anomalous Eras - Best Evidence: Best Theory', Toronto Conference, The Invented Middle Ages, June 28, 29, 30, 2005
- Iorio, L., 'On the anomalous secular increase of the eccentricity of the orbit of the Moon', 2011, *Mon. Not. R. Astron. Soc.*
- IPCC, *Third Assessment Report - Climate Change 2001*, Working Group I: The scientific basis
- Jago, L., *The Northern Lights*, Alfred a Knopf, NY, 2001.
- Janiskee B., 'Shenandoah National Park Ranger Roy Sullivan Set the World Record for Being Hit by Lightning', 1 August 2008.

- Janke, Paul R., *A Correlated History of the Universe*, Pan Terra Inc, 2002
- Jermacans, K. & Laws, K.. 'Coalescence of Raindrops in an Electrostatic Field', *The Physics Teacher,* vol. 37, April 1999
- Jolley D. *et al.*, 'Two large meteorite impacts at the Cretaceous-Paleogene boundary', *Geology*, April 2010
- Jones, C.D., 'Large electric fields due to industrial chimney stack plumes', *Atmospheric Environment, 1967, 11(12),p. 1197*
- Kaldellis, Anthony, *Procopius of Caesarea: Tyranny, History, and Philosophy at the End of Antiquity*; 2004, University of Pennsylvania; p. 130.
- Kamer, F. 'Hedge Farm! The Doomsday Food Price Scenario Turning Hedgies into Survivalists', *observer.com*, 18 May 2011
- Kamra, A., 'Measurements of the electrical properties of dust storms', *Journal of Geophysical Research*, vol. 77, no. 30, pp. 5856-5869
- Kanter, J., 'Carbon trading: Where greed is green', *New York Times*, 20 June 2007
- Karlsson, G., *The History of Iceland*, University of Minnesota, 2001
- Keay, C. & Ostwald's Laboratory, 'Test of the Production of Electrophonic Sounds', *Journal of the Acoustical Society of America*, 89(4) Pt(1), 1823-1824, 1991
- Keay, C. 'Continued progress in electrophonic fireball investigations', *Earth, Moon, and Planets*, 1995, vol. 68(1), pp. 361-368
  _________, 'Progress in Explaining the Mysterious Sounds Produced by Very Large Meteor Fireballs', *Journal of Scientific Exploration*, 7, No 4, 337-354, 1993
  _________, 'The 1978 New South Wales Fireball', *Nature*, 285, 464-466, 1980
- Keay, C. S. L., 'Audible fireballs and geophysical electrophonics', *Astronomical Society of Australia*, vol. 11, no. 1, p. 12-15
- Kelly, J. *The Great Mortality: An Intimate History of the Black Death*
- Keppler J., 'A View Behind the Scenes of Matter Provides Valuable Clues for the Development of a Theory of Consciousness', *Talk at the TSC 2012 Conference*, Tucson, April 2012
- Kirkby, J. *et al.*, 'Cloud formation may be linked to cosmic rays', *Nature,* 476, pp. 429-433 (2011)
- Kivelson, M. G., *et al.* 'Ganymede's magnetosphere: Magnetometer overview.' *Journal of Geophysical Research: Planets* (1991–2012) 103.E9 (1998): 19 963-19 972.
- Knight-Jadczyk, L., *Comets and the Horns of Moses: Secret History of the World Vol. 2.,* Red Pill Press 2013
  _________, *The Secret History of the World*
- Kobayashi, M. *et al.*, 'In vivo imaging of spontaneous ultraweak photon emission from a rat's brain correlated with cerebral energy metabolism and oxidative stress', *Neuroscience Research*, July 1999, 34(2):103-13
- Koehli, H., 'The Cs Hit List 06: Let's Do the Planetary Twist to the Tune of the Brothers Heliopolis', *Signs Of The Times*, 29 February 2012
- Kohsen, Anita, *Experiments in Mental Suggestion*, 1963
- Kramer, L., 'Strange Sounds Maps', *Seektress*, 21 January 2012
- Krasinsky, G. A.; Pitjeva, E. V.; Vasilyev, M. V.; Yagudina, E. I., (July 2002). 'Hidden Mass in the Asteroid Belt'. *Icarus* 158 (1): 98–105.
- Krasnikov, S., 'Traversable wormhole', *Physical Review D*, 2000, vol. 62(8)
- Kriesa, C., *The Influence of Distance on Psychokinesis as Measured by Yeast Metabolism*, Sonoma State University, 1978
- Kropotkin *et al.*, 'Baron Von Toll on New Siberia and the Circumpolar Tertiary Flora', *Geogr. J.*, vol. XVI, pp. 95-98.

- Kross B., 'How many atoms are in the human body?', *Jefferson Lab*
- Kumar P., 'information in a photon', DARPA, Defense Science Office.
- Laine U. K., Turunen E., Manninen J., Nevanlinna H., 'Measurements and analysis of sounds during active aurorae in Finland 2000-2001', *URSI General Assembly*, Maastricht, August 17-24, 2002.
- LaMarche, V. 'Paleoclimatic Inferences from Long Tree-Ring Records', *Science* 15 March 1974: vol. 183, no. 4129, pp. 1043-1048
- Langel, R., *Geomagnetism Vol. 1,* (1987).
  __________, Hinze, W. J., *The Magnetic Field of the Earth's Lithosphere: The Satellite Perspective*, 1998
- Lappa, Marcello, *Rotating Thermal Flows in Natural and Industrial Processes*, 2012,
- Lard, L., Paull, C., & Hobson, B., (1995). 'Genesis of a submarine sinkhole without subaerial exposure'. *Geology* 23(10): 949–951.
- Laštovička J. *et al.*, 'Global Change in the Upper Atmosphere', *Science*, 24 November 2006, vol. 314, no. 5803, pp. 1253-1254
- Lazlo, E., *The Interconnected Universe: Conceptual Foundations of Transdisciplinary Unified Theory*, World Scientific Pub Co Inc, 01 November 1995, p.100
- Le Bon, Gustave, *The Crowd: A Study of the Popular Mind*; Aristeus Books, 1895
- Lee C., 'The insanely weird quantum wave function might be 'real' after all', *Ars Technica,* 22 November 2011
- Leinert C *et al.*, 'Interplanetary Dust', *Physics and Chemistry in Space.* Springer-Verlag
- Leroy S., 'Natural hazards, landscapes, and civilizations' in: Shroder J. *et al.*, *Treatise on geomorphology*, Academic Press, San Diego, CA, vol. 13
- Leslie, R., 'First Sighting of a Noctilucent Cloud', *Nature*, 32, 245, 16 July 1885
- Lewin, Roger, 'Is your brain really necessary?', *Science*, Dec 12, 1980, vol. 210, p. 1232(3)
- Lide, D., (2004). *CRC Handbook of Chemistry and Physics*
- Lippman Walter, *Public Opinion*; NuVision, 2007.
- Livingston, Ian, 'Polar vortex delivering D.C.'s coldest day in decades, and we're not alone', *Washington Post*, 7 January 2014.
- Lockwood, M, Rouillard, A. P. and Finch, I. D., 'The rise and fall of open solar flux during the current grand solar maximum, *Astrophys. J.*, 2009
- Lomborg, B., 'Bjørn Lomborg on Haiyan/Yolanda', *wattsupwiththat*, 10 November 2013.
- Longo, M., 'Does the Universe Have a Handedness?', *arXiv:0812.3437 [astro-ph],* 2008
- Louie, J., *Earth's Interior,* 1996, University of Nevada, Reno.
- Louzada, K., Stewart, S. & Weiss, B., 'Shock Demagnetization of Pyrrhotite', *Lunar and Planetary Science* 36, 2005
- Lund, R. 'Study in the changes in the tropical cycle record in the North Atlantic between 1851 and 2008', *Journal of the American Statistical Association,* 2009
- Lunsford, R.; 'Meteor activity from Comet ISON', *American Meteor Society.*
- Lyall S., 'Heat, Flood or Icy Cold, Extreme Weather Rages Worldwide', *The New York Times*, 10 January 2013.
- Lykawka, P. S. and Mukai, T., 'An Outer Planet Beyond Pluto and the Origin of the Trans-Neptunian Belt Architecture', *Astronomical Journal,* 2008
- Madrigal, A., 'Mysterious, Glowing Clouds Appear Across America's Night skies', *Wired,* 16 July 2009
- Mahaney, W. *et al.* 'New Evidence from a Black Mat Site in the Northern Andes Supporting a Cosmic Impact 12,800 Years Ago', *Journal of Geology,* vol. 121, No. 4 (July 2013), pp. 309-325

- Mahapatra, L. 'Mount Sinabung is just one of 35 currently erupting volcanoes', *International Business Times*, 26 November 2013.
- Mandelkehr Moe M., 'An Integrated Model for an Earthwide Event at 2300 BC', *Chronology & Catastrophism*, Vol. X, 1988
- Mango, C., *Byzantium: The Empire of New Rome*, Scribner, 1980
- Mann, M., 'Medieval climatic optimum', *Encyclopedia of Global environment change*, Wiley, 2001
- Martin Wild, M., (2012) 'Enlightening Global Dimming and Brightening', *Bull. Amer. Meteor. Soc.*, vol. 93
- Master, J., 'Arctic sea ice loss tied to unusual jet stream patterns', *Weather Underground,* 2 April 2012.
- Master, S., 'A possible Holocene impact structure in the Al 'Amarah marshes' - *Meteoritics & Planetary Science*, vol. 36, Supplement, p.A124; 2001.
- Maté, G, *When the Body Says No: Understanding the Stress-Disease Connection,* 2011
- Matese, John J.; Whitmire, Daniel P., 'Persistent evidence of a jovian mass solar companion in the Oort cloud'. *Icarus,* 2011
- Matheus, P. & Jenniskens, M., *Meteor Showers and Their Parent Comets*, Cambridge University Press, 2006
- McCanney, J.M., *Atlantis To Tesla, The Kolbrin Connection*, Jmccanneyscience Press, 2003
  __________, *Planet X, Comets & Earth changes*, Jmccanneyscience Press, 2007
- McCarthy, D. *et al.*, 'The Physical Basis of the Leap Second', *Astronomical Journal,* vol. 136 (2008), pp. 1906–1908
- McKay, Brett & Kate, 'How to Survive a Plane Crash: 10 Tips That Could Save Your Life', *Manliness,* 30 July 2013
- McTaggart, L., *The Field: The Quest for the Secret Force of the Universe,* Harper Perennial, 2008
- Melott A. & Bambach R., 'Nemesis Reconsidered', *Monthly Notices of the Royal Astronomical Society Letters* 407
- Michael the Syrian, *Chroniques Universelles,* Académie des Inscriptions et Belles Lettres, 1905
- Mitchell, T., 'Arctic Oscillation time series', *JSAO.*
- Moore, Patrick, *Guide to the comets*, Lutterworth Press, 1977
- Mudur, G. 'Seed safety vault in Ladakh', *The Telegraph* (India), 16 May 2010.
- Muller, R., *Nemesis: The Death Star*, Wiedenfeld and Nicholson, 1988
- NASA, 'Atmospheric aerosols: what are they and why are they so important', August 1996.
  __________, 'Night Time Artificial Cloud Study Using NASA Sounding Rocket', 19 September 2009.
- Nash, C.B., 'General and comparative study of the psychokinetic effect on a fungus culture', *J. Barry Journal of Parapsychology*, 1968
- Natural Environment Research Council, 'Soluble rocks', *British Geological Survey.*
- Naudiet, A., 'Das Geheimnis der Präzession', *Efodon Synesis* n° 9, 17-23, 1995
- Nelson, B., 'Scientists baffled to discover that Venus' spin is slowing down', *MNN*, 17 February 2012
- Nelson, J., 'Major Fires Since 2001', *UX blog*, 13 July 2012.
- Nelson R. D. *et al.*, 'Global resonance of consciousness: Princess Diana and Mother Teresa' *Electronic Journal of parapsychology,* 1998
- Nelson R. D., 'Field REG anomalies in group situations', *Journal of Scientific Exploration,* 1996, 10(1):111-41
  __________, 'Terrorist disaster, September 11, 2001', *Global Consciousness Project*
- Newcomb, S., *Tables of the Four Inner Planets,* 2nd ed., 1898
- Newton, *Atmospheric circulation systems: their structure and physical interpretation*, 1969

- Niemitz Hans-Ulrich, *Did the early middle age really exist*, Dragos Iulian, 1997
- Nissen V., *Masters of Manipulation*, 2010
- Nodon, A, *Action électrique du Soleil*, Gauthier Villars, 1910
- North, G., *Astronomy in Depth*, Springer, 2002
- Oak Ridge National Laboratory, *Technical Basis for Regulatory Guidance on Lightning Protection in Nuclear Power Plants*, U.S. Nuclear Regulatory Commission, p.6
- Orndorff, R., 'The Science of Sinkholes', *WAMU*
- Ortega, X., 'More Unexplained Sounds, Tremors & Lights from the Skies', *ghosttheory.com*, 20 August 2011.
- Ouspensky, P., *Fragments of an unknown Teaching*, Harcourt, Brace and Company, 1940
- Ovid, *Metamorphoses*, Oxford University Press, 2009
- Pajola, M., *et al.*, (2013), 'Phobos as a d-type captured asteroid, spectral modeling from 0.25 to 4.0 μm', *The Astrophysical Journal*, vol. 777, no. 2.
- Panel on Atmospheric Effects of Aviation, *Atmospheric Effects of Aviation: A Review of NASA's Subsonic Assessment Project*
- Pawar, S., 'Anomalous electric field changes and high flash rate beneath a thunderstorm in northeast India', *Journal of Earth System Science*, October 2010, vol. 119, no. 5, pp. 617-625
- Peabody, A. W., 'Considerations of telluric current effects on pipeline, in Solar System', *Plasma Physics*, pp. 349-352.
- Pearce, F., 'Is weird winter weather related to climate change?' *Yale Environment 360*, 24 Feb 2014.
- Pearson, M. & Zarrella, J. 'A loud crash, then nothing: Sinkhole swallows Florida man', *CNN*, 5 March 2013.
- Peiser, B. J., 'Catch a falling star', *Jewish Chronicle*, 6 March 1998
  ______, 'collapse of early bronze age civilizations: has the smoking gun been found?', *Cambridge Conference Correspondence*
- Peratt, A. L., 'Advances in Numerical Modeling of Astrophysical and Space Plasmas', *Astrophysics and Space Science,* 1996, vol. 242, no. 1-2, pp. 93-163
- Pérez, J., *The Spanish Inquisition: A History*, Profile Books, 2006
- Pesnell, Dean, 'You would expect it to get brighter and brighter but unfortunately it got dimmer and dimmer as it got close to the sun', *The Globe And Mail*
- Petric, Boris, 'The land rush - Transnational strategies for land grabbing', *Revues transcontinentales*, 10 November 2011
- Philips, T., 'Meteor Smoke Makes Strange Clouds', *NASA*, 7 August 2012
- Pinker, R.T., *et al.*, 'Do Satellites Detect Trends in Surface Solar Radiation?', *Science*, 2005, vol. 380, 2005
- Plotkin, V., 'Earth Charge Produced By Thunderstorms', *Russian Geology and Geophysics*, vol. 44, no. 3, pp. 260-267, 2003
- Pogge, R., *The moons of Jupiter*, Ohio State University, 17 November 2007
- Potter, M., 'History, III The Renaissance', *Lectures on the Harvard Classics.* The Harvard Classics, 1909–14
- Prescott, George B., *History, Theory and Practice of the Electric Telegraph*, 1866
- Pribram, K., *Brain and Perception*
- Procopius, *The Secret History*, translated by Richard Atwater, 1927
- Prölss, Gerd, *Physics of the Earth's Space Environment: An Introduction*, Springer, 2004
- Purdey, M., 'Chronic barium intoxication disrupts sulphated proteoglycan synthesis: a hypothesis for the origins of multiple sclerosis', *Medical Hypotheses,* 2004, 62(5):746-54
- Quénot, K., *Cathares. Le massacre oublié,* Hugo et Comapgnie, 2012
- Quinn, J., 12 October 2012, 'Meteorite Impacts Earth in Minden, Louisiana - Media and Government Cover It Up', *Signs of the Times.*
  ______, 'Strange Noises in the Sky: Trumpets of the Apocalypse?', *Signs Of The Times. 29* January 2012

- Radin, D. I., & Ferrari, D. C., 'Effects of consciousness on the fall of dice: A meta-analysis', *Journal of Scientific Exploration*, 5(3), 61-84
- Radin, D. I., *Lunar Correlates Of Normal, Abnormal And Anomalous Human Behavior,* Subtle Energies, 1994, vol. 5, no. 3
  ______, *The Conscious Universe*, Harper One, 2009
- Rajeev Syal, 'Meteor showers blotted out man's first civilisations', *The Sunday Times*, 14 December 1997
- Ramanathan V. *et al.*, 'Cloud-Radiative Forcing and Climate: Results from the Earth Radiation Budget Experiment', *Science*, vol. 243, No. 4887. (Jan. 6, 1989)
- Randerson, J. 'Sea change: why global warming could leave Britain feeling the cold', *The Guardian*, 27 October 2004
- Rankin, I., *This Erratic Planet: What Happens When the Earth Changes Its Axis of Rotation*
- Raup, D. & Sepkoski, J., 'Mass extinctions in the marine fossil record', *Science*, vol. 215, no. 4539
- Ravaud M., 'Alain Aspect: Shedding new light on light and atoms', *CNRS International Magazine*.
- Redd, Nola Taylor, *What is a Wormhole,* Space.com.
- Renter, E., 'Another Study Finds GMO Compounds in 100% of Pregnant Women and Fetuses', *Activist Post*, 27 October 2012
- Reynolds, J., *An Introduction to Applied and Environmental Geophysics*
- Rhine, L. E., & Rhine, J. B. (1943). 'The psychokinetic effect. The first experiment'. *Journal of Parapsychology*, 7, 20–43.
- Rigby, E. *et al.*, 'A comet impact in AD 536?', *A&G* (2004) 45 (1): 1.23-1.26
- Roble, R., *The Earth's Electrical Environment, National Academies*, 1986
- Roth, A., 'What's behind mysterious booms? Phenomena produce theories, but no answers' *San Diego Union-Tribune*. 23 April 2006.
- Runcorn, S., 'The Earth's magnetism', *Scient Am*, 1955, pp. 152-162
- Russell C.T. & Vaisberg O. 'The interaction of the solar wind with Venus', *Venus*, pp. 873-940, 1983
- Russell III, J.M., 'Analysis of low latitude Noctilucent Cloud occurrences using satellite data and modeling', *Aeronomy of Ice in the Mesosphere,* March 2013
- Saklani, A., 'Psychokinetic effects on plant growth', *Research in Parapsychology*, 1989, 1990 37-41
- Sample, I., 'Scientists reveal the full power of the Chelyabinsk meteor explosion', *The Guardian,* 7 November 2013
- Sarovar *et al.*, `Quantum entanglement in photosynthetic light-harvesting complexes', *Nature Physics*, 2010, 6:462–467
- Schilling, A. *et al.*, 'Superconductivity in the Hg-Ba-Ca-Cu-O system', *Nature*, 1993
- Schmeidler, G. R., & Murphy, G., 'The influence of belief and disbelief in ESP upon individual scoring level', *Journal of Experimental Psychology*, 1946, 36 (3): 271–276
- Schmidt, H., 'Mental Influence on Random Events', *New Scientist and Science Journal*, June 1971, 757-768.
  ______, 'PK effect on pre-recorded targets', *J Am Soc Psychical Research,* 1976, 70(3):267-291
- Schmüser, P. *The Strange Features of Quantum Mechanics in the Light of Modern Experiments,* pp. 24-37
- Schneider, H. S. *et al.*, *Encyclopedia of Climate and Weather*
- Schneider, H. S., *Encyclopedia of Climate and Weather*, Oxford University Press, 1996
- Schwabl, H. *et al.*, 'Spontaneous ultraweak photon emission from biological systems and the endogenous light field'. *Forsch Komplementarmed Klass Naturheilkd*, April 2005 (2):84-9
- Schwartz, P. and Randall, D. *An Abrupt Climate Change Scenario and Its Implications for United States National Security,* October 2003
- Schwartz, S., *The Secret Vault of Time: Psychic Archaeology and the Quest for Man's Beginnings*, Grosset & Dunlap, 1978
- Schwarz Domink, 'Is the low-l microwave background cosmic?', *Astrophysics Journal*, November 24, 2004
- Scott, D.E., 'Electric cosmology – Stellar Evolution', *The Electric Sky*, online version.
  ______, *The Electric Sky,* Mikamar, 2006

- Scott, Susan & Duncan, Christopher, *Return of the Black Death,* Wiley, 2004
- Scuderi L., 'A 2000-Year Tree Ring Record of Annual Temperatures in the Sierra Nevada Mountains', *Science,* 5 March 1993: vol. 259 no. 5100, pp. 1433-1436
- Seager, R., 'Climate mythology: The Gulf Stream, European climate and Abrupt Change', *Lamont-Doherty Earth Observatory of Columbia University*
- Sédir, *Initiations*, 1920
  ______, Letters sent to the *Bulletin des Amitiés Spirituelles*, 1919
- Segalstad, T., 'Correct Timing is Everything - Also for CO2 in the Air', *CO2 Science*, vol. 12, no. 31: 5 August 2009.
- Segura, Gabriela, *New Light on the Black Death: The Viral and Cosmic Connection*; The Dot Connector Magazine, no. 13, vol. 1/2011.
- Shah, D., 'South Korea develops Electromagnetic Pulse Bomb', *fareastgizoms*, 6 September 2011.
- Shamah, D., 'An electromagnetic pulse attack — the 'other' Iranian nuclear threat', *Times of Israel*, 14 October 2012
- Sharp, T., 'What is Earth Made Of?', *Space*, 26 September 2012.
- Sheldrake R., *Morphic resonance: The Nature of Formative Causation,* Park Street Press, 2009
- Sheppard, Scott S., 'The Jupiter Satellite and Moon Page', *Scott Sheppard's homepage*.
- Shiller B., *Origin of Life: The 5 Option*, Trafford (2004), pp. 305-315
- Short, T., *A General Chronological History of the Air, Weather, Seasons, Meteors, Etc.*, 1749, London
- Silagadze, Z.K., 'Tunguska genetic anomaly and electrophonic meteors', *Astronomy & Astrophysics,* 2008
- Simpson, R. & Hope, J., 'Atlantic hurricane season 1971', *NOAA*, 1971
- Sincell, Mark, 'Hubble Snaps Martian Hurricane', *Science*, 19 May 1999
- Solanki, S K, Usoskin, I G, Kromer, B, Schussler, M and Beer, J, 'Unusual activity of the Sun during recent decades compared to the previous 11,000 years', *Nature* 431 1084–7
- Solfvin, G. F., 'Psi expectancy effects in psychic healing studies with malarial mice', 1982, *European Journal of Parapsychology*
- Solfvin J., *Mental Healing, in Advances in Parapsychological Research 4*, McFarland and Company, Jefferson, NC, 1984, pp. 31-63
- Sommerville, J., 'The Divine Right of Kings', Wisconsin University.
- Soon, W. & Yaskell, S., *Maunder Minimum And the Variable Sun-Earth Connection,* WSPCI, 2004
- Stanhill, G. & Cohen, S. 'Global dimming: a review of the evidence for a widespread and significant reduction in global radiation with discussion of its probable causes and possible agricultural consequences', *Agricultural and Forest Meteorology*, vol. 107, no. 4, 19 April 2001
- Stathakopoulos, D. *Famine and Pestilence in the Late Roman and Early Byzantine Empire: A Systematic Survey of Subsistence Crises and Epidemics*, Ashgate, 2004
- Steiger, B. & Steiger, S., *Conspiracies and Secret Societies: The Complete Dossier*, Visible Ink Press, 2012
- Stone, R., 'Preparing for Doomsday', *Science*, 7 March 2008
- Stonier, T., *Information and the Internal Structure of the Universe*, Springer, 2013
- Svensmark, *et al.*, 'Experimental evidence for the role of ions in particle nucleation under atmospheric conditions', *Proc. R. Soc. A*, February 2007
- Svensmark H & Calder N., *The Chilling Star*, 2007
- Svensmark, H *et al.*, 'Variation of cosmic ray flux and global cloud coverage: A missing link in solar climate relationships'. *J. Atmos. Solar-Terr. Phys.,* 59, 1225– 1232.
- Swanson B. 'Wind and Jet Stream', *USA Today*. June 27, 2007
- Takeuchi, K. & Chavoshian, A. *Quick Report of Great Eastern Japan Earthquake and Tsunami*

- Talbot, M. *The Holographic Universe,* Harper Perennial, 2011
- Tauxe, L., *Paleomagnetic Principles and Practice*, Modern Approaches in Geophysics, vol. 18, 1998.
- Taylor, J., 'Mars Is Warming, NASA Scientists Report', *The Heartland Institute*, 1 November 2005
- Taylor, K., 'Rapid climate change', *American Scientist*, vol. 87, no. 4, July-August 1999
- Thompson, A., 'Earth's Upper Atmosphere Cooling Dramatically', *Space*, 17 December 2009.
- Thompson, W. *The Historical Evolution of World-Systems*, p.22-35
- Thomson, W. (1862) 'On the age of the sun's heat', *Macmillan's Mag.* 5, 288–93; PL, 1, 394–68
- Thornhill, W. & Talbott, D., *The Electric Universe*, Mikamar Publishing, 2007
- Thornhill, W., 'Newton's Electric Clockwork Solar System', *Holoscience*
- Tomkin, J., 'What gas lies behind greenhouse effect?', *Phys. Today*, December 1992, pp. 13 - 15.
- Toomer, G. J., *Ptolemy's Alemagest*, Princeton University Press
- Toupin, L. P., 'Riding the electromagnetic wave', *Houghton Mifflin Science*
- Tracy, L. *Torture and Brutality in Medieval Literature: Negotiations of National Identity*
- Tressoldi P. *et al.*, 'Extrasensory Perception and Quantum Models of Cognition' *NeuroQuantology*, 2010; 4: S81-87
- Trower, W.P., '*Luis Walter Alvarez – A biographical memoir'*, 2009
- Tsytovich, V.N., *Elementary Physics of Complex Plasmas*, Springer, 2008
- Tuttle, Robert J., *The Fourth Source: Effects of Natural Nuclear Reactors*, Universal Publishers, 2012
- Tyerman, C., *God's War: A New History of the Crusades*, Penguin, 2007
- Ulanowsky, Z. *et al.*, 'Alignment of atmospheric mineral dust due to electric field', *Atmos. Chem. Phys.*, 7, 6161–6173, 2007
- Ullman, F., *Encyclopedia of Industrial Chemistry*, Wiley-VCH, 2011
- Ulmer, Gregory L., *The Florida Landscape: Revisited*, The Polk Museum, 1992
- URS, 'Évaluation de l'impact environnemental et social de l'usine d'engrais de Port-Gentil', *EIES*, p. 94
- Usoskin, I.G., *et al.*, 'Cosmic ray-induced ionization in the atmosphere: spatial and temporal changes', *Journal of Atmospheric and Solar-Terrestrial Physics*, vol. 66, no. 18, December 2004, Pages 1791–1796
- Valet, J.-P., Meynadier, L. & Guyodo, Y. 'Geomagnetic field strength and reversal rate over the past two million years', *Nature*, vol. 435, 802-805 (2005)
- Van der Sluijs, R., 'Venus' Tail of the Unexpected', *Thunderbolts*
- Van Dyke, D. 'The CFC Ban: Global Warming's Pilot Episode', February 4, 2010, *americanthinker*
- Vaughan A., 'England and Wales hit by wettest winter in nearly 250 years', *The Guardian,* 27 February 2014.
- Vedral, V., Quantifying entanglement in macroscopic systems, *Nature*, vol. 453, 19 June 2008
- Villiger, M., 'The Arctic – Our Global Thermostat', *Scientific American,* 15 June 2004.
- Vinkovic D. *et al.*, 'Global Electrophonic Fireball Survey: a review of witness reports, *WGN, the Journal of the IMO*, 11 November 2002
- Wang C. *et al.*, 'Spontaneous and visible light-induced ultraweak photon emission from rat eyes', *Brain Research*, 19 January 2011;1369:1-9
- Ward-Perkins, Bryan, *The fall of Rome: And the End of Civilization,* Oxford University Press, 2007
- Warlow, P., 'Does pole-flipping account for earth magnetism?', *New Scientist*, 9 November 1978, p.1224
- Wei, J., 'A one world seed bank vault', *Jerilee Wei Blog*
- Weltecke, D., The World Chronicle by Patriarch Michael the Great (1126-1199): Some reflections, *Journal of Assyrian Academic Studies.*
- Werthmann, M., 16 October 2012. 'Thousands reported bang, rumble of Maine quake in Massachusetts', *Boston.*

- Wheeler, J.A., *Superspace and Quantum Geometrodynamics*
- White Jr., Lynn, *Die mittelalterliche Technik und der Wandel der Gesellschaft.* München, 1968
- Whitmire, Daniel P. & Jackson, Albert A., 'Are periodic mass extinctions driven by a distant solar companion?', *Nature* 308, 713–715 (19 April 1984)
- WHO Regional Office for Europe, *Air Quality Guidelines, Second Edition*, Copenhagen, Denmark, 2000
- Wickramasinghe, C., et al 'Fossil Diatoms In A New Carbonaceous Meteorite', *Journal of Cosmology*, vol. 22, 2013
- Wild, Martin; et al (2009-05-16). 'Global dimming and brightening: An update beyond 2000'. *Journal of Geophysical Research Atmospheres*
- Wittke, J., *et al.* 'Evidence for deposition of 10 million tons of impact spherules across four continents 12,800 y ago', *PNAS*, 2013
- Wogan, Tim, 'Coherent Schrödinger's cat still confounds', *IOP*, 23 November 2011)
- Zeilinger, B., 'Young's experiment and the finiteness of information', *Philosophical Transactions of the Royal Society,* 360: 1061–1069 (2002)
- Ziegler, P., *The Black Death,* Harper Perennial Modern Classics, 2009
- Zillmer, R., *Irrtümer der Erdgeschichte*, Langen Müller, 2001
- Zuang, H.C., *et al.*, 'The influence of the interplanetary magnetic field and thermal pressure on the position and shape of the magnetopause', *Journal of Geophysical Research, (*1981) vol. 86, n. A12, pp. 10 009-100021

Index

Other books and DVDs from Red Pill Press

“From Paul To Mark”

PaleoChristianity

Laura Knight-Jadczyk

Nearly two thousand years ago the seeds of a new religion were sown in the eastern fringes of the Roman empire. An apostle named Paul wrote letters to his small congregations offering support, rebukes, and the outline of the gospel that would come to be known as Christianity. In the decades after came the Gospel of Mark, followed by more letters and more Gospels, controversies and debates, factions and infighting, until finally, Christianity became an empire.

But what if nearly everything you thought you knew about early Christianity was wrong? When read without preconceptions, the available contemporary sources tell a very different story, filled with ‘colorful’ characters, hardened revolutionaries, political maneuvering, and ideological conflict. In this groundbreaking study, Laura Knight-Jadczyk strips away centuries of assumptions and dogma to reexamine the fundamentals of what we can truly know of the early Christians, how we know it, and how that changes our picture of what was really happening in first-century Judea.

Why are there no historical references to Jesus and Christianity until decades after the events of the Gospels were supposed to have occurred? Why do the first non-Christian historians who mention Jesus seem dependent on the Gospels? Why does Paul make no unambiguous references to the Gospels’ Jesus of Nazareth? What was Paul talking about? Laura Knight-Jadczyk’s answers to these questions are revolutionary. After reading this book, you’ll never see the origins of Christianity the same way again.

“What will happen to you if you read this book? I’ll be glad to tell you. Your paradigm will begin to shift, perhaps only gradually at first. Your assumptions, even your axioms, will be challenged, and this time you will no longer be able to nervously default to the familiar. And all this will happen because you will be seeing the emergence of an exciting new stage of biblical criticism. Laura Knight-Jadczyk has here synthesized the work of a new generation of scholars who are not afraid to venture beyond convention and consensus. She has shown that the work of Wells, Doherty, Doughty, Carrier, Detering, Pervo, and myself are not merely isolated fireworks displays but rather gleams of a new, rising dawn. And in that light she presses on to her own striking advances. Won’t you join her?”—Robert M. Price, host of The Bible Geek podcast, author of *Jesus Christ Superstition* and *The Amazing Colossal Apostle.*

“Quite a delight, well written, well researched.”—Russell Gmirkin, author of *Plato and the Creation of the Hebrew Bible* and *Berossus and Genesis, Manetho and Exodus*

ISBN 978-1-7349074-1-4

"Cometary Encounters"

Flash-Frozen Mammoths, Mars–Earth Discharge, Comet Venus and the 3,600-Year Cometary Cycle

Pierre Lescaudron

What flash froze the woolly mammoths in a matter of minutes? Was Earth inundated by water from Mars? What series of catastrophic events caused the advent of the Younger Dryas? Is there evidence to support the existence of an ongoing 3,600 year Earth-threatening cometary cycle? Was Venus a comet that made several destructive earthly flybys? These are just some of the questions explored in this book that takes the reader on a fascinating and hair-raising journey through time and space that marshals abundant evidence to show that the official view of the history of our planet, on which mainstream science obtusely insists, is deeply problematic. Rather than a uniformitarian evolutionary process, we are confronted with a traumatic picture of relatively frequent global cosmic catastrophes that profoundly altered our planet's geologic and social landscape. Of all the alarming data presented by the author, undoubtedly the most pertinent and pressing is that these events cannot and should not be consigned to the past,. but viewed as an urgent wake-up call about our possible future.

ISBN 978-1-7349074-2-1

"The Secret History of the World", Book 1

... and how to get out alive

Laura Knight-Jadczyk

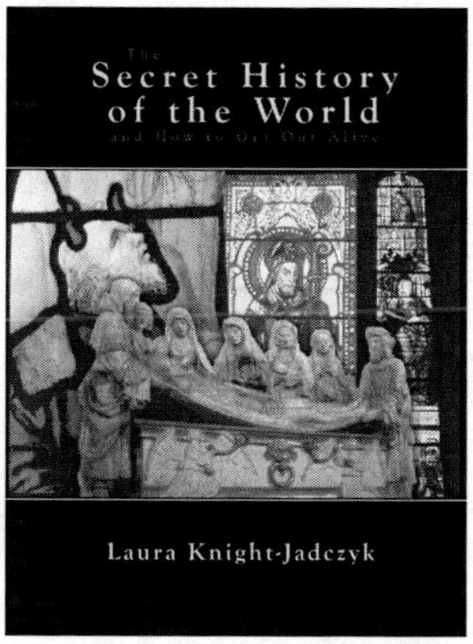

If you heard the Truth, would you believe it? Ancient civilisations. Hyperdimensional realities. DNA changes. Bible conspiracies. What are the realities? What is disinformation?

The Secret History of The World and How To Get Out Alive is the definitive book of the real answers where Truth is more fantastic than fiction. Laura Knight-Jadczyk, wife of internationally known theoretical physicist, Arkadiusz Jadczyk, an expert in hyperdimensional physics, draws on science and mysticism to pierce the veil of reality. Due to the many threats on her life from agents and agencies known and unknown, Laura left the United States to live in France, where she is working closely with Patrick Rivière, student of Eugene Canseliet, the only disciple of the legendary alchemist Fulcanelli.

With sparkling humour and wisdom, she picks up where Fulcanelli left off, sharing over thirty years of research to reveal, for the first time, The Great Work and the esoteric Science of the Ancients in terms accessible to scholar and layperson alike.

Conspiracies have existed since the time of Cain and Abel. Facts of history have been altered to support the illusion. The question today is whether a sufficient number of people will see through the deceptions, thus creating a counter-force for positive change - the gold of humanity - during the upcoming times of Macro-Cosmic Quantum Shift. Laura argues convincingly, based on the revelations of the deepest of esoteric secrets, that the present is a time of potential transition, an extraordinary opportunity for individual and collective renewal: a quantum shift of awareness and perception which could see the birth of true creativity in the fields of science, art and spirituality. *The Secret History of the World* allows us to redefine our interpretation of the universe, history, and culture and to thereby navigate a path through this darkness. In this way, Laura Knight-Jadczyk shows us how we may extend the possibilities for all our different futures in literal terms.

With over 850 pages of fascinating reading, *The Secret History of The World and How to Get Out Alive* is rapidly being acknowledged as a classic with profound implications for the destiny of the human race. With painstakingly researched facts and figures, the author overturns long-held conventional ideas on religion, philosophy, Grail legends, science, and alchemy, presenting a cohesive narrative pointing to the existence of an ancient techno-spirituality of the Golden Age which included a mastery of space and time: the Holy Grail, the Philosopher's Stone, the True Process of Ascension. Laura provides the evidence for the advanced level of scientific and metaphysical wisdom possessed by the greatest of lost ancient civilizations - a culture so advanced that none of the trappings of civilization as we know it were needed, explaining why there is no 'evidence' of civilization as we know it left to testify to its existence. The author's consummate synthesis reveals the Message in a Bottle reserved for humanity, including the Cosmology and Mysticism of mankind Before the Fall when, as the ancient texts tell us, man walked and talked with the gods. Laura shows us that the upcoming shift is that point in the vast cosmological cycle when mankind - or at least a portion of mankind - has the opportunity to regain his standing as The Child of the King in the Golden Age.

If ever there was a book that can answer the questions of those who are seeking Truth in the spiritual wilderness of this world, then surely *The Secret History of the World and How to Get Out Alive* is it.

ISBN 9781897244166

Comets and the Horns of Moses

"The Secret History of the World", Book 2

Laura Knight-Jadczyk

The Laura Knight-Jadczyk's series, The Secret History of the World, is one of the most ambitious projects ever undertaken to provide a cogent, comprehensive account of humanity's true history and place in the cosmos. Following the great unifying vision of the Stoic Posidonius, Laura weaves together the study of history, mythology, religion, psychology and physics, revealing a view of the world that is both rational and breathtaking in its all-encompassing scope. This second volume, Comets and the Horns of Moses, (written in concert with several following volumes soon to be released) picks up the dangling threads of volume one with an analysis of the Biblical character of Moses – his possible true history and nature – and the cyclical nature of cosmic catastrophes in Earth's history.

Laura skillfully tracks the science of comets, revealing evidence for the fundamentally electrical and electromagnetic nature of these celestial bodies and how they have repeatedly wreaked havoc and destruction on our planet over the course of human history. Even more startling however, is the evidence that comets and cometary fragments have played a central role in the formation of human myth and legend and the very concept of a 'god'. As she expertly navigates her way through the labyrinth of history, Laura uncovers the secret knowledge of comets that has been hidden in the great myths, ancient astronomy (and astrology) and the works of the Greek philosophers. Concluding with a look at the political and psychological implications of cyclical cometary catastrophes and what they portend for humanity today, Comets and the Horns of Moses is a marvel of original thought and keen detective work that will rock the foundations of your understanding of the world you live in, and no doubt ruffle the feathers of the many academics who still cling to an outdated and blinkered view of history.

ISBN: 9781897244838

The Wave 1 – "Riding the Wave"

... The Truth and Lies about 2012 and Global Transformation

Laura Knight-Jadczyk

As 2012 fast approaches, opinions about what to expect on this much-anticipated date are sharply polarized. Will humanity experience a global, spiritual transformation? Cataclysmic Earth Changes? Or both? Or nothing? If Earth and its inhabitants are scheduled for some life-changing or life-ending event, we should ask ourselves what we know and how we know it, and how to prepare for our future.

Drawing on decades of research into history, religion, and the esoteric, Laura Knight-Jadczyk introduces the concept of "the Wave" to describe the possible phenomena behind all the hype surrounding global transformation. *Riding the Wave* not only collects the most probable scenarios we may face in the near future – it provides the context to make it all intelligible.

With roots in the science of hyperdimensions made popular by physicist Michio Kaku and the Fortean theories of the late John Keel, *Riding the Wave* suggests that many of the noticeable changes to our world in the last century are symptoms of the approaching Wave. From climate change, extreme population growth and technological development, as well as novel social and political movements, to the advent of UFO sightings, crop circles, and a variety of otherworldly experiences, something is up on the Big Blue Marble, and it all seems to be leading to a sea change in the way we see and interact with the world. The only question is, will it be for the better or the worse?

An intimate blend of science and mysticism, this volume of Laura Knight-Jadczyk's Wave Series initiates the process of unveiling the truth about life on Earth, and the man behind the curtain...

ISBN: 9781897244500

The Wave 2 – "Soul Hackers"

... The Hidden Hands Behind the New Age Movement

Laura Knight-Jadczyk

Why are we here? Why do we suffer? If this is an infinite school, what are we here to learn? And why do our efforts at "fixing" our lives often do exactly the opposite? As mystic and researcher Laura Knight-Jadczyk writes in this volume of her expansive Wave Series: "when you ask a question – if the question is a burning one – your life becomes the answer. All of your experiences and interactions and so forth shape themselves around the core of the answer that you are seeking in your soul. In [my] case, the question was: 'How to be One with God,' and the answer was, 'Love is the answer, but you have to have knowledge to know what Love really is.'"

Soul Hackers is a deeply personal and insightful account of this very process – of burning questions and transformative answers. Through the story of her own struggle with mainstream and alternative religion and the solutions they claim to offer, Knight-Jadczyk lays bare the problems inherent in the New Age movement as a whole – from Reiki, Wicca, and the phenomenon of channeling, to the very real problems of spirit attachments, mind control, and otherworldly predators posing as benevolent beings. She asks what it really means to "create your own reality." Is it merely self-hypnosis, or is something more hidden in this New Age truism?

The answers lie in the very nature of the Wave – the cosmic force and fabric of personal and collective evolution. For anyone wishing to understand the deeper meaning and reality of the human experience, and what our very near future may very well have in store for us, *Soul Hackers* provides a map to our symbolic reality and the knowledge necessary to weather the approaching storm.

ISBN: 9781897244517

The Wave 3 – "Stripped to the Bone"

... The Path to Freedom in the Prison of Life

Laura Knight-Jadczyk

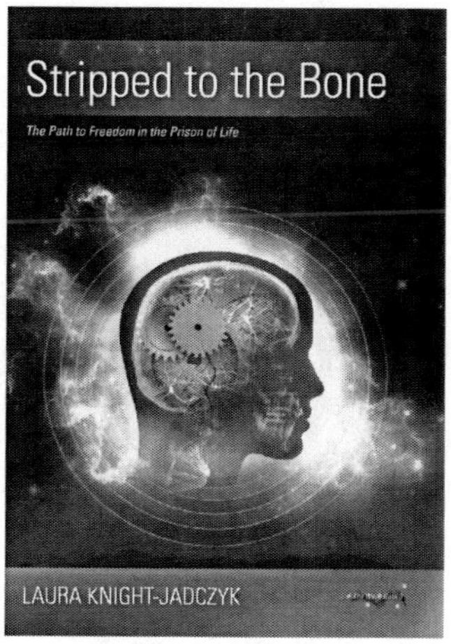

Media propaganda. Official cover-ups. Dishonest science. "Non-lethal" weaponry. Mind control technology. Racial stereotypes. Social engineering. Religious programming. The cold pursuit of profit. And the unrelenting pull of materialism... In a world where "freedom" is exported at the barrel of a gun, true freedom seems more like a distant fairytale, blocked for us in more ways than we can imagine.

In *Stripped to the Bone*, author Laura Knight-Jadczyk lays bare the forces seeking to keep humanity in a prison of its own creation. She lucidly describes evil's place in the cosmos, from the dark world of political conspiracy and government mind control to the reality behind the UFO phenomenon. But in response to the grim state of affairs on the Big Blue Marble, she also asks: Is there a solution? What can we learn from those who came before us? *Stripped to the Bone* suggests that this knowledge was not only known and widely practiced in humanity's prehistory, but that it can be rediscovered.

Through her extensive reading on all things esoteric, Knight-Jadczyk maintains that by knowing our limitations, we may overcome them. In this volume of her acclaimed Wave Series, she tears down our illusions about freedom and the idea that it can be won in any war. Rather, the path to freedom is an inner battle against the many limitations placed on our ability to choose by official culture, our own beliefs, and the forces behind the reality of our everyday experience. By showing us our own limitations she also succeeds in presenting anew the real possibilities and true potential of a free humanity.

ISBN: 9781897244524

The Wave 4 – "Through A Glass Darkly"

... Hidden Masters, Secret Agendas and a Tradition Unveiled

Laura Knight-Jadczyk

Behind the surface of everyday life lie secrets that have been kept from the eyes of the humanity. In every field of knowledge, we seem to take a wrong turn, coming to conclusions that are diametrically opposed to the truth of the matter. It seems that true science, history, the purpose and aim of human life, our past and potential futures are all off limits to public consumption. How can this be the case, and can these truths come to be known?

In *Through a Glass Darkly*, Laura Knight-Jadczyk continues to make it clear that nothing is what it appears to be. From the stories stitched together to make up our own personal identities to the myths of history on which nations are founded, we live in a sea of lies and half-truths. Just as we lie to ourselves and each other about who we really are, often putting ourselves in the best light possible, there are those who manufacture, manipulate, and shape current and past events to suit their own vested interests. And the current events of today will become the history of tomorrow, erroneously shaping our notions of who we are as a people, just as those of the past have done before.

But behind this sorry state of affairs, the truth awaits discovery. In this fourth volume of her series *The Wave or Adventures with Cassiopaea*, Knight-Jadczyk follows the trail of the hidden masters of our planet, exposing the agenda behind the alleged secret society, the "Priory of Sion", and that mystery's connections with alchemy, Oak Island, and the Kabbalists of old. In the process she reveals aspects of the tradition kept under wraps by these very groups. By exposing the agendas and conspiracies of the elite, we can come to know the truth about ourselves, and why it is has been kept hidden.

ISBN: 9781897244562

The Wave 5 & 6 – "Petty Tyrants & Facing the Unknown"

... Navigating the Traps and Diversions of Life in the Matrix

Laura Knight-Jadczyk

From the myths of romance to the tales of the hero's journey, the quest for knowledge and being has always been portrayed in terms of struggle. Far from home, the hero faces obstacles and tests of his or her courage, will, and cunning. But how do the labyrinths and monsters of these 'messages in a bottle' from our remote ancestors relate to our lives in the 21st century? In an age of mass media, the worldwide web, and multinational corporations, how do these archetypal dramas play themselves out?

In these two volumes of her revolutionary series, *The Wave or Adventures with Cassiopaea*, Laura Knight-Jadczyk continues her project of laying bare the nature of our reality. Through her own experiences and interactions over the course of the Cassiopaean Experiment, many of which just go to show that truth is stranger than fiction, Laura describes the real-life dynamics only hinted at in myth. Most importantly, she gives the tools and clues necessary to actually read the symbols of reality: the theological substrate in which our ordinary psychological motivations are embedded.

With these stunning revelations, Shakespeare's famous words take on a whole new meaning: "All the world's a stage, and all men and women merely players."

First published on her groundbreaking website cassiopaea.org, *Petty Tyrants & Facing the Unknown* have now been fully revised and packaged together in one attractive volume. For anyone interested in the world of esoteric knowledge, studies in the paranormal and everything 'alternative', or even just curious about life in general and its possible significance and meaning, these volumes are a must-read.

ISBN: 9781897244630

The Wave 7 – "Almost Human"

... A Stunning Look at the Metaphysics of Evil

Laura Knight-Jadczyk

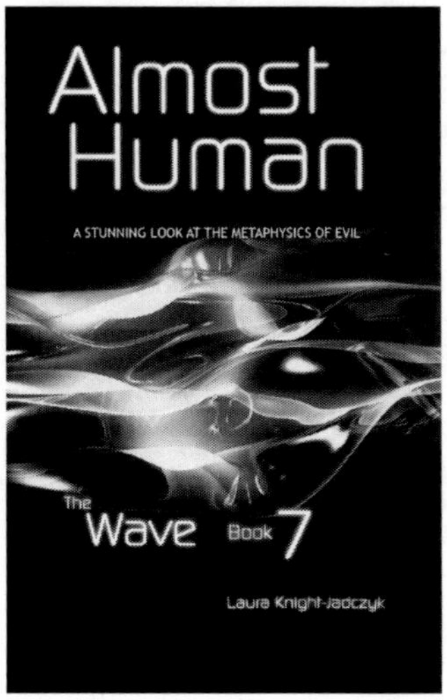

In this volume of her prescient Wave series, Laura Knight-Jadczyk brings order to the chaotic and labyrinthine world of murder, conspiracy, and the paranormal. In a unique and probing synthesis of science and mysticism she presents a detailed series of case studies and application of her hypothesis of hyperdimensional influence.

From interpersonal relationships and their expression of archetypal dramas to the vectoring of human behavior to achieve hyperdimensional purposes, *Almost Human* reveals the mechanics of evil, how it creeps into our lives, and what we need to be aware of in order to avoid it.

The case studies of John Nash, the schizoidal creator of Game Theory, and Ira Einhorn, the New Age psychopath who murdered his girlfriend, are the window through which Knight-Jadczyk unravels the intricate web of deception, aims, and counter-aims of the Powers That Be.

Almost Human is essential reading for anyone wondering why our world is becoming increasingly controlled and our freedoms more restricted.

ISBN: 9781897244463

The Wave 8 – "Debugging the Universe"

... The Hero's Journey

Laura Knight-Jadczyk

The Path of the Fool, the Hero's Journey, the Great Work – by whatever name it takes, the path of self-development and growth of knowledge is one fraught with difficult lessons and intense struggle. But what exactly is the nature of those lessons, and what insights can the latest advances in modern science provide for us along the way?

Debugging the Universe takes us into the heart of what it means to be human, from the molecules of our DNA to our life purpose and true place in the universe, and everything that separates us from embodying that higher potential. Explored within are real-life applications of the Hero's archetype, the relevance of neuroscience and the 'molecules of emotion', the hidden meaning behind the enigmatic symbols of esoterica, and what it means to live inside a complex system: the universal breath of chaos and order.

This volume concludes the publication in print of Laura Knight-Jadczyk's unparalleled and controversial magnum opus: *The Wave or Adventures with Cassiopaea.* Originally published online at www.cassiopaea.org, The Wave is a fully modern exposition of the knowledge of the ancients, with subjects ranging from metaphysics, science, cosmology, and psychology to the paranormal, UFOs, hyperdimensions and macrocosmic transformation.

ISBN: 9781897244654

"High Strangeness"

... Hyperdimensions and the Process of Alien Abduction

Laura Knight-Jadczyk

High Strangeness: Hyperdimensions and the Process of Alien Abduction is an enlightening attempt to weave together the contradictory threads of religion, science, history, alien abduction, and the true nature of political conspiracies. With thorough research and a drive for the truth, Laura Knight-Jadczyk strips away the facades of official culture and opens doors to understanding our reality.

The Second Edition includes additional material that explains the hyperdimensional mechanisms by which our reality is controlled and shaped by 'alien' powers. The self-serving actions of unwitting puppets – psychopaths and other pathological types – who may have no knowledge that they are being used, become the portals through which an agenda that is hostile to humanity as a whole, is pushed forward.

High Strangeness takes the study of ponerology into a whole new dimension!

ISBN: 9781897244340

“The Cassiopaea Experiment Transcripts”

Volume 1, 1994

Laura Knight-Jadczyk

The Cassiopaea Experiment is unique in the history of channeling, mediumship, and parapsychology. For years prior to the first Cassiopaean transmission, Laura Knight-Jadczyk went to great lengths to study the channeling phenomenon, including its history, its inherent strengths, weaknesses, dangers, and the various theories and methods developed in the past. After having exhausted the standard literature in search of answers to the fundamental problems of humanity, Laura and her colleagues (including her husband, mathematical physicist Arkadiusz Jadczyk) have held regular sittings for more than twenty years.

With the goal of applying true scientific standards and critical thinking, Laura began her experimentation with the spirit board, chosen for the optimum conditions of conscious feedback it offers. After two years of working through various levels of phenomena, including alleged discarnate entities and denizens of the ‘astral realms’, a new source came through in 1994. Unlike previous contacts, the Cassiopaeans came through strong and clear, often delivering full paragraphs’ worth of complex material at one to two letters per second, with a high density of content and impeccable orthography. After telling Laura, “we are you in the future,” the C’s have been covering a wide range of topics by transmitting more than one million letters over the last twenty years, answering questions from practically all fields of knowledge, including physics, mathematics, psychology, philosophy, parapsychology, esotericism, history, politics, health, and astronomy. Where the material can be verified, the C’s have proven to have an amazing track record, especially when compared to the variable quality of most examples of channeling from the past 150 years.

For the first time in print, this volume includes complete and extensively annotated transcripts of 36 experimental sessions conducted in 1994, beginning with the first contact with the Cassiopaeans on July 16. The sessions of this year introduced many of the themes that would recur in more detail over the next twenty years, including such topics as cyclical cometary bombardment of the Earth, the solar companion hypothesis, ancient history, metaphysics, the hyperdimensional nature of reality, and the possibility of evolution of humanity.

ISBN: 978-1-897244-99-9

"The Cassiopaea Experiment Transcripts"

Volume 2, 1995

Laura Knight-Jadczyk

The Cassiopaea Experiment is unique in the history of channeling, mediumship, and parapsychology. For years prior to the first Cassiopaean transmission, Laura Knight-Jadczyk went to great lengths to study the channeling phenomenon, including its history, its inherent strengths, weaknesses, dangers, and the various theories and methods developed in the past. After having exhausted the standard literature in search of answers to the fundamental problems of humanity, Laura and her colleagues (including her husband, mathematical physicist Arkadiusz Jadczyk) have held regular sittings for more than twenty years.

With the goal of applying true scientific standards and critical thinking, Laura began her experimentation with the spirit board, chosen for the optimum conditions of conscious feedback it offers. After two years of working through various levels of phenomena, including alleged discarnate entities and denizens of the 'astral realms', a new source came through in 1994. Unlike previous contacts, the Cassiopaeans came through strong and clear, often delivering full paragraphs' worth of complex material at one to two letters per second, with a high density of content and impeccable orthography. After telling Laura, "we are you in the future," the C's have been covering a wide range of topics by transmitting more than one million letters over the last twenty years, answering questions from practically all fields of knowledge, including physics, mathematics, psychology, philosophy, parapsychology, esotericism, history, politics, health, and astronomy. Where the material can be verified, the C's have proven to have an amazing track record, especially when compared to the variable quality of most examples of channeling from the past 150 years.

For the first time in print, this second volume includes complete transcripts of 51 experimental sessions conducted in 1995. Questions and answers have been annotated extensively, giving unprecedented insight into the background and interpersonal dynamics of the early Cassiopaea Experiment.

In this year, the dialog with the Cassiopaeans revolved around topics of 'New Age' and historical disinformation, reincarnation and past lives, karma, soul evolution and the purpose of humanity, the 'Golden Age' and ancient civilizations, weather and Earth changes, 'The Wave' and coming turmoil on our planet, the 'New World Order', hidden government and the all-encompassing control system, higher 'density' experience and the 'fluid' nature of the UFO and alien phenomenon, time and the nature of our universe. Also, interspersed with these topics, the sessions of this year document the conflicting interests arising within the group undertaking the Cassiopaea Experiment and the struggle against sometimes obvious skewing of the information and attempted sidetracking of the communications.

ISBN: 978-0-692484-51-7

"The Cassiopaea Experiment Transcripts"

Volume 3, 1996

Laura Knight-Jadczyk

The Cassiopaea Experiment is unique in the history of channeling, mediumship, and parapsychology. For years prior to the first Cassiopaean transmission, Laura Knight-Jadczyk went to great lengths to study the channeling phenomenon, including its history, its inherent strengths, weaknesses, dangers, and the various theories and methods developed in the past. After having exhausted the standard literature in search of answers to the fundamental problems of humanity, Laura and her colleagues (including her husband, mathematical physicist Arkadiusz Jadczyk) have held regular sittings for more than twenty years.

With the goal of applying true scientific standards and critical thinking, Laura began her experimentation with the spirit board, chosen for the optimum conditions of conscious feedback it offers. As Laura writes, "We take the approach of a sort of scientific mysticism where mystical claims are submitted to rational analysis and testing, and the required scientific proofs are modified to allow for the nature of evidence from theorized realms outside of our own where ordinary scientific proofs might not apply."

The dynamics and content of channeling are in general tightly entangled with current global events and the lives of the participants and their interpersonal relationships. For this reason, clarifying context has been added, and questions and answers have been annotated extensively, giving unprecedented insight into the background of the Cassiopaea Experiment.

In this year, the dialogue with the Cassiopaeans revolved around the further exploration of various theories that were promulgated at that time in books and the internet, the solar companion hypothesis, the theorized states of being above our own, hypnosis, mind programming, false prophecies, the rediscovery of a possible ancient technology utilized in the pre-history of mankind, quantum principles and possibilities for their macroscopic application, as well as the process of gaining more protection by assimilating and applying knowledge and networking with like-minded people.

For the first time in print, this volume includes complete transcripts of 44 experimental sessions conducted in 1996.

ISBN: 978-0-692725-73-3

"The Cassiopaea Experiment Transcripts"

Volume 4, January – May 1997

Laura Knight-Jadczyk

Following the various clues given by the Cassiopaeans, Knight-Jadczyk includes email correspondence documenting the early stages of her research leading to the publication of her book *The Secret History of the World*, as well as illuminating the rationale behind the questions and answers in this year – a year that demonstrates the material manifestations of how reality shifts begin.

For the first time in print, this volume includes complete transcripts of 17 experimental sessions conducted between January and May 1997.

The Cassiopaea Experiment is unique in the history of channeling, mediumship, and parapsychology. For years prior to the first Cassiopaean transmission, Laura Knight-Jadczyk went to great lengths to study the channeling phenomenon, including its history, its inherent strengths, weaknesses, dangers, and the various theories and methods developed in the past. After having exhausted the standard literature in search of answers to the fundamental problems of humanity, Laura and her colleagues (including her husband, mathematical physicist Arkadiusz Jadczyk) have held regular sittings for more than twenty years. With the goal of applying true scientific standards and critical thinking, Laura began her experimentation with the spirit board, chosen for the optimum conditions of conscious feedback it offers. As Laura writes, "We take the approach of a sort of scientific mysticism where mystical claims are submitted to rational analysis and testing, and the required scientific proofs are modified to allow for the nature of evidence from theorized realms outside of our own where ordinary scientific proofs might not apply."

ISBN: 978-1-548314-87-3

"9/11 – The Ultimate Truth"

Joe Quinn and Laura Knight-Jadczyk

In the years since the 9/11 attacks, dozens of books have sought to explore the truth behind the official version of events that day - yet to date, none of these publications has provided a satisfactory answer as to **why** the attacks occurred and who was ultimately responsible for carrying them out.

Taking a broad, millennia-long perspective, Laura Knight-Jadczyk's *9/11: The Ultimate Truth* uncovers the true nature of the ruling elite on our planet and presents new and ground-breaking insights into just how the 9/11 attacks played out.

9/11: The Ultimate Truth makes a strong case for the idea that September 11, 2001 marked the moment when our planet entered the final phase of a diabolical plan that has been many, many years in the making. It is a plan developed and nurtured by successive generations of ruthless individuals who relentlessly exploit the negative aspects of basic human nature to entrap humanity as a whole in endless wars and suffering in order to keep us confused and distracted to the reality of the man behind the curtain.

Drawing on historical and genealogical sources, Knight-Jadczyk eloquently links the 9/11 event to the modern-day Israeli-Palestinian conflict. She also cites the clear evidence that our planet undergoes periodic natural cataclysms, a cycle that has arguably brought humanity to the brink of destruction in the present day.

For its no nonsense style in cutting to the core of the issue and its sheer audacity in refusing to be swayed or distracted by the morass of disinformation that has been employed by the powers that be to cover their tracks, *9/11: The Ultimate Truth* can rightly claim to be **the** definitive book on 9/11 - and what that fateful day's true implications are for the future of mankind.

The new Second Edition of *9/11: The Ultimate Truth* has been updated with new material detailing the real reasons for the collapse of the World Trade Center towers, the central role played by agents of the state of Israel in the attacks, and how the arrogant Bush government is now forced to dance to the Zionists' tune.

ISBN: 9781897244876

JFK: The Assassination of America

Laura Knight-Jadczyk

Anyone who has taken the time to study the facts about that fateful day in Dallas, TX, November 22, 1963, will already know that John F. Kennedy was deliberately murdered by a cabal of psychopathic warmongers who were opposed to his plans for a more peaceful world. This ebook written by Laura Knight-Jadczyk brings into focus how the convergence of greed and the power-mad forces of big oil, organized crime, and the military-industrial complex brought about the destruction of JFK. Drawing on an early analysis of Kennedy's assassination, *Farewell America*, which was produced by a French intelligence group, Mrs. Knight-Jadczyk brings a deeper understanding of this tragic event by placing it in the light of the psychopathic motivations of these criminal elements. *JFK: The Assassination of America* shows a world that could have been, and a great man silenced by forces who will stop at nothing to keep that world from becoming a reality.

ISBN: 9781897244883 (Amazon Kindle only)

Amazing Grace

Laura Knight-Jadczyk

Laura Knight-Jadczyk has lived intimately – and mysteriously – with the world of spirit. In *Amazing Grace*, Laura takes us back to her beginnings in a Gulf Coast Florida childhood, mapping the first decades of her extraordinary search for an objective reality of spirit, of the play of forces that exist as a subtext to the lives of all human beings – a journey toward knowledge and understanding.

From her first experiences with a terrifying Face at the Window in childhood, to her work as an exorcist, chronicled by Pulitzer Prize–winning journalist Thomas French in the *St. Petersburg Times*, Laura relates the many experiences in her search for the existence of truth about our reality, which forced her to recognize the validity of perceptions beyond those of materialism.

This is also the story of how the Cassiopaeans came to be a part of her life. Their channeled messages, which include important concepts of physics and the underlying nature of reality, have drawn the attention of intellectually advanced yet spiritually hungry people from all over the world. This is not just the story of one woman's experience with personal quantum jumps from one reality to another, but is also the greater story of the potential that exists in every seeker. We have the potential to discover the genuine existence of spirit and the play of the archetypal forces of the world, and to connect with them in a dynamic way. Amazing Grace, or Quantum Future, can be a reality in our lives.

ISBN: 9781897244814

The Apocalypse: Comets, Asteroids and Cyclical Catastrophes

Laura Knight-Jadczyk

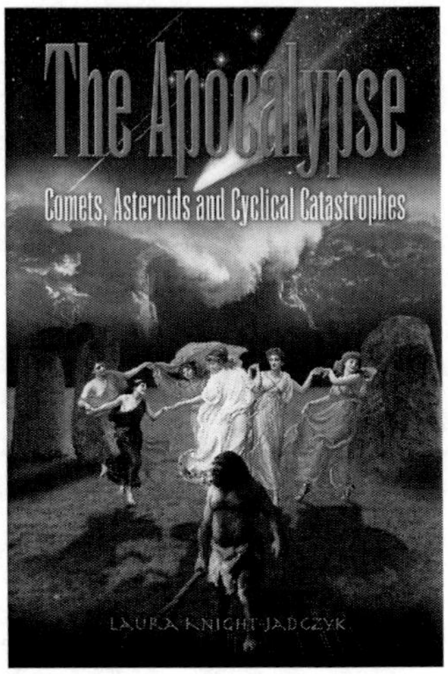

For untold millennia, comets and asteroids have struck fear into the hearts of humankind. Their stark radiance was observed everywhere with a sense of impending doom, interpreted as signs of the gods' judgment, omens of plague, mass destruction and the end of time. Astronomers recorded their appearance the world over, building large scale observatories to track their movements and predict their ominous arrival. What was it about these majestic wonders of the heavens that inspired such dread? Was it simply a product of mere superstition and social hysteria?

The latest scientific analysis and historical analysis strongly suggest otherwise. Our ancestors knew something we have since forgotten, their secrets deeply embedded in the archaeological record and the myths passed on throughout generations. And we have only begun to unravel their mysteries ...

Spurred on by the discovery of a little known letter of warning to the European Office of Aerospace Research and Development by astrophysicist Victor Clube, author Laura Knight-Jadczyk began an in-depth research project to get to the bottom of the very real threat to humanity posed by these celestial visitors. In *The Apocalypse: Comets, Asteroids, and Cyclical Catastrophes*, Knight-Jadczyk shares what she found: historical evidence for mass destructions, comet-borne plagues, and repeated cover ups littering our past, as well as clues that a similar fate may be fast approaching.

ISBN: 9781897244616

The Noah Syndrome

Laura Knight-Jadczyk

"As it was in the days of Noah ..."

Technological progress married with moral decay. A people enraptured by the trivial and superficial, entrenched in a culture of materialism and endless warfare. A civilization whose time has come. If a phrase defines the condition of our era, it is this: The Noah Syndrome.

After twenty-six years, Laura Knight-Jadczyk's unpublished book is now in print for the first time. And it's more relevant than ever. Drawing on prophecies ancient and new - from biblical narratives to modern-day visionaries - yet grounded in cutting-edge scientific discoveries about earth's cataclysmic history, this book presents a remarkable vision of humanity's dramatic past and extremely hazardous future.

The Noah Syndrome also introduces the concept of quantum cosmic metamorphosis - the spiritual ark that may carry us through the coming catastrophe. If our past is the key to our future, as Laura suggests, heeding the counsel in these pages could mean the difference between transformation and destruction.

ISBN: 9781897244791

Personality-Shaping Through Positive Disintegration

Kazimierz Dabrowski

For psychologist and psychiatrist Kazimierz Dabrowski, personality is not a given – it must be consciously created and developed by the individual. In his second English-language book, *Personality-Shaping Through Positive Disintegration*, first published in 1967, Dr. Dabrowski presents a comprehensive treatment of personality that is still relevant, perhaps more so today than when it was first written. Here Dabrowski describes personality's individual and universal characteristics, the methods involved in shaping it, and case studies of famous personalities (including Augustine and Michelangelo) demonstrating the empirical and normative nature of personality development.

Included in this edition are the original introduction, written by former APA president O. Hobart Mowrer, an appendix detailing a study on gifted children and outstanding abilities conducted by Dr. Dabrowski, as well as previously unpublished biographical pieces analyzing the personalities of Beethoven, Kierkegaard, and Unamuno. Grounded in Dabrowski's theory of positive disintegration, Personality-Shaping introduces the concepts at the heart of the theory and at the heart of human potential, creativity, social service, inner conflict, mental illness, and personal growth. Dabrowski's all-embracing perspective is at once a fresh alternative to the one-dimensional theories and trends pervasive in the field of psychology, and a full statement in its own right of all those aspects of human nature too often marginalized, ignored, or denied – a revolutionary and heartfelt product of Dr. Dabrowski's incisive observations and all-embracing vision.

ISBN: 978-0-692427-49-1

Fear of the Abyss

Aleta Edwards, Psy. D.

Most self-help or personal growth books help their readers learn to "cope" better with one specific problem that troubles them. In contrast, *Fear of the Abyss* offers healing for a wounded core by providing guidance to those with a certain constellation of issues, which Dr. Edwards calls the "PCS" personality. While perfectionism, control, and shame may be the issues that readers most easily relate to, they form part of a pattern that also includes such issues as rigidity, black-and-white thinking, indecisiveness, and relationship difficulties caused by a need to be needed.

Dr. Edwards sees this entire constellation as rooted in a core fear that varies from person to person. It is the object of this fear that she calls the "abyss." Through a series of clinical vignettes drawn from her own experience as well as exercises accompanying most chapters, she helps the reader develop self-awareness and compassion for him- or herself, as well as a greater sense of authenticity.

"This book is a true gift. Aleta Edwards is ... a genius in simplifying how to get to the root of the issues that trouble us – how to embrace the shadow and transcend to love, compassion, and honor. *Fear of the Abyss* is a must read for every psychotherapist, not to mention anyone who suffers from shame and perfectionism – which I'd say is most if not all all of us." –Lisa Ortigara Crego, PhD, LCSW, CEDS, CAP, NBCCH, Clinical Psychotherapist, Doctor of Addictions Psychology, Nat. Bd. Cert. Hypnotherapist

"In this engaging book, Dr. Aleta Edwards sheds new light on the issues of perfectionism, control, and shame. Breaking from the traditional model of diagnosis, she explains the role of trauma as a major factor in the development of the 'PCS' personality and emphasizes the adaptive nature of these issues. This book is recommended for its thoughtful text and helpful exercises, as well as the spirit of hope." –Nina Savelle-Rocklin, PsyD, Psychoanalyst, award-winning writer of the "Make Peace with Food" blog and host of the "Win the Diet War" podcast

"This is a great book for individuals who struggle with shame and perfectionism. I've recommended it to several clients and they have found it very helpful. I believe this book is a good tool as an addition to therapy or by itself as a self-help book." –Ana Maria Aluisy, MA, LMFT, LMHC, CRC, relationship specialist, featured on CNN Latino, Super Q Radio and ESPN Deportes Radio

ISBN: 978-0-692717-0-59

Evidence of Revision

Quantum Future Group

Evidence of Revision is a six part documentary containing historical, original news footage revealing that the most seminal events in recent American history have been deeply and purposefully misrepresented to the public. Footage and interviews provide an in-depth exploration of events ranging from the Kennedy assassinations to the Jonestown massacre, and all that lies between.

The footprints left in this archival footage reveal the coordinated, clandestine sculpting of the America we know today. Evidence of Revision proves once and for all that history has been revised even as it was written!

Part 1: The Assassinations of Kennedy and Oswald Part 2: The why of it all referenced to Vietnam and LBG Part 3: LBJ, Hoover and others: what so few know even today Part 4: The RFK assassination as never seen before Part 5: RFK assassination, MKULTRA and the Jonestown massacre Part 6: The assassination of Martin Luther King

6 part, 3 DVD set. Region-Free DVDs – Watch on any DVD player or computer anywhere in the world.

Newly subtitled in English, Spanish, French and Polish – the documentary is filmed in English, the subtitles are for clarity due to archival footage on which the audio is, at times, unclear.

Duration: 10 hours 25 minutes

Cover Design: Humberto Braga

ISBN: 9781897244784

Éiriú Eolas, An Amazing Stress Control, Healing and Rejuvenation Program

Laura Knight-Jadczyk

Are you stressed? Do you suffer from chronic fatigue, conditions that your doctor cannot diagnose or that he thinks are "all in your head"? Are you in physical pain more often than not? Is your system toxified from living in today's polluted environment? Do you wish you could face life's challenges with greater calm and peace of mind? Would you like to actually feel healthy, happy and pain-free every day?

"Introducing Éiriú Eolas (pronounced "AIR-oo OH-lahss"), the amazing scientific stress-control, healing, detoxing and rejuvenation program which is THE KEY that will help you to change your life in a REAL and immediately noticeable way:

Proven benefits of the Éiriú Eolas Program include: instantly control stress in high energy situations, detox your body resulting in pain relief, relax and gently work through past emotional and psychological trauma and regenerate and rejuvenate your body/mind.

Éiriú Eolas will enable you to rapidly and gently access and release layers of mental, emotional and physical toxicity that stand between you and a healthy, younger feeling and younger looking body!"

The Éiriú Eolas technique grew out of research conducted by the Quantum Future Group under the direction of Laura Knight-Jadczyk and Gabriela Segura, M.D. The practice has been thoroughly researched and proven to work by the thousands of people who are already benefiting from this unique program. The effects are cumulative and results and benefits can be seen in only a very short time, sometimes after just one session!

Disc 1, PART 1: In this video lecture, you will receive a detailed overview of the science behind the breath and the simple action of breathing by program developer, Laura Knight-Jadczyk. She outlines the importance of the breath for our physical, emotional, and spiritual well being. You will also learn the primary breathing technique that will be used in the exercises on Discs 1, part 2, and disc 2: Pipe Breathing. This is the Magic Key to instant stress reduction and control. Pipe breathing brings the entire body-mind system into balance. You can use it any place at any time for instantaneous results; control fears and phobias; recover quickly from shocks or use in high energy situations where decisive action is required and stress control is essential.

PART 2: This disc is a video demonstration of the Éiriú Eolas warm-up exercises and the breathing techniques that comprise the Éiriú Eolas program.

Disc 2 is an audio CD for you to use to practice the entire Éiriú Eolas breathing program itself, culminating in the powerful Prayer of the Soul, a non denominational, non-sectarian, scientific prayer - a truly unique achievement in our time.

Disc 3: The audio from the CD with subtitles, set against a backdrop of relaxing nature scenes. Use this disc if Éiriú Eolas CD is not available in your language or if you are hearing impaired. This disc also contains a copy of the Éiriú Eolas guide book in languages in PDF format that can be accessed from your computer along with MP3 versions of the CD in English and other languages.

Subtitles available in: English, Danish, German, Spanish, Greek, French, Croatian, Italian, Dutch, Polish, Russian, Serbian, Turkish and Vietnamese!

ISBN: 9782916721187

Red Pill Press
info@redpillpress.com
www.redpillpress.com

Made in United States
North Haven, CT
24 October 2026

21056081R00220